Climate in Motion

Climate in Motion

SCIENCE, EMPIRE, AND THE PROBLEM OF SCALE

Deborah R. Coen

The University of Chicago Press CHICAGO AND LONDON

The University of Chicago Press, Chicago 60637
The University of Chicago Press, Ltd., London

Published 2018
Paperback edition 2020
Printed in the United States of America

29 28 27 26 25 24 23 22 21 20   1 2 3 4 5

ISBN-13: 978-0-226-39882- 2 (cloth)
ISBN-13: 978-0-226-75233-4 (paper)
ISBN-13: 978-0-226-55502-7 (e-book)
DOI: https://doi.org/10.7208/chicago/9780226555027.001.0001

Library of Congress Cataloging-in-Publication Data

Names: Coen, Deborah R., author.
Title: Climate in motion : science, empire, and the problem of scale / Deborah R. Coen.
Description: Chicago ; London : The University of Chicago Press, 2018. | Includes bibliographical references and index.
Identifiers: LCCN 2017054497 | ISBN 9780226398822 (cloth : alk. paper) | ISBN 9780226555027 (e- book)
Subjects: LCSH : Climatology—Research—Austria. | Climatology—History—19th century. | Science—Austria—History—19th century. | Science—Political aspects—Austria. | Austria—Intellectual life—19th century.
Classification: LCC QC 857. A 92 C64 2018 | DDC 551.509—dc23
LC record available at https:// lccn.loc .gov/ 2017054497

♾ This paper meets the requirements of ANSI/NISO Z39.48-1992 (Permanence of Paper).

For Amalia and Adam, and in memory of my sister Gwendolyn Basinger (1965–2017)

A barometric low hung over the Atlantic. It moved eastward toward a high-pressure area over Russia without as yet showing any inclination to bypass this high in a northerly direction. The isotherms and isotheres were functioning as they should. The air temperature was appropriate relative to the annual mean temperature and to the aperiodic monthly fluctuations of the temperature. The rising and setting of the sun, the moon, the phases of the moon, of Venus, of the rings of Saturn, and many other significant phenomena were all in accordance with the forecasts in the astronomical yearbooks. . . . In a word that characterizes the facts fairly accurately, even if it is a bit old-fashioned: It was a fine day in August 1913.

ROBERT MUSIL, *The Man without Qualities*

There is no affair that a priori and according to general principles could be called large or small; matters are only large or small in comparison to and in relation to other things.

ATTRIBUTED TO EMPEROR FRANCIS I OF AUSTRIA

CONTENTS

ILLUSTRATIONS

FIGURES

PLATES

Following page 210

INTRODUCTION

Climate and Empire

In 1869, at the age of thirty, Julius Hann seemed poised for a brilliant academic career. Holding two degrees in physics from the University of Vienna, Hann had earned an international reputation three years earlier by explaining the cause of a mysterious wind. Foehn blows warm and dry on the northern side of the Alps, resembling what those living east of the Rockies call chinook. In the Alps it strikes most often in the cold months, producing incongruously warm temperatures and allegedly unleashing a variety of human woes, from heart conditions to epileptic seizures. In Hann's day, because of foehn's warmth and dryness, it was assumed to originate in the Sahara. Indeed, it was common at the time to ascribe the health effects of a local climate to the prevailing air current, which in turn was understood to be determined by its place of origin. By contrast, Hann's "On the Origin of Föhn" of 1866 zoomed in to explain foehn as a local effect. Through years of hiking in the Alps, notebook in hand, he had tracked the movements of clouds and deduced the patterns of wind. At university, he had learned the new science of thermodynamics, the physics of heat and motion. This gave him the tools he needed to interpret foehn in terms of the transformations undergone by a parcel of air as it crosses the mountains, losing moisture as it rises and warming as it descends the northern side. The resulting wind will feel hot and dry—not because it arrives from an African desert, but because of the physical properties of the atmosphere and the relationship between heat and motion.[1]

This was a foundational moment in the history of an international research program that I will refer to as *dynamic climatology*: the application of the physics of heat and fluid motion to explain climatic conditions in their past and present distribution across the surface of the earth. Dynamic climatology deserves a prominent place in the history of the Anthropocene, for it identified and began to address the problem of conceptualizing interactions across scales of space and time, from the human to the planetary.[2]

Dynamic climatology flourished in Austria for as long as the Habsburgs ruled in Vienna. It was, as we will see, a science geared to the needs of this multifarious empire, just then in the throes of modernization, and Hann stood to be rewarded handsomely for his service to the state. He was on his way to acquiring the authority of an imperial-royal (*kaiserlich-königlich, k.k.*)[3] scientist, that is, an expert on the territory of the Habsburg Monarchy as a whole—which covered nearly seven hundred thousand square kilometers of central, eastern, and southern Europe in the nineteenth century, an area nearly twice as large as the state of Germany today. Already he had begun to synthesize data from across the imperial lands, forging a continental-scale window onto the workings of the atmosphere.

Why then was Hann plagued by self-doubts as he entered the summer of his thirty-first year? As his personal diary reveals, he felt his future in Vienna to be "uncertain." He was tormented by homesickness for the very mountains that had set him on his path to success, longing for the "peace" and "mercy" of their forests, "as in the days of youth."[4] He hovered on the verge of renouncing his scientific career and returning to the seclusion of the monastery where he had been educated.

To be modern, in the words of the geographer Yi-Fu Tuan, is to experience the conflicting pulls of "cosmos" and "hearth": the yearning to venture beyond known frontiers and the craving for the stability of the familiar.[5] *Climate in Motion* explores the consequences of this tension for the production of climatology as a global science in the nineteenth century. It argues that essential elements of the modern understanding of climate arose as a means of thinking across scales of space and time, in a state—the multinational Habsburg Monarchy, a bricolage of medieval kingdoms and modern laws—where such thinking was a political imperative.

Until the twentieth century, theories of the global circulation of the atmosphere ignored motions smaller and shorter-lived than a major tropical cyclone. The circle of scientists around Hann bridged that gap, developing a picture of the interaction of small scale and large that still underwrites the climate models of today. Arguably, what makes modern climate science modern is its integration of phenomena of radically different dimensions. This requires methods appropriate to each scale; satellites are ideal for tracking hurricanes, for instance, but of little use for following the minute fluctuations in temperature and humidity that matter to the growth of a seedling. What's more, atmospheric phenomena at different scales do not proceed independently of each other. The breeze that, out of nowhere, ruffles the pages of your book may be an indirect effect of a storm brewing off the coast. As the sun's energy warms air at the equator and causes it to rise, it drives a hemispheric churning motion, which in turn generates smaller eddies, from which even smaller whirls spin off. These cross-scale exchanges of energy and even matter are what make global warming so all-encompassing and so hard to predict. One of the greatest challenges scientists face today is to model the very small-scale process by which water condenses around an atmospheric particle and begins to form a cloud, in relation to the very large-scale effects that clouds have on the global radiative energy balance. In order to predict, say, the earth's average temperature a century from now, scientists cannot ignore processes that unfold far more quickly and on far smaller scales, such as cloud formation. They deal with this problem by means of what they call "scale analysis," meaning that they estimate which are the significant phenomena for the purposes at hand and work with those to produce a reasonable approximation. The history of dynamic climatology is the story of this multiscalar, multicausal way of engaging with the world.[6]

When I began the research for this book in 2001, the field of climate science looked rather different than it does today, in 2017. The questions being put to climate models were almost exclusively about the *global* impacts of increasing concentrations of greenhouse gases. The spatial and temporal resolution of most models was too low to generate predictions that could be of use to communities planning for the medium term; nor did most models factor in regional-scale forcing mechanisms. Only recently has the question of regional impacts gained the attention of a significant portion of the modeling community.[7] Meanwhile, stalemates over international agreements on climate change have confirmed the need for action at the local and regional levels—and thus for reliable predictions at these spatial scales. Back in the early 2000s, one of the few research groups modeling regional impacts was CLEAR, or Climate and Environment in the Alpine Region. This group emphasized the value of alternating between local and global perspectives, or *downscaling*.[8] Based in Switzerland, they pointed to the Alps as a region that exemplified the benefits of small-scale climate analysis alongside global modeling. Prescient as their program may have been for the twentieth century, it struck me as an echo of the lessons of Habsburg climatology.

Working without the aid of digital computers, Habsburg scientists invented creative ways to detect, model, and represent atmospheric motions. These included turning plants into measuring instruments, imagining riverbeds as models of atmospheric waves, and inventing new literary and cartographic genres. Although you may not have heard their names before, their ingenuity was widely celebrated in their own day. Hann was nominated for a Nobel Prize in 1910, and his publications were said by an American colleague to tower "over other books as the pyramids of Gizeh tower over the valley of the Nile."[9] Others have observed that the research of Hann and his colleagues was "eclectic": whereas most climate scientists today specialize in research at a single scale, "the range of phenomena considered [by Austrian scientists] spanned the space-time scales from the global circulation to boundary layer turbulence,"[10] from the planetary scale to the dimensions of agriculture and human health. Indeed, these scientists produced a novel conceptual apparatus for tracking the transfer of energy from hemispheric flows down to molecular motions. What's more, they found creative ways to make their insights vivid to a nonscientific audience. This new way of thinking about local winds in relation to a hemispheric circulation in turn breathed life into the Habsburg idea of transnational interdependence.

ENVIRONMENT AND EMPIRE

A twenty-first-century reader might be surprised to open the first volume of the *Journal for Austrian Ethnography* (1895) and find an article titled "The Importance of Collecting Popular Names for Plants."[11] What did the study of human cultures have to do with botany? How was the emergence of human science in this multiethnic state tied to the study of the nonhuman environment?

Over the past two decades, historians have demonstrated the many ways in which the making of environmental knowledge in the long nineteenth century was inextricably bound to European empire-building. This is true, first, in the sense that imperial expansion transformed colonial environments in far-reaching and irreversible ways. Environmental historians have shown how empires depleted natural resources, disrupted ecosystems, and introduced invasive species, often leading to the extinction of native ones. Europeans justified their empires by pointing to their superiority as "masters of nature." But such mastery counted both humans and nonhumans among their victims. In some cases, the nonhuman environment has avenged itself against the imposition of control by means of epidemics, floods, earthquakes, and other catastrophes.[12] In other cases, empires eventually instituted policies of conservation, and yet these have often come at the expense of indigenous rights. This is the tragic side of the relationship between environment and empire.[13]

At the same time, the environmental history of modern empires is also a history of learning. This is not to argue, as George Basalla did in the 1960s, that the explosion of natural historical knowledge that began in the sixteenth century was the achievement of intrepid European naturalists penetrating the wilds of other continents.[14] Nor is it to forget that some intellectual traditions were *unlearned*, even forcibly suppressed, as a result of colonization. Rather, over the past two decades, historians have shown precisely how the celebrated discoveries of metropolitan elites rested on knowledge appropriated from colonial naturalists and native informers.[15] The knowledge produced in this way included strains that we would now call ecological: in particular, knowledge of the distribution of plants and animals in relation to conditions of climate and soil. Indeed, the classification of new species has surged along the edges of expanding empires, even if closely followed by the disappearance of those newly cataloged species.[16] In other words, there is an uncomfortably intimate relationship between the growth of environmental knowledge and the environmental destruction wrought by imperialism.

What's more, some of the modern era's most damning critiques of imperialism and its ecological costs have come from scientists in the service of empires, from Johann Forster in the eighteenth century to Arthur Tansley in the twentieth. Twenty years ago, Richard Grove argued that the origins of modern environmentalism lie in the responses of colonial naturalists to the environmental degradation they witnessed on Pacific islands and for which they knew Europeans to be responsible. More recently, Helen Tilley has shown how twentieth-century scientists working for the British Empire in Africa developed their own environmentalist ideas by contrasting European land-use practices with the sustainable methods of local agriculture.[17] In some cases, imperial science proved sufficiently independent of imperial ideology to generate genuinely new knowledge. No less significant for the history of environmental science and politics was the *scale* of empire in the nineteenth century. Its networks of information exchange laid the basis for the global science of the late twentieth century and have been linked suggestively to the emergence of a "planetary consciousness."[18] From this angle, empires have been experimental sites for exploring ties of interdependence among far-flung humans, nonhumans, and the inorganic world.[19]

DEFINING CLIMATE

This book inquires into the role of nineteenth-century empires in producing knowledge specifically about climate. In the nineteenth and early twentieth centuries, the meaning of climate in European languages was in flux. Since ancient times, climate had signified the latitude-dependent effect of solar radiation, *klima* being the Greek word for inclination or the angle of incidence of sunlight. Three climate zones neatly banded each hemisphere: frigid at the poles, temperate at midlatitudes, and torrid at the equator. By the nineteenth century, however, geographic explorers and medical and agricultural researchers were drawing attention to the complex spatial variability of atmospheric conditions. Thus late nineteenth-century scientists contrasted the "solar climate," meaning the distribution of heat to be expected on the basis of solar radiation and the known composition of the earth's atmosphere, with the "physical" or "terrestrial climate," meaning climate as modified by any number of factors, such as the shape of the earth's surface and the distribution of land, water, and vegetation. This terminological confusion reflected a methodological problem of scale. Scientists of the nineteenth century possessed rudimentary physical models of the planetary circulation of the atmosphere, but they had no way to integrate these models with accumulating evidence of the fine-grained variability of climate across the surface of the earth. As one medical writer in the Silesian spa town of Kudowa expressed the difficulty in 1908, "Even if this locally restricted meaning of climate is not common usage, the phrase 'weather or atmospheric conditions' [*Wetter oder Witterung*] wouldn't always be accurate in what follows, and I must apologize in advance for the lack of an appropriate expression."[20] Thus climate was a fundamentally ambiguous concept in the nineteenth century, in ways that have escaped the attention of historians.

The temporal dimensions of climate were just as ambiguous as its spatial extent. The relevant time scales ranged from the seasonal, in the case of medical and agricultural climatology, to the geological, in the case of theories of the Ice Ages. Indeed, these temporal perspectives were proving impossible to reconcile. How could the evidence unearthed in the 1830s and 1840s of an ancient Ice Age be squared with the apparent stability of the earth's climate since the start of human records?[21] If climate implies a time-averaged state, it was not clear how long an interval was appropriate to its definition.

Finally, the meanings of climate also varied in their proximity to human interests. Alexander von Humboldt had famously defined climate as all the changes in the atmosphere that perceptibly affect our sense organs.[22] This anthropocentric definition reflected the fact that climate was recognized as an influence on almost every aspect of human life: health, agriculture, labor, trade, even the psyche. Indeed, the historical roots of climate science lie largely in practical efforts to improve everyday life. Much of our knowledge of climate has come from everyday experience with the many aspects of nature that stand in mutual dependence with the atmosphere, such as vegetation, water supply, and soil conditions. As historians have noted, professional scientists often built on the traditional knowledge of those whose livelihoods depended on climatic knowledge, such as farmers, sailors, and fishermen.[23] At the same time, climatological knowledge was being produced by new kinds of experts of many stripes, including botanists, foresters, agricultural scientists, geologists, paleontologists, mining officials, medical geographers, pharmacists, and balneologists. Nineteenth-century botanists, for instance, referred to vegetation as an ideal climatic record, "everywhere a reflection of the local climate," while geologists recognized glaciers as the best possible "climatic measuring instruments" (*Klimamesser*).[24] Just as numerous were the institutions that produced knowledge relevant to climatology. At European universities, climatology might be studied and taught in any or all of three different faculties: geography, physics, and medicine. Climate knowledge was also produced far beyond the walls of universities: at public observatories, by academies of forestry or agriculture, by voluntary associations like Austria's popular Meteorological Society, and by European explorers and travelers the world over.[25] Thus the study of climate mattered in practical ways to people of all kinds. At the same time, scientists were increasingly interested in atmospheric phenomena unfolding in the upper layers of the atmosphere. These were accessible only to intrepid mountaineers and balloonists—and they bore no obvious relation to human life.[26]

When did people begin to speak of "climate" in the singular, in the sense of a property of the earth as a whole? As we will see in chapters 7 and 9, the concept of the global climate entered popular discourse in central Europe in the 1870s, in the midst of a heated debate over anthropogenic climate change. But that is only half the story. Just as important was the rigorous definition of a "local climate" that entered with it. A modern, multiscalar concept of climate was born from encounters between scientists and others with wide-ranging claims to climate expertise.

VARIETIES OF EMPIRE

European empires of the nineteenth century seized on climatology as a tool for adapting newly acquired colonial environments to imperial ends. Climate was studied in relation to a swiftly changing global distribution of production and trade, as well as shifting theaters of military conflict. It was with the aid of such research that the global spread of capitalism was understood at the time to "make the best of the diverse endowments of the world by encouraging the most efficient use of each country's particular resources."[27] Eighteenth-century savants had typically regarded the fit between a society and its natural environment as the result of a divine plan.[28] Nineteenth-century geographers saw it instead as the result of a process of biological and cultural adaptation, one in which imperialists intended to intervene. Theories of acclimatization supported imperial plans to settle Europeans permanently in the "torrid zone."[29] In principle, if not always in practice, climatology could inform decisions about where to locate agricultural settlements and colonial towns. It also inspired the construction of high-altitude medical rehabilitation centers for sick and weary colonists in the tropics, taking advantage of cooler, drier mountain air.[30]

One historian has characterized nineteenth-century climatology as a "sleepy backwater,"[31] a charge that held true in certain parts of the world. In many parts of Europe's overseas empires, climatology developed as a scattershot study of local climates, paired with the self-serving theory that "the tropics," whatever they were, were unsuitable for European labor.[32] Across Europe and North America in the mid-nineteenth century, governments and private societies organized networks of meteorological observers. But this data was mined primarily for the purposes of storm forecasting. As the Habsburg climatologist Alexander Supan explained in 1881, "Climate has retreated with the advance of weather."[33] In 1908 Napier Shaw complained of the "waste" of the British Meteorological Office, from which he had recently retired as director. Data was streaming in, but to what end? Shaw lamented as well the discontinuities of the imperial network—"four observatories and over four hundred stations of one sort or another in the British Isles; an elaborate installation of wind-measuring apparatus at Holy head; besides other ventures squandered abroad; an anemometer at Gibraltar, another at Saint Helena; a sunshine recorder at the Falkland Isles, half a dozen sets of instruments in British New Guinea, and a couple of hundred on the wide sea."[34] "Waste" in this sense was a function of the inability or unwillingness of scientists of *overseas* empires to mediate between local observations and global models.[35]

Indeed, histories of science and empire tend to assume a divergence of interests and epistemic commitments between imperial scientists and local collectors. Imperial scientists figure in this scholarship as intellectual "lumpers," proponents of generalization, against the "splitters" who would insist on nature's local variability. Imperial science in this sense was embodied by the London botanist Joseph Hooker, the director of the imperial botanical garden at Kew, who claimed that his colonial correspondents paid too much attention "to minute differences that when long dwelt upon . . . assume minute value."[36]

According to this account, the global perspective of modern science emerged at the expense of the local.

However, historians have paid far less attention to the geography of knowledge production in the continental empires of nineteenth-century Eurasia: Habsburg, Romanov, Ottoman, and Qing. In these cases the dividing line between metropolitan and provincial knowledge was far less clear. These states had in common the capacity to govern multiple cultures and to wield power from multiple centers. They took up the tools of modern science in part in the interest of developing new imperial "languages of self-description" that could serve as alternatives to the simplifications of nationalism. Undermining reified categories of nation and race, these sciences traced histories of migration, mixing, and cultural transfer. In other words, these late imperial sciences established the hybridity of their populations and territories as empirical facts.[37]

As this suggests, modern states do not all render society "legible" in the same way.[38] Making a territory and its population legible need not mean abstracting away local particularities. "Seeing like a state" might instead mean bringing imperial diversity into sharper focus. Historians of cartography recognize that maps do not simply reflect the world; they actively contribute to the creation of a territory by visualizing its features selectively, thereby encouraging some modes of interaction with the environment and discouraging others. In this way, maps that accentuate heterogeneity may help to preserve local differences of environment and culture.[39]

In the Habsburg Monarchy, in particular, "local knowledge" in fields like ethnography, medicine, and physical geography was often produced as "imperial science" and funded by the imperial state. Recent research by Pieter Judson and others has shown how the nineteenth-century Habsburg state created a legal space in which even former serfs could assert their rights as citizens and pursue economic development. Far from inhibiting modernization, the state's decentralized structure allowed grassroots politics and cultural pluralism to flourish.[40] Already in 1817, the Habsburg archduke Johann, a great patron of natural history, articulated the principle that provincial patriotism could go hand in hand with loyalty to the dynasty: "Austria's strength rests on the diversity of its provinces . . . which one should carefully preserve. . . . Austria has recovered her strength after every misfortune, because each province stood on its own and considered its survival independently of the others, and yet contributed loyally to the common goal."[41] Previous research hints at the implications of this ideology in fields like ethnography, medicine, and physical geography. In these cases, research conducted in the service of nationalism—perhaps even the "national schools of thought" celebrated by post-Habsburg historians—was sponsored by the ministry of education and articulated, often in German, by Habsburg-loyal scholars as contributions to the ideology of "unity in diversity."[42]

So it was that the Austrian Society for Ethnography instructed its members in the collection of botanical knowledge from peasants, women in particular.[43] Knowing the popular names for plants in multiple dialects enabled the botanist and the ethnographer jointly to track the spatial distribution of a given species and the diffusion of a given dialect. Research in this vein was of particular significance "in our multilingual Austria . . . since the popular names are not only linguistic but also cultural-historical records." As we will see in chapter 10, the discovery that in some cases the *same* name was being used for *different* plants would have profound repercussions for interpretations of the relationship between climate and living things. By the same token, ethnographers took it upon themselves to record names for characteristic local winds and the rituals associated with them. Thus Hann was familiar with the adage of Alpine peasants: *Steigt man im Winter um einen Stock, so wird es wärmer um einen Rock* ("If you climb in winter up a stair, one less coat you will need to wear").[44] To Hann, this meant that inversions of the usual decrease of temperature with height must be a common phenomenon, which provided a clue to atmospheric conditions of thermal equilibrium. In short, the fine-grained ethnic geography patronized by the Habsburg state often involved tracking a given natural kind across the Monarchy, in order to extract differences of language and culture. By attaching value to natural diversity, such research promoted the integration of "indigenous places" into "imperial space."[45]

SEEING VARIABILITY IN NATURE

Climatology in imperial Austria merits comparison to related enterprises in other nineteenth-century imperial states. The royal-imperial Central Institute for Meteorology and Geomagnetism (Zentralanstalt für Meteorologie und Geomagnetismus, later Geodynamik; ZAMG) was founded in Vienna in 1851, four years after the opening of the Royal Prussian Meteorological Institute. The British Meteorological Office evolved out of a committee established in 1854, while the French meteorological service was founded in 1855 and the US Weather Service in 1870. Like these other meteorological offices, the ZAMG was charged with taking advantage of the new technology of telegraphy in order to issue advance warnings of approaching storms. Indeed, Austrian scientists played an active role in the development of an international storm-warning system in the 1860s, including hosting the first-ever International

Meteorological Congress in 1873, where the conventions of meteorological telegraphy were negotiated. And yet issuing forecasts was work that most ZAMG scientists disdained. Kreil, its first director, was distressed to find that not only was he expected to predict the weather, he was even held responsible for it—in certain circles he became known as the "weather maker." Given the state of meteorological knowledge at the time, Kreil believed that reliable weather forecasting was a long way off. For now, as he tried to convince the public in his popular writings, the most useful knowledge would not seek to "organize agriculture according to the daily fluctuations of the atmospheric conditions, but once and for all according to the average climatic character of specific places and regions."[46] Kreil and his associates genuinely believed that their research would benefit the state. But those benefits would come from knowledge of long-term regularities, not short-term predictions. They would come, in other words, from climatology, with its practical value for agriculture, medicine, tourism, trade, and military operations.

Austria seems to have been unusual in this respect, as far as one can tell from the sparse historical literature on nineteenth-century atmospheric science. At the state-sponsored observatories of Berlin, London, Paris, and Washington, DC, it seems that climatological research tended to be viewed as little more than a quaint relic of the eighteenth century, a worthy pastime for provincial amateurs or peasant farmers but of little relevance to modern science. Only in rare cases did meteorological officers in North America and western Europe choose to invest in the study of climate—that is, to shift their priority from short-term forecasts to long-term planning.[47]

Climatology thrived in particular in the great land empires of Eurasia: Russia, India, and Austria. Presented with an expansive and *continuous* field for geoscientific research, scientists in these states began to think in new ways. Tasked with the goal of stably and efficiently exploiting the territory's resources, they set out to map its "natural regions" and "transition zones." One historian of imperial Russia has aptly described this enterprise as a scientific project of "regionalization." The regionalizing sciences of the nineteenth century were holistic enterprises intent on analyzing local differences within a larger, integrated context.[48] The baseline definition of these regions was climatic: average temperatures and rainfall were the fundamental data on which a mental map of imperial difference was being constructed. Scientists thus came to appreciate the *heterogeneity* of climates and to seek the causes of their differentiation.

The choice to emphasize or de-emphasize the diversity of an empire's human and nonhuman subjects—the choice between pursuing imperial climatology as a coherent overview or as a patchwork of local details—was always a strategic political decision. From the perspective of Henry Francis Blanford, chief of the Indian Meteorology Department (founded 1875), the subcontinent furnished an ideal site for climatological research: a single, contained landmass displaying in miniature nearly all the variations in climate to be found on the earth's surface, with an unobstructed view of the formation of tropical storms. To speak of "the climate of India" in the singular, he argued, "would be as misleading as if we were to speak of its inhabitants in terms implying that they are a homogeneous race, alike in ethnic and social characters, culture and belief."[49] Yet others expressed India's suitability for climatological study in language with quite a different political valence. They emphasized the order and regularity of the Indian climate (now in the singular) and the ease with which British science could master it.[50]

In Russia, as Wladimir Köppen pointed out in 1895, climatology followed the tsar's troops, as the Russians pushed into the Crimea, the Caucasus, Siberia, and beyond. "Russia's conquests to its west have quickly expanded our knowledge. For the Russian soldier is followed with remarkable speed by the Russian [scientific] observer, and out of the impenetrable dens of thieves . . . have for a long time now been appearing practicable, uninterrupted meteorological data series."[51] The work of Russia's foremost climatologist, Alexander Voeikov (1842–1916) reflected the ambitions of a modernizing and expanding empire. He produced theories of the global climate system and of the influence of climate on social life, and he argued that unfavorable environments could be improved through rational water use.[52] Meanwhile, a more cautious research program in soil science under Vasilii Dokuchaev was investigating charges of human-induced climate change in the "black earth" region of the Russian steppe.[53] As Catherine Evtuhov shows, the "sheer bulk and diversity of information" gathered by Dokuchaev "rendered it virtually unusable by the central authorities."[54] Thus enterprising Russian scientists and citizens were hard at work developing scientific methods of agricultural assessment and improvement—methods in which the imperial government took little if any interest at all.

The situation was different still in the United States. Lorin Blodget's *Climatology of the United States* (1857) explicitly set out to counter the claim that the nation's climate was especially "variable or extreme."[55] East of the one hundredth meridian, Blodget insisted that the climate was "uniform"; and where mountains intruded, they "do not shelter or expose either side, nor cause any contrasts in the character of productions respecting them." Significantly for the present study, Blodget favorably compared his nation's climatic uniformity with "the abrupt transitions, and the predominance of local changes in southern and central Europe."[56] He concluded that whatever local variability the climate might display was not worth scientific attention. Climatic uniformity, not diversity, was his ideal. It is therefore unsurprising to learn that the federal government dedicated few resources to the investigation of local climates.[57]

"K. UND K." CLIMATOLOGY

In Austria, too, climatology was closely tied to empire-building. From the perspective of the Habsburg state, climatology was not about exploring a dark continent but about reimagining a familiar one in the throes of change. Over the course of seven centuries, through strategic marriages and battles, the Habsburg dynasty had acquired a motley yet contiguous set of kingdoms, duchies, and principalities. Extending through nine degrees of longitude, the dynasty's lands were crisscrossed by mountains: one chain running east from the tall peaks of the Alps in the westernmost province of Vorarlberg, then turning south along the Adriatic into the craggy towers of the Dinaric Alps; another rising to the east from the Bohemian massif up into the Carpathians, a largely unmapped range that isolated Galicia, the Habsburgs' newest acquisition. Before the coming of the railways, the Monarchy was largely dependent on its main river way, the Danube, for transport. Its waters originated in melting glaciers in the Alps, flowed east through Vienna, Bratislava (German: Pressburg; Hungarian: Poszony), and Budapest (German: Ofen-Pest), into lands still controlled by the Ottomans in the nineteenth century, and emptied into the Black Sea. Across much of the Habsburg world—the grassy plains of the Hungarian steppe; the rocky, porous ground along the Adriatic; and the high-altitude pastures of the Alps and Carpathians—agriculture was possible only during a brief season when adequate rainfall and mild temperatures prevailed. A swing of the climate could easily spell hunger. On the other hand, excess rainfall could trigger devastating floods along the Danube and its tributaries, especially in deforested areas, as well as rockslides in the mountains. In the marshy lands of the Carpathian basin, eighteenth-century attempts at water regulation had wreaked havoc with the ecosystem and threatened floodplain agriculture. Transport and communication across the Habsburgs' lands was also highly dependent on climate. During Europe's Little Ice Age, from roughly the thirteenth through the early nineteenth centuries, portions of the Danube often froze, and snow blocked certain mountain passes year round. With the period of warming that set in circa 1860, baseline expectations from the past were not necessarily useful guides to the future.[58]

FIGURE 1. *Karst Landscape in Kupreško polje* (Bosnia-Herzegovina), by Zygmunt Ajdukiewicz, 1901. The "karst" lands of the Dinaric Alps were known for their sinkholes, caves, and underground waterways.

The changes afoot in central Europe in the nineteenth century were socioeconomic as well as climatic. "For now a little rural village and its surroundings can shut itself in with all that it has, all that it is, and all that it knows," wrote the novelist and naturalist Adalbert Stifter at midcentury. "Soon, however, this will no longer be so, and [the village] will be caught up in the dealings of the world beyond it."[59] The nineteenth-century Habsburg Empire was a geographically extensive but semiclosed economic system, having been excluded from the German *Zollverein* in the 1830s. It was out of the race for overseas colonies and had minimal prospects for expansion in southeastern Europe. Within its borders, capitalism was already beginning to produce its signature geography of unevenness. Between the end of the Napoleonic Wars and the stock market crash of 1873, industrialization transformed the geography of production in the Habsburg lands. Regions of the Monarchy that had once been economically self-sufficient were thrust into the role of imperial "periphery," generating a flood of migrants to the Monarchy's new industrial centers. After the crash, capital investment shifted to the periphery in order to take advantage of cheaper land, raw materials, and labor.[60]

In this context, the new genre of *climatography* (chapter 6) played in part the role of advice manual for repurposing natural environments to fit the new

FIGURE 2. *Angyalháza puszta*, eastern Hungary, 1891. Note the draw well in the background, typical of the region.

economy. Possibilities included introducing new crops (beets for sugar, potatoes for spirits), launching a tourist industry, or building spas or sanatoriums to provide "climatotherapy." At the same time, imperial climatologists were called on to adjudicate between proponents of development and those of conservation. Deforestation or swamp drainage was blamed for a deterioration of the climate within the span of historical memory in several areas, above all Bohemia, the Hungarian plains, and the karst regions of Carniola, Dalmatia, Croatia, and Bosnia.[61] Everywhere it was argued that restoring forests would create a moister, more fertile climate. Inquiries into suspected climatic shifts in the course of human history were thus highly public and politically fraught.

In all these ways, climatology's history in Austria sheds light on this science's close ties to nineteenth-century imperialism more generally. At the same time, the scientific institutions of the supranational state formed a unique lens onto the natural world. Unlike administrators in Washington or Saint Petersburg, the emperor's ministers in Vienna saw good reasons, both practical and ideological, to support the study of climate down to the details of the smallest scales. And unlike British scientists, with their insistence on the fundamentally orderly atmosphere of India, Habsburg climatologists were far more ready to grapple with the true complexity of their data. As we will see, their day-to-day focus on minute fluctuations and statistical subtleties was the flip side of their quest for an all-encompassing overview.

Julius Hann made this point most effectively. After admitting that he had spent an entire week determining how best to standardize measurements of the distribution of air pressure in central and southern Europe, he cited Francis Bacon in his defense. Bacon had once remarked that the natural philosopher who neglects details is like a haughty prince who ignores the petition of a poor woman: "He who will not attend to things like these as being too paltry and minute, can neither win the kingdom of heaven nor govern it."[62] The Baconian metaphor of the "empire of nature" underscored the affinity of this scientific precept with the logic of the multinational state. In fact, Francis I (1804–35) had been praised in nearly identical language for his comprehensive knowledge of the Habsburg realm. The first ruler to call himself "Emperor of Austria" and to envision the dynastic lands as a unified state, Francis was known to occupy himself with what might seem to be insignificant local matters. Particularly on his frequent tours of the provinces, he was said to display "his extensive, one could even say all-encompassing knowledge of the laws, customs, and morals of all parts of his great kingdom. . . . A phrase that was often on his lips was this: 'There is no affair that a priori and according to general principles could be called large or small; matters are only large or small in comparison to and in relation to other things.'"[63]

TOWARD A HISTORY OF SCALING

The history of climate science needs to be seen, then, as part of a history of *scaling*: the process of mediating between different systems of measurement, formal and informal, designed to apply to different slices of the phenomenal world, in order to arrive at a common standard of proportionality. In the natural and social sciences, scaling (upscaling, downscaling) refers to the process of adapting models to apply at larger or smaller dimensions of space and time. But scaling is also something we all do every day. It is how we think, for instance, about how one individual's vote might influence a national election, or whether buying a hybrid car might slow global warming. It can also be a way of situating the known world in relation to times or places that are distant or otherwise inaccessible to direct experience. Scaling makes it possible to weigh the consequences of human actions at multiple removes and to coordinate action at multiple levels of governance. It depends on causal factors that are likewise of varying dimensions, from an individual's imagination to translocal infrastructures, institutions, and ideologies. A focus on scaling is timely for the field of history today, at odds as it is over "going big." Rather than framing this as an either/or choice, we might think historically about what is involved in working between and across different scales of observation, analysis, and action.[64]

Scaling is a necessary step in the production of what John Tresch calls "cosmograms": the representations we use to situate ourselves in relation to the rest of the universe, to map our interconnections and mutual influences. The history of scaling thus calls for a "materialized" intellectual history on Tresch's model. As he writes, "Cosmological ideas take on realist force when they are anchored, housed, and transmitted in objects, technical networks, routine practices, and social organizations."[65] Histories of scaling must therefore attend to the tools and practices of commensuration, which are not limited to measuring instruments in the traditional sense. For instance, Benedict Anderson pointed out decades ago that the novel and the newspaper were key early tools for creating a new scale of thought in the nineteenth century: the nation.[66] More recently, Richard White has traced the role of the railroads in creating a new, politicized space in nineteenth-century North America, for which transport rates provided the new metric of proximity and distance.[67] When it comes to climate change, some of the most important tools of scaling—for understanding the weather of the here and now in relation to large-scale, long-term processes—have been found, not made. These include migratory animals, tree rings, ice cores, fossils, rocks, and living plants.

In the nineteenth century, the scalar imagination was restless. A host of technical developments—including the improvement of microscopes, telescopes, photography, telegraphy, and electrical clocks—allowed human observation to push down into the smallest dimensions of space and time, as well as outward, across the earth, to the solar system, the galaxy, and beyond. Some in the nineteenth century spoke of the annihilation of time and space. In fact, time and space were measured more carefully than ever before. New intervals of space and time were introduced: a tenth of a second, an electron-width, a light-second, the height of the stratosphere, the depth of the earth's crust. Meanwhile, new political entities were emerging in an unprecedented variety of sizes and forms. Circa 1850 may well have been a high point for the diversity of state forms, after the upheavals of the Napoleonic Wars and before nation-states became ubiquitous.[68] Thus the political imagination was not confined to the spatial scale of nation-states and the temporal scale of their historical memory. In all, there was a wide variety of ways to envision relations between the individual and the state, between nation and empire, between the small scale and the large. Today, we have been so conditioned by statistical reasoning that we tend to view the micro solely as an instantiation of, or an exception to, the macro. Alternatives proliferated in the nineteenth century, proposing relations between micro and macro that were, for instance, emblematic, metonymic, or ecological. In the Habsburg lands, as we will see in chapter 1, dynastic iconography maintained the vitality of Renaissance cosmologies.

Meanwhile, new modes of representation juxtaposed these radically disparate dimensions to dramatic effect. Part 2 of this book traces the emergence in the nineteenth-century Habsburg lands of new techniques in a range of media and disciplines—from landscape painting to geography, fiction, and atmospheric physics—all dedicated to representing precise local detail within a large-scale overview. Elsewhere at the time, writers and artists were developing related techniques. In the United States, for instance, the landscape paintings of Frederic Church shifted away from Romanticism and toward what Jennifer Raab calls an "aesthetic of information," in which a proliferation of natural details vies with and ultimately triumphs over any overall impression of unity. This new orientation grew out of Church's sense of the interdependence of living things and culminated in his reinvention of himself as a landscape architect intent on revealing nature as a "living system."[69] Likewise, literary scholars have shown how certain Victorian novels integrated their human stories into grand visions of earth history. As Anna Henchman has recently put it, "Victorian writers move restlessly back and forth between self and universe, part and whole."[70] These literary exercises in scaling were often responding directly to developments in the sciences. "The novel," write Jesse Oak Taylor, "helps reconcile the expansive timescales of evolution, climate, and geological change with those of human history and everyday life."[71] Thus the multiscalar vision of nineteenth-century climatology arose interdependently with a host of new ways of seeing and representing the human and natural world.

To be sure, some of these aesthetic trends proved more mystifying than illuminating. Turning attention from human events to cosmic ones could be a strategy for obfuscating the relationship between humans and nature back on earth. In this vein, Victorian scientists liked to compare British industry's consumption of energy with the thermodynamic "dissipation" of energy on a cosmic scale. In doing so, they made an unsustainable system of production appear natural and inevitable.[72] Hence the nineteenth century saw a growing need for representational approaches that might discipline a scalar imagination run wild.

It was in this spirit that many in the Habsburg world undertook the work of scaling. They insisted that nature could not be measured solely according to a scale derived from human concerns; other measures of significance were needed when studying nature on the very small scale or the very large. As the physiologist and leader of the Czech national awakening J. E. Purkyně wrote in the preface to the first issue of his scientific journal *Živa*: "We do not suppose that anyone will take exception to our discussing this or that seemingly insignificant matter in greater detail and more meticulously; in infinite nature nothing is insignificant, nor are the needs of man its only measure."[73] Likewise, the geologist and liberal statesman Eduard Suess insisted that "the planet may well be measured by man, but not according to man." As he explained,

> the standard for small and big as well as for the duration and the intensity of a natural phenomenon is in many cases based on the physical organization of man. . . . When we speak of a thousand years, we introduce the decimal system and, with it, the structure of our extremities. We often measure mountains in feet and we distinguish long and short periods of time according to the average length of human life and therefore based on the frailty of our bodies; and we unconsciously borrow the standard for the expressions "intense" or "less intense" from our personal experience.[74]

This theme was echoed by the pioneers of climatology in the Habsburg lands. Karl Kreil, the founding director of the ZAMG in Vienna, insisted that "everywhere there is a macrocosm and a microcosm, a world on the large scale and on the small—the latter just as important, often more important than the former." He urged his colleagues to pursue the "interaction," *die Wechselwirkung*, between large-scale atmospheric phenomena and the "organic and inorganic shell of the earth."[75] We might think of such comments as the self-defense of Habsburg scientists against the accusation of drudgery that would haunt them into our own day.

By distinguishing the multiple scales of climatic processes, and by devising methods of observation and analysis appropriate to each, dynamic climatology operationalized Suess's principle that nature must be measured "by man but not according to man." His injunction to think across dimensions of space and time rested on an implicit distinction between "lived" and "absolute" scales of measurement. In this respect, dynamic climatology's ambition of studying the small scale in relation to the large resonates with quite a different tradition of central European thought: the philosophy of phenomenology, as it was developed in the early twentieth century by Edmund Husserl, Jan Patočka, and Ludwig Landgrebe, all born into the Habsburg Monarchy. These philosophers contrasted the experience of the realm of "self-extension," of human "work," with the experience of the "absolute" sphere of the earth, "the global context on which we depend."[76] It seems fitting, therefore, to describe the history of Habsburg climatology as what Landgrebe called "a history of the transformation of the horizons of the world."[77] From Husserl this book also adopts the goal of restoring "the link between our knowledge in physics and our intuitive experience of things in our 'surrounding world of life.'"[78] It accepts phenomenology's challenge to the historian of science to re-embed inherited scientific knowledge in the existential context in which it originated.

Yet scaling is a more uncertain and imperfect process than the phenomenologists supposed. Not even the natural sciences have access, in practice, to an "absolute" measurement scale. Every measurement depends on an agreed-upon definition of a standard unit and its instantiation in an exemplary object. These standards are social conventions and recent research has shown that behind many such conventions lurk hidden histories of contention.[79] It is interesting to note that contemporary English offers no verb to denote the process of negotiation that produces a measurement standard. The use of "commensurate" as a verb seems to have fallen by the wayside in the nineteenth century, when the work of commensuration had apparently been completed once and for all by boards of experts. The term "scaling" fills this linguistic gap and so reminds us of the work that goes into mediating between different ways of measuring the world. As we will see, such work is not only cognitive; it may challenge the body, put social relationships to the test, and expose a mediator to the pull of conflicting desires.

AN OUTLINE OF THE BOOK

Part 1, "Unity in Diversity," analyzes the mutual development of imperial ideology and empire-wide institutions of environmental science. Its sources include state-institutional archives and distinctive compilations of knowledge such as the encyclopedic Austro-Hungarian Monarchy in Word and Image, also known as the Kronprinzenwerk. Chapter 1, "The Habsburgs and the Collection of Nature," takes a broad and deep view of the production of climatic knowledge in the dynasty's lands. It reveals the motivations behind the long-term collection and preservation of information about the physical and biological diversity of this territory. Chapter 2, "The Austrian Idea," recasts the long debate over the late Habsburg Monarchy's ideological underpinnings or lack thereof. It calls attention to the spatial character of emerging justifications for the empire and to the empirical research programs they rested on, which shaped the human and natural sciences alike. Chapter 3, "The Imperial-Royal Scientist," introduces the figure of the *kaiserlich und königlich* scientist—like Hann, an expert on the territory of the Monarchy as a whole. Chapter 4, "The Dual Task," charts the construction in the 1840s and 1850s of an empire-spanning geophysical observing network and a central observatory in Vienna—an institution whose responsibilities were described as "dual," serving both knowledge of particular places and of universal processes.

Part 2, "The Scales of Empire," focuses on the problems of scale facing the scientific servants of the Habsburg Monarchy and the representational techniques they developed in response. Chapter 5, "The Face of the Empire," traces the rise of cartographic and painterly efforts to achieve a synthetic overview of the Monarchy. Chapter 6, "The Invention of Climatography," introduces a nineteenth-century genre that aimed to make atmospheric data meaningful to readers of many backgrounds. Chapter 7, "The Power of Local Differences," tracks the spread of the metaphor that linked Habsburg ideology to the physics of the atmosphere. Chapter 8, "Planetary Disturbances," analyzes the physical-mathematical description of climate produced by Hann and his colleagues as the fruit of their practices of scaling.

Part 3, "The Work of Scaling," relies on scientists' unpublished letters and diaries to reconstruct the social and personal dimensions of the process of scaling. In the Austrian press and parliament in the 1870s and 1880s, arguments raged over the climatic consequences of deforestation and swamp drainage. Chapter 9, "The Forest-Climate Question," illustrates scaling as a social process by showing how imperial-royal scientists intervened to *rescale* this debate. Chapter 10, "The Floral Archive," considers plants as tools of temporal scaling, showing how botany became a crucial source of knowledge of climatic history. Finally, chapter 11, "Landscapes of Desire," turns to the private side of scaling. Comparing scientists' intimate accounts with published sources, this chapter explores the emotional experience of the imperial-royal scientist as he reoriented his sense of near and far. The conclusion considers the legacies of Habsburg climatology for twentieth-century central Europe and the present climate crisis.

* 1 *

Unity in Diversity

CHAPTER 1

The Habsburgs and the Collection of Nature

In 1867 Anton Kerner von Marilaun made a historic discovery in the mountains northwest of Innsbruck, where he was professor of botany. To explain its significance, he was obliged to turn back three hundred years. It was then, in the mid-sixteenth century, that the cultivation of auriculas (see figure 3) had caught on in Holland and spread to England and beyond, triggering a demand for the flowers second only to the craze for tulips. The auricula was, to Kerner's knowledge, the only Alpine flower to have become a "widespread ornamental plant in gardens."[1] High-born ladies had enjoyed an abundant selection of these beauties at Viennese markets. But the geographical origin of the auricula was unknown, and Kerner set out to uncover its story.

Kerner's historical detective work led him to the writings of Carolus Clusius (1526–1609), perhaps the most celebrated European naturalist of the sixteenth century. Clusius, from Flanders, had come to Vienna in 1573 at the bidding of Emperor Maximilian II, who sought to collect everything that was rare, beautiful, or useful in nature.[2] The Flemish botanist was tasked with developing a medical garden for the imperial palace. As Kerner noted, gardeners proved eager to send Clusius plants. Many varieties came from the Mediterranean, others from Turkey and the Banat.[3] Clusius, meanwhile, had an uncommon passion for Alpine flowers. In fact, he took great pains to grow these high-altitude plants in Vienna. Even his failures were instructive, as Kerner explained. It appeared that many Alpine varieties were unhappy with the warmer conditions in Vienna, even if some managed to thrive in the garden's shadier spots. In the end, Clusius was able to cultivate two varieties, which Kerner identified by their Linnaean names, *Primula auricula L.* and *Primula pubescens Jacq.* The latter,

FIGURE 3. Illustration of an auricula by Clusius, 1601.

according to Kerner, was a hybrid of the former and another variety. The hybrid, which came to be known simply as the auricula, reproduced vigorously and with dazzling variations.[4] It found its way from Vienna to merchants in Antwerp and soon became the basis of the new botanical craze.

But where had the flower come from? Clusius had first encountered and described auriculas in the Vienna garden of the physician Johann Aichholz.[5] All Aichholz knew was that the plant had been a gift from a noblewoman. Hearing that the auricula was common "in the Alps near Innsbruck," Clusius set out to find it. This was a remarkable undertaking in an age when mountains were still regarded as godforsaken wilderness.[6] Yet Clusius searched "the highest passages of the Austrian and Styrian Alps in vain."[7] To later generations of botanists, the origins of the auricula remained a mystery. Many searched the Alps for it, but it was nowhere to be found in the wild. That is, until 1867, when Kerner first discovered the plant (*Primula pubescens Jacq.*) growing amid limestone and slate on a steep hill above the village of Gschnitz, at an elevation of seventeen hundred to eighteen hundred meters. This discovery meant so much to Kerner that he chose to build a summer villa for his family on that very site. In 1874, when he was ennobled by Emperor Franz Josef in recognition of his scientific service to the state, the auricula became the basis for his family's coat of arms.

One might wonder why the flower's origin had remained a mystery all this time. As we will see in chapter 10, Kerner would glimpse a possible answer to that question as he began to decipher the climatic history of the eastern Alps. For the present chapter, the question is this: why would a nineteenth-century scientist have worked so hard to reconstruct the efforts of a plant collector who had been dead for three centuries? As director of the botanical gardens in Innsbruck in the 1860s and later in Vienna, Kerner believed he could not reform these institutions without understanding their historical development. In order to chart a course for the future, one needed to appreciate "how, in the course of time, botanical gardens had come to possess the collections that we find in them today." The director needed to know not only what the garden contained, but why. He had to learn to read the garden for clues to the "state of botany at the time. . . . The dominant ideas of an age in the sciences are like the air we must breathe. They have a refreshing and invigorating influence not only on the intellectual lives of individuals, but also on all our institutions."[8] Kerner's point was that the garden was an archive for the study of both natural history and the history of natural knowledge.

This argument signals Kerner's intention to contribute to the writing of imperial history. The discipline of imperial or "whole-state" (*Gesamtstaat*) history had first emerged at Habsburg law faculties in the early nineteenth century, when its goal had been to piece together the legal basis of the dynasty's "historic rights" to the crown lands.[9] By the 1860s, however, this narrow vision had expanded. In the words of the historian and imperial adviser Joseph Chmel, the historian now had a "far more difficult, but all the more valued responsibility": to explain the Austrian Empire as "one of the most remarkable natural phenomena, as a practical solution to a daunting problem of nature"—that is, the problem of uniting "the most varied nationalities and education levels in one state."[10] This new imperial history would recover the development of the arts and sciences in the Habsburg lands since 1526; it would promote "the intellectual union of the Austrian lands."[11] Documenting the natural world region by region was central to this initiative. "Is it not part of a region's history to know how it has come to be and how it has gradually taken shape; to know on what kind of ground we are standing . . . ? . . . The oldest regional history is furnished by the work of the geologist, the physicist, the geographer; their research must teach us how Austria gradually took form."[12] Imperial history would have to include both a history of natural science in Austria and an account of the development of the natural environment itself.[13]

This chapter considers the consequences of this project for the production of climatological and related knowledge in nineteenth-century Austria. The long rule of the Habsburg dynasty in central Europe ensured that environmental knowledge accumulated in repositories of many kinds, including botanical gardens, libraries, mineral collections, herbariums, weather diaries, and map collections. In these institutions, nineteenth-century scholars found material for their own histories of nature, science, and empire.

THE BIRTH OF THE IMPERIAL IDEA

Anton Kerner looked back at the late sixteenth century and saw a crucial turning point in the history of natural knowledge. Not only did the wellborn begin to devote their leisure time to travel and collecting. They also began to take a scholarly interest in "domestic"—or what we might call "indigenous"—nature.[14] These trends were manifest in the reorientation of princely gardens from narrowly medical purposes to the collection of "wonders," whether native or exotic. This quest for nature's wonders was fueled by the ambitions of Habsburg princes.

In 1526 the Habsburgs met with the good fortune that would define their political predicament for the next four hundred years. Hungary had been defeated by the Ottomans at the Battle of Mohács. The death of the Hungarian king on the battlefield and a complicated set of marriage agreements left the house of Habsburg with both the Hungarian and Bohemian crowns. In this way, the Habsburg Monarchy became a strange new beast, "a complex composed of different and overlapping historico-political, ethnic, and later, also administrative units."[15] In what were known as the Austrian hereditary lands (including Upper Austria; Lower Austria; the duchies of Styria, Carinthia, and Carniola; the principalities of Istria, Gorizia and Trieste; and the more recently acquired county of Tyrol), the power of the dynasty was already established. Its authority was much looser in the remaining lands of the Holy Roman Empire, which the Habsburgs ruled continuously (with the exception of one five-year gap in the eighteenth century) from 1438 until its dissolution in the midst of the Napoleonic Wars in 1806. Thus the Habsburg lands resembled a Venn diagram of partially overlapping dominions: Hungary and Croatia lay outside the Holy Roman Empire, while many of the German principalities lay outside the Austrian hereditary lands.

The territorial windfall that the Habsburgs received in 1526–27 was a mixed blessing from the start. The Ottoman Empire was still in control of parts of Hungary and poised to expand further north and west. Concurrently, the Protestant Reformation was threatening to rip apart the Holy Roman Empire. Emperors Ferdinand I (king of Bohemia, Hungary, and Croatia from 1526 and Holy Roman Emperor from 1558 to 1564), Maximilian II (1564–76), and Rudolf II (1576–1612) all attempted to stave off a direct confrontation between Catholic and Protestant rulers, while presenting themselves as defenders of Christendom against the Ottomans. Not until the early seventeenth century would the Habsburgs abandon this policy of irenicism and embark on the forceful suppression of Protestantism.[16] The sixteenth-century Habsburg rulers styled themselves as heirs to a universalist legacy—that of the ancient Roman Empire and of the "Holy Roman Empire" founded by Charlemagne in the year 800.

NATURE AND EMPIRE

In Kerner's own day, in the face of burgeoning nationalist movements, the house of Habsburg was revisiting older ideas and symbols. The Renaissance iconography of Habsburg power got a new lease on life. Architects, for instance, crowned countless buildings with female figures meant to personify Austria, an allegorical motif that dated back to the late sixteenth-century reign of Rudolf II. Likewise, nineteenth-century sculptors designed fountains to illustrate the convergence of the Monarchy's four principal rivers, mimicking a famous water feature built for Rudolf's father, Maximilian II (see figure 4). The 1860s and 1870s even saw a renaissance of *tableaux vivants*, in which actors enacted allegories of imperial unity, just as princes and princesses had done at the Habsburg courts of the sixteenth century.[17] As we will see, allusions and artifacts like these were part of a modern visual and material culture that preserved an older way of thinking about nature and empire, in terms of an intimate relationship between microcosm and macrocosm.

It was in the sixteenth century that Habsburg rulers had begun to link the careful observation and representation of the natural world to their political ideal of universal harmony. Like many other European rulers of their day, the Habsburgs sought to demonstrate their power by displaying collections of rarities. From the late sixteenth century, in northern Europe and Italy, it became common for princes to house cabinets of curiosity or *Wunderkammer*. The category of "wonders" included rare specimens of animate and inanimate nature (*naturalia*), marvelous works of art (*artificialia*), and ingenious scientific instruments (*scientifica*). Many Renaissance natural philosophers believed

FIGURE 4. *Wien, Freyung mit Austriabrunnen*, by Rudolf von Alt, 1847. Von Alt's watercolor illustrates the newly erected fountain, featuring allegories of the four major rivers of the Austrian territory at that time, the Danube, Elbe, Po, and Vistula.

that objects in nature contained hidden, symbolic meanings. Things pointed beyond themselves to a web of interrelations, which might link a given species to seemingly unrelated objects and, ultimately, to the cosmos as a whole. Individual objects might therefore provide symbolic or perhaps even magical control of the world at large. Paracelsus, a physician and alchemist at the court of Maximilian I, used the skills of artisans to show how power over the world could be derived from the imitation of nature's own creative processes.[18] Thus *Wunderkammer* conveyed the message that power over nature constituted power over the human world.[19]

As Kerner noted, the sixteenth century also saw the rise of a new, empirical approach to natural knowledge. In fact, more recent historians have argued that collections of wonders played an essential role in stimulating the close observation of natural specimens. Accordingly, the meaning of scientific "experience" began to shift from an Aristotelian ideal of knowledge of the common course of nature, to a modern emphasis on knowledge of particular "facts."[20] This was a global phenomenon, in the sense that such collections could eventually be found in many parts of Europe, and they in turn depended on networks of exchange with Africa, Asia, and the New World. Within this global historical transition, the collections of the Habsburgs were of special significance. In the judgment of historian Bruce Moran, "it was especially at the courts of the imperial Hapsburgs that collections grew to unprecedented proportions."[21]

The Habsburgs sought to illustrate the universality of their rule by expanding their collections to a truly encyclopedic scope. Three were particularly noteworthy. The cabinet of Ferdinand II of Tyrol (1529–95) in Ambras Castle, outside of Innsbruck, was famous for its variety of natural objects.[22] The gardens and menageries of Ferdinand II's brother, Emperor Maximilian II, were renowned for their rare plants and animals, including medicinal plants from the Americas, tulips from Turkey, and an elephant that arrived in Vienna in 1552.[23] Even more remarkable than these two was the collection of Maximilian II's eldest son, Emperor Rudolf II, who moved his court to Prague.

Rudolf's collection included works of art, beautifully crafted scientific instruments, such as terrestrial and celestial globes, but also minerals, plants, and animals.[24] Thomas Kaufmann, the leading interpreter of Rudolf's collections, argues that the emperor prized his collections as a microcosm of the world he sought to rule. His castle in Prague contained a wing specially designed to house the collection, with an anteroom decorated with cosmic motifs: Jupiter, the four elements, the twelve months of the year.[25] Likewise, his gardens were designed with mathematical precision and according to classical architectural theory, such that garden and museum could "serve as a key to the understanding and study of the harmony of the creative universe."[26] Other princes may have harbored similar aspirations, but Rudolf turned the dream into a systematic enterprise, recruiting an entire team of naturalists to seek out natural wonders and learn their powers. What's more, historians consider Rudolf's collection to mark the origin of the institution of the research museum, where naturalists could study specimens in person. Unlike most *Wunderkammer*, these displays were not arranged to dazzle the viewer, but to invite patient attention.[27]

Like Rudolf II's *naturalia*, works of art on display at Habsburg courts often carried a political message. Paintings and sculptures represented the dynasty's universal dominion by evoking subtle interconnections between microcosm and macrocosm. For instance, a fountain designed by Wenzel Jamnitzer for Maximilian II and completed in 1578 was an ingenious incarnation of the Habsburgs' aspiration to govern a universal realm. Symbolized by an eagle, their empire soars over figures representing aspects of the natural world: the four elements, the rivers of the dynasty's lands (Rhine, Danube, Elbe, and Tiber), and the four seasons, all crowned by a celestial globe.[28] The eagle is thus the instrument of union among parts of the natural world existing at different scales, spatial and temporal. Or consider the famous portrait of Rudolf II as Vertumnus, the Roman god of the seasons, painted circa 1590 by the Italian artist Arcimboldo, who served three consecutive Habsburg emperors. Rudolf's face is composed of plants that grow in every season of the year, from spring flowers to fall gourds. In Kaufmann's reading of this painting, it is an "allegory of the imperial dominion over the macrocosm." Like a garden or a naturalia collection, the painting is an assemblage of "the world in small." "As the emperor rules over the body politic, so he may be seen to rule over the microcosm of his collection—which in turn mirrors the greater world, or macrocosm, that Arcimboldo's paintings also reflect."[29] One last example illustrates this motif particularly vividly. In 1571 a "tournament" was staged to celebrate the marriage of Archduke Karl to Maria, the daughter of the Duke of Bavaria. Choreographed by Arcimboldo and performed by members of Maximilian's court, it featured Maximilian himself in the role of "winter" and others as "personifications of the rivers of Europe, metals, planets, European countries, continents, seasons, elements, and the liberal arts. In short the tournament represented the macrocosm or greater world through the microcosm of man." As Kaufmann concludes, "the tournament obviously represented the cohesion of Europe under the Habsburgs, and in turn the Habsburgs' dominion over the world."[30]

It's worth pausing here to note that these examples illustrate a very different understanding of the relationship between part and whole than is familiar in the sciences today. The relationship is not statistical: the part does not instantiate or stand as an exception to the whole. Nor is the relationship causal in the modern sense of being linked by a chain of intermediate physical actions and reactions. Rather, these Mannerist creations presume that part and whole are aligned by the actions of hidden forces. On this view, individual human bodies and souls respond directly to the motions of heavenly bodies or the cycle of the seasons. This cosmological outlook was not uncommon in Europe at the time, but it seems that Rudolf II welcomed it like no other ruler. It also provided a framework for thinking about weather and climate in the sixteenth century.

METEOROLOGICAL THEORY AND OBSERVATION

The age of Maximilian II and Rudolf II was a time when learned men were contemplating competing cosmologies. Natural philosophers were questioning the Aristotelian doctrine of the celestial spheres and attempting to come to terms with Copernicus's heretical thesis. Theories of weather and climate were likewise in flux. In the Aristotelian tradition of natural philosophy that dominated the Renaissance universities, meteorology was the study of the *causes* of "meteors," a category that included storms, floods, earthquakes, and comets. These were explained in terms of the "exhalations" of the earth and atmosphere, which were understood to act on different forms of matter in different ways. This branch of knowledge was little concerned with observation in a modern sense, because learned men believed that causal explanation must derive from knowledge of the common course of nature, not from experience of particular instances. A second tradition of weather knowledge in the medieval tradition was astrometeorology, which flourished at Renaissance courts. Astrometeorology, unlike Aristotelian meteorology, was observational and predictive. It sought to forecast weather on the basis of the positions of celestial bodies. Here in all likelihood lies the motivation behind the observations of weather that were occasionally recorded in the margins of almanacs and ephemerides in the early Middle Ages. On the basis of such observations, predictions could be tested and discrepancies recorded between predicted and actual weather.[31]

At the Habsburg courts, learned men resolved these contradictions in eclectic ways. Tycho Brahe, court astronomer to Rudolf II, borrowed from Stoic cosmology and Paracelsan alchemy. In contrast to Aristotelian cosmology, which assumed that the medium pervading the heavens was separate from the terrestrial atmosphere, the Stoics conceived of this celestial substance as fluid and continuous with the earth. From this perspective, the influence of the planets on earthly weather could be understood as a direct action by this universal fluid. Brahe believed, moreover, that this same fluid served as the vital substance of living things.[32] Astrometeorology was therefore an essential aid to the practice of medicine.

Johannes Kepler, Brahe's successor at the Prague court, had been responsible for producing astrological calendars while provincial astronomer in Graz from 1594 to 1600. These calendars specified, for instance, when a physician should bleed a patient or even a particular part of the patient's body; they even dictated when a patient should bathe. To Kepler, astrology was fundamentally like medicine, in that both required generating prognoses based on causes that remained hidden. Kepler went so far as to suggest that the physician and the astrologer studied phenomena that were causally analogous. Just as a human soul at birth receives light rays from the planets that influence its future course, so the soul of the earth reacts to light from the heavens and responds with corresponding meteorological phenomena. For Kepler, as for Brahe before him, the astrological principle of a resonance between microcosm and macrocosm validated attempts to apply knowledge of human-scale forces to theories of the cosmos.[33]

It was undoubtedly to test the predictions of astrometeorology that Brahe and Kepler embarked on some of the earliest systematic weather records.[34] In Europe, systematic weather observations have survived from as early as the late fifteenth century, recorded by the astronomers Peurbach and Regiomantanus, both of whom served the Habsburg emperor Frederick III.[35] According to historical climatologists, the practice of keeping a weather diary spread from the University of Kraków through much of central and northern Europe in the late fifteenth century.[36] This habit became increasingly common in the humanist circle around Frederick's son Maximilian I, often in connection with astrometeorology.[37]

Other practical motivations for weather observing came from medicine, botany, and agriculture. For instance, the first long-term systematic weather records in the Austrian hereditary lands were launched in the 1540s by the physician Johann Aichholz, in whose medical garden Clusius first encountered the auricula.[38] Aichholz was then at the start of a stellar career. In the plague epidemic of 1558, he was appointed "Magister sanitatis" and two years later sent to Hungary to tend to an ailing Count Nádasdy. Although he had converted to Protestantism, the emperor allowed him to teach as the youngest member of the medical faculty at the University of Vienna, where he was elected dean five times and rector in 1574. In 1581, when the emperor fell sick, Aichholz was called to Prague to treat him. Alone among the weather diarists of his circle, he showed no sign of interest in astrometeorology, and may instead have been trying to track conditions favorable to gardening.

Aichholz and his contemporaries may also have been motivated by their experiences of climatic variability. The late sixteenth century marked a period of cooling in Europe that has been linked to harvest failures and food shortages, and which seems to have played a significant role in the political upheavals of the first half of the seventeenth century.[39] Thus, what is known today of the European climate circa 1600 comes in part from weather diaries kept by scholars patronized by the Habsburgs.[40]

MICROCOSM AND MACROCOSM IN THE NINETEENTH CENTURY

The multiple cosmologies of the sixteenth century did not die a sudden death, neither with the advent of Newtonian science in the seventeenth century nor with the battle of the *Aufklärer* against "superstition" in the eighteenth. How individual minds and bodies might respond to the vicissitudes of earthly weather—and to the more distant oscillations of celestial bodies—remained a live question in the nineteenth century.

Kepler, for instance, was celebrated by nineteenth-century Habsburg scientists as a model of free inquiry. Despite his Protestantism, his statue adorned a landing of the stairwell leading up to the observatory of Kremsmünster Abbey, where Julius Hann trained as a gymnasium student.[41] Thinkers as different as the Vienna physicist Andreas Baumgartner and the Prague aesthetician Josef Durdík acknowledged the central role of Kepler's cosmic mysticism in his elucidation of the planetary orbits.[42] Perhaps the most insightful commentary on Kepler came from the Moravian-born astronomer Norbert Herz, director of a private observatory in Vienna. Herz drew attention to the causal model that underlay Kepler's astrometeorology, his theory of the earth-soul or "sensitive earth." Are we justified, Herz asked, in viewing this as an error on Kepler's part? In answer, Herz pointed to the ignorance of the science of his own day, at a time when Sigmund Freud was still employing hypnosis to treat his neurotic patients in the Berggasse. Herz observed that "alongside certain impulses that influence and move our mind, our entire being, for which, in the course of time . . . we believe to have found an apparently sufficient explanation," there were others that had not yet "reached that stage of knowledge" and which were still designated as "marvelous," "mysterious," "mystical," or "still unexplained."[43] While the psychiatry of his day tracked the "mystery" of human emotions inward, to innate drives, Herz instead followed it outward, with Kepler, to the "effect of inorganic nature on the organic being."[44] The foundation of Kepler's astrology lay in the "mysterious power" of the natural environment, or what Kepler called the "affinity between heaven and earth." Whether or not Kepler's astrology was a form of superstition was not for the modern-day scientist to judge. Herz pointed out that the meteorology of his own day had "likewise postulated certain connections between terrestrial and cosmic phenomena," by which he meant the correlation of climate with sunspots. "Belief! Superstition! Where is the line that divides them?"

Herz was not alone in urging research into the sensitivity of the human psyche to geophysical conditions.[45] One of the first writers to use the term "unconscious" (*das Unbewußte*) in the modern sense of "inaccessible to consciousness" (rather than "innocent" or "artless") was Romuald Lang, an instructor of geography and history at the gymnasium of the Kremsmünster Abbey, where Julius Hann was among his students. In "The Unconscious in Man" (1858), Lang insisted on "the great influence of climate, the atmosphere, nutrition, and so forth on the constitution and health of the body." He sought to understand this influence in terms of an alternative cosmology, one that acknowledged the interdependence of human and nonhuman nature. "For the human microcosm is an organic part of the macrocosm. . . . The history of an individual finds its foundation and its explanation only in relation to the cosmos as a whole." It was not long after taking Lang's classes that the young Julius Hann began to explore his own susceptibility to the natural environment, as we will see in chapter 11.

Astrometeorology survived as well in popular weather almanacs, which offered forecasts for a year or even an entire century, based on the positions of the moon and planets. These continued to sell well in the nineteenth century throughout the Habsburg lands, as in much of Europe and North America.[46] In fact, in Habsburg languages other than German, this latter-day astrometeorology made up the bulk of the meteorological literature. At the start of the nineteenth century, approximately eight annual weather almanacs were published in Czech per year; by the 1850s, this number had risen to twelve to fifteen. In addition, Czech readers could choose from a whole series of hundred-year calendars.[47] Bohuslav Hrudička, an early twentieth-century meteorologist in Brno, found these nineteenth-century publications to be remarkably similar to the earliest printed almanacs in Czech from the fifteenth century, down to the repetition of some of the same mnemonic verses. At the same time, newer versions of astrometeorology laid claim to scientific authority, such as the disaster prophecies of the Styrian ex-theologian Rudolf Falb, which were based on the tidal influence of the moon's phases.[48]

Modern astrometeorology also took more reputable forms. Recognition of the periodicity of sunspots in the mid-nineteenth century inspired a wave of studies of correlations between these solar cycles and the fluctuations of the terrestrial climate. Historians have suggested that this line of research was particularly appealing to British scientists, who sought methods of long-term weather prediction after famines wreaked havoc in India in the last quarter of the nineteenth century.[49] Other theories promised to explain and predict longer term variations in the terrestrial climate. In Britain, James Croll proposed an astronomical theory for the Ice Ages, arguing that variations in the earth's orbit around the sun could cause significant shifts in the earth's climate. Croll's language invited comparisons to astrometeorology, as he claimed that the "true cosmical cause" of climate change "must be sought for in relations of our earth to the sun," and that "geological and cosmical phenomena are physically related by a bond of causation."[50] In the early twentieth century, Croll's theory provided inspiration to Milutin Milanković, a native of Habsburg Croatia and a graduate of Vienna's Technical University. Milanković's calculations of the variation in solar radiation due to shifts in the earth's orbit finally turned "cosmic causation" into a serious, if radical, climatological hypothesis—one that has since become a standard element of long-term climate models.[51]

What can these echoes of Renaissance cosmology within modern climatology tell us about scaling as a historical process? They suggest that the global standardization of measurement in the late nineteenth and early twentieth centuries did not necessarily displace other ways of tracking and imagining space and time. Older notions of proximity, simultaneity, and relatedness persisted, notions that had nothing to do with industrial efficiency. Vanessa Ogle has shown, for instance, that the imposition of Greenwich mean time after 1884 sparked resistance that briefly served to strengthen alternative traditions of time keeping and time management.[52] That is to say that processes of scaling need not result in the occlusion of one measurement framework by another. Commensuration is not always homogenization.

NATIVE EXOTICA

The quest to gather nature's secrets at the early modern Habsburg court also generated vital records of ecological change. As R. J. W. Evans noted in his seminal history of the early modern Habsburg Empire, the "precise fidelity to animate and inanimate objects" that was characteristic of sixteenth-century art also extended to the study of local provinces.[53] The passion that Maximilian II and Rudolf II displayed for the "exotica" of the New World, Africa, and Asia has captured the most attention from historians, but the emperors also had an enormous appetite for "wonders" to be found closer at hand.

Aichholz, for instance, "grew many native [*einheimische*] plants" in his Vienna garden, as his nineteenth-century biographer would later note.[54] The humanist Martin Mylius and the theologian Caspar Schwenckfeld each published on the flora of their native Silesia. The painter Georg Hoefnagel, who traveled throughout Europe, made exquisite drawings for Rudolf of many species of flora and fauna, both foreign and native, along with sketches of provincial towns and local dress. The landscape artist Roelant Savery was inspired by Rudolf's fondness for the Tyrol to study its wildlife and included many Tyrolean birds and other animals in his paintings.[55] Arcimboldo, too, drew common species, such as sheep, cats, deer, hares, and sparrows. What the historian Paula Fichtner says of Clusius holds for many of the humanists active at the Habsburg courts in the sixteenth century: one cannot read his "descriptions of plants and stones from all the corners of the Habsburg imperium without appreciating how much his work added to understanding of the natural environment of east-central Europe."[56]

Part of the motivation for such work was to overturn the demeaning image of the German lands that had been propagated in classical literature.[57] Tacitus and other classical authorities had portrayed Germany as a backward, barbaric land. By contrast, sixteenth-century central European humanists emphasized the fertility and natural riches of their home region—a quality that made their accounts of special interest to nineteenth-century provincial historians.

Humanists in the patronage of Maximilian and Rudolf pursued nature's wonders up mountains, along rivers, and even underground. In search of alpine plants, Clusius and the astronomer Johannes Fabricius scaled the Ötscher, an 1,893-meter peak near the border between Lower and Upper Austria. Leonhard Thurneysser studied the waters of the Danube and its tributaries and discussed their properties in his *Ten Books on Mineral and Metallic Waters*, while his *Magna Alchemia* treated mining in Bohemia and Hungary. Šimon Tadeáš Budek, who called himself "His Majesty's prospector for treasures, metals, precious stones, and all hidden secrets in the whole of nature," collected both geognostic information and folk legends in northwestern Bohemia.[58] Anselm Boethius de Boodt, imperial physician under Rudolf II, traveled through Bohemia and Moravia to gather information for his 1609 treatise on gems; along the way, he collected plants to send to Clusius.[59] The information these men recorded laid the foundations for the natural histories of the Habsburg lands written in the nineteenth century.

The lands newly conquered from the Ottomans proved particularly rich in "exotica" for Habsburg collectors. Stolen in battle, cultural objects from the Ottoman Empire held an important place in Habsburg collections from the early sixteenth century.[60] The same seems to have been true of *naturalia* from the Ottoman lands. The naturalist Ogier Ghiselin de Busbecq returned from Constantinople to Vienna in 1562 "with the tulip and other plants that had been uncultivated in Europe outside Turkey."[61] Clusius too traveled to Hungary, where he encountered botanical "rarities" that included many types of mushroom, a particular passion of his. Most remarkably, he recorded the names of 346 plants in both Latin and Hungarian, the first compilation of Hungarian plant names.[62] Many of these species were included in Clusius's 1601 *History of Rare Plants*, which treated plants he had found in Spain, Austria, and Hungary.

In these ways, rarities came from near and far. As Kerner von Marilaun recognized, the auricula and the tulip were equally exotic wonders in the eyes of the sixteenth-century Viennese, although one species originated within the borders of the empire and one without. The line between "native" and "exotic" was blurred from the start, in ways that set a course for the subsequent history of science in central Europe.[63]

FROM PRINCELY COLLECTIONS TO PROVINCIAL MUSEUMS

What became of Habsburg collections of natural knowledge in the nineteenth century? Many were opened to the public, thanks to the wave of provincial museums founded in the wake of Napoleon's defeat. In the course of the first half of the nineteenth century, patriotic museums sprang up in towns across the Monarchy: Pest, L'viv, Graz, Brno, Opava, Prague, Ljubljana, Innsbruck, Salzburg, and Sankt Anton. In stark contrast to the collections of Rudolf II and his contemporaries, which were accessible only to an elite few, the new provincial museums of the post-Napoleonic era embraced the goal of educating the public. While they invited experts to use the collections for their research, they also encouraged the public to come and appreciate the variety of natural resources to be found in their home province.[64] What distinguished these from other museums, founded elsewhere in Europe at the time, was the multiplicity of forms of knowledge they sought to preserve. Anything and everything was of interest that related to the local province, its human and nonhuman inhabitants, and its physical environment. While the new Imperial-Royal Natural History Museum in Vienna set its sights on rarities from distant lands, the provincial museums made space for far less exotic specimens. These museums constituted *Wunderkammer*, library, and archive under one roof. Indeed, had the dynasty ever achieved its plan to centralize its archival collections, no doubt much local environmental information would have been disposed of or lost.[65]

One of the earliest models for a provincial museum was the Joanneum in Graz.[66] Founded in 1811, it was named for Archduke Johann (1782–1859), a pivotal figure for the history of the geosciences in Austria. He took a passionate interest in the folk culture and landscapes of the valleys of the eastern Alps. Fittingly, he commissioned a portrait of himself in hunting gear in the hills of his beloved Styria (see figure 5). This was no mere hobby. Rather, it reflected his ambition of crafting a new, more modern ideology of Habsburg rule: namely, to shift the object of patriotic attachment from the dynasty to the lands of their dominion. What he envisioned was a territorialization of Habsburg authority, yet in a different spirit than that of the enlightened absolutists of the eighteenth century. Johann sought to bind Austrian patriotism to an appreciation of the uniqueness of local landscapes. The museum he founded in Graz gave pride of

FIGURE 5. Archduke Johann (1782–1859).

place to mineralogical and zoological specimens, such as stuffed birds replete with eggs and nests. It served as the model for other provincial museums that opened across the Habsburg lands in the following decades.

The displays and publications produced by these museums helped to crystallize distinct regional as opposed to national identities. In Bohemia and Moravia, for instance, the provincial museums published research reports in both German and Czech. This was a way to support the linguistic demands of Czech nationalists, while also aligning the museum with a regional—and thus nonnational—identity. At the time, such regional identities were understood as parts of a greater whole. The Joanneum, for instance, highlighted its debt to the house of Habsburg in its very name. As Werner Telesko argues, these museums did not aim for "a provincial and autonomous self-conception, but rather presented the idea of a *supranational space and identity*, which also manifested itself in their collections." They emphasized both "regional patriotism" and "trans-regional inter-connections," a duality resolved by the identification of the regions themselves with the supranational state and dynasty.[67]

This system of provincial museums preserved records of local natural history and made them available to nineteenth-century researchers. In the wake of the Napoleonic Wars, Habsburg scholars mined these resources to assemble both patriotic regional histories and inventories of local natural resources. For instance, Kaspar von Sternberg, the founder of the Bohemian National Museum in Prague, composed his *Outline of a History of Bohemian Mines* (1836) and its accompanying map on the basis of extensive research in regional archives. Much as Kerner would argue in the case of botany, Sternberg believed that the effective management of mining depended on historical knowledge. This was true "not simply in order to orient oneself in the field, but also in order to get to know the reasons why mining was led into decline, in order to identify and avoid these." Moreover, Sternberg saw the history of mining as an inextricable part of the history of Bohemia. His historical research relied so heavily on esoteric archival documents that he found it impossible to knit it together into a narrative, for which he apologized. "Indeed there could not be many researchers who would have the desire and patience to undertake such a painstaking effort in the dust of old, rarely visited archives and go to such trouble to procure the oldest certificates and reports."[68]

REINVENTING THE PAST

As nineteenth-century authors like Sternberg constructed their provincial natural histories, they rediscovered, reprinted, and reinterpreted sixteenth-century accounts of the natural world. So, for instance, the 1877 *Topographie von Niederösterreich* traced attempts to construct a mathematically precise map of the region back to Fabricius. It memorialized Fabricius's ascent of the Ötscher in 1574, in the company of Clusius and Aichholz, in order to determine the position and height of the peak using astronomical instruments.[69] Eduard Suess's pioneering contributions to historical seismology made use of accounts of the earthquake of 1590 that he found in the Lower Austrian provincial archive. In Czech-speaking circles, scientists turned back to sixteenth-century almanacs as the starting point for histories of astronomy and meteorology.[70] In some of these cases, sixteenth-century sources proved genuinely useful to nineteenth-century science. In other cases, modern scientists cited these forerunners in order to establish their work as part of a centuries-old Habsburg tradition.

Seen through a nineteenth-century lens, Brahe, Kepler, and Clusius appeared as pioneers of a patriotic tradition of *Landeskunde*, the study of the region as a natural, social, and cultural unit. We find Clusius, for instance, cast as a practitioner of *Landeskunde* despite his cosmopolitan life. Analogously, Tycho Brahe and Johannes Kepler were embraced as part of a Czech intellectual tradition. Both were cited in Czech publications as pioneers of meteorology and astronomy in Bohemia. Likewise, the maps drawn by Fabricius and Jan Amos Komenský (Comenius) in the late sixteenth and early seventeenth centuries were reprinted as foundations of a Moravian *vlastivěda*, or regional science.[71] In this way, descriptions of nature's rarities that had shown little regard for their distribution in space were reinvented as early forms of provincial patriotism.

It was for the sake of *Landeskunde*, for instance, that a local archivist in Bohemia published a trove of Renaissance weather diaries in the 1880s and 1890s. Wenzel Katzerowsky was a gymnasium instructor and the archivist for the city of Litoměřice/Leitmeritz in Bohemia. When he was a child, his parents' house had been struck by a small meteor, which crashed right through the roof. This was the start of his lifelong fascination with meteorology, in the broad Aristotelian sense. Throughout his career as an archivist, Katzerowsky worked assiduously to reconstruct the climatic history of the area around Litoměřice/Leitmeritz, from the earliest records to his own instrumental observations. He combed local archives and even private collections in search of weather diaries, which he resourcefully combined with all manner of municipal records. Since the observations came from different sources, Katzerowsky helped to produce a continuous record by including proxy information for the later period, such as extreme weather events, harvest dates, and crop prices. In this way, he managed to publish a more or less continuous record of observations in Litoměřice/Leitmeritz covering the period 1454 to 1892.[72] These records have since been used repeatedly by historical climatologists, and the director of the local museum in Litoměřice today carries on this research.[73]

Sixteenth-century naturalists also figured prominently in the new *imperial* histories of the nineteenth century. In these cases, however, they appeared as "Austrian" naturalists in the new whole-state sense. Thus Clusius was portrayed by the director of the imperial-royal herbarium as an expert on "Austrian" plants and plant names. He was "the first scholar of the flora of Austria," and the first to report "the occurrence of more than a hundred varieties of mushroom in *our empire*." It was only fitting that every subsequent scholar of Austria's flora should "linger affectionately" on his work.[74] Similarly, the geologist Franz von Hauer, later the director of the imperial Museum of Natural History, identified Renaissance scholars such as Georg Agricola and Rudolf II's physician Anselm de Boodt as "the first founders of geological knowledge of Austria."[75]

CONCLUSION

When von Hauer addressed the Academy of Sciences in Vienna at a ceremonial meeting in 1861, he presented a historical narrative of the rise of geology that argued for the patriotic orientation of this science since the time of Agricola. Saluting the military officers in his audience two years after Austria's defeat at Solferino, he described the historical role of the imperial scientist as analogous to theirs. "Intellectual progress" was the surest way to "replace the bonds that have weakened in this state" and to "establish a more secure base for the great power of our empire."[76] Indeed, nineteenth-century scientists stressed their debt to dynastic institutions and traditions of research dating back to the sixteenth century. These include the princely collections that were to become the provincial museums of the nineteenth century, with their rich stores of *naturalia* and historical documentation. These also include visual techniques developed by naturalists in the service of the Habsburg state for depicting precise detail in relation to a systematic overview. In the sixteenth century, answers to this challenge included Mannerist paintings and sculpture, books of flora and fauna, the design of botanical gardens, the organization of the *Kunstkammer*, and the staging of spectacles in which human players embodied the harmonious union of the parts of the empire and of the cosmos. After 1848, genres like these were revived and others invented to illustrate the theme of unity in diversity, now with a new urgency in the face of the challenge that modern nationalism posed to the unity of the Habsburg lands.

CHAPTER 2

The Austrian Idea

The late Habsburg Monarchy has been said to have suffered from an ideological deficit. The dynasty had traditionally justified its rule as divinely ordained, portraying their lands as the fulfillment of the medieval dream of universal empire, as the "bulwark of Christendom" against the "infidel" Ottomans, and as a mediator between "West" and "East." Those arguments faltered with the waning of the Ottoman threat to Europe in the eighteenth century. Thereafter, the Habsburgs styled themselves as patrons of the arts and sciences, as incubators of enlightenment, and, most ambitiously, as the protectors of the panoply of cultures under their rule. This idea of "unity in diversity" has looked so weak in retrospect, viewed through the scrim of twentieth-century nationalism, that it has hardly seemed to merit the attention of intellectual historians. As A. J. P. Taylor put it in his classic history, "Francis Joseph was an Emperor without ideas; this was his strength and enabled him to survive. Yet, by the end of the nineteenth century, ideas made a state and kept it going. . . . An 'Austrian idea' had to be found. The phrase was everywhere; the translation into practice never took place."[1]

It is ironic that Taylor, defender of the "small nations," was echoing a position common among nineteenth-century *German* nationalists. To them, too, the Habsburg dynasty seemed to have been depleted of a legitimating ideology. The anti-Semitic Prussian conservative Paul de Lagarde, who hoped to see all of central Europe unified under Prussian domination, wrote in 1853 that "Austria has at present no idea holding it together."[2] In the absence of such a unifying "idea," Lagarde warned that Austria was destined to "descend into materialism." Austria, in Lagarde's view, was the massive animal body to which Prussia would supply the rational mind.

But perhaps Taylor, like Lagarde, was blinded by a narrow preconception of what an "idea" is and where to find one. Taylor had been looking for glimmerings of a new idea in high politics, in the transnational ideologies of political parties like the Christian Socials or Social Democrats. More generally, historians have identified nineteenth-century justifications for Austria-Hungary's existence associated with four domains: military defense, politics, culture, and economics. Let us briefly survey each in turn.

From the perspective of much of western Europe in the nineteenth century, Austria remained a military necessity. Even after the decline of Ottoman power, it formed a defense against Russia. As the Habsburg-loyal Hungarian statesman Count Gyula (Julius) Andrássy argued in 1897, Austria's expansion in the eighteenth and nineteenth centuries had made Hungary even more vital to its security. He likened Galicia, Bohemia, and Tyrol to *Vorwerke*, in the medieval sense of fortified castles outside the walled city, while Hungary was the *Festungskern*, the heart of the monarchy's fortifications. "Austria consists of provinces that protrude and extend into foreign territories and are only barely connected to each other," and an enemy could "easily cut them off from each other." Hungary, he argued, made "Austria's defense system into an enclosed whole."[3] This military justification for Austria-Hungary's existence foundered, however, on the question of the organization of its military force.

The Austro-Hungarian Compromise of 1867 had divided the empire into two halves, ruled independently of each other, sharing only financial, foreign, and military policy. The Habsburg Common Army was the only institution that served both halves. Because the army introduced universal conscription at this time, it also became the only institution with a potentially formative influence on most male citizens of the empire.[4] However, conscripted soldiers had the right to receive training in their native language, in keeping with the principle of the formal equality of the empire's official languages. This meant that the army was obliged to form separate regiments for up to twelve Habsburg languages.[5] In the eyes of many critics at home and abroad, this multilingual military seemed anachronistic in the age of what the Napoleonic general Antoine-Henri Jomini called "national war." These critics suspected that the efficacy of universal conscription in countries like Prussia and France hinged on the motivating force of nationalism.

A second category of argument for the supranational empire viewed it as a means to political ends—that is, as a vehicle for defending the civic rights of its citizens, or more often, the rights of one particular class. Consider the position of Austro-Slavism, most closely associated with the Bohemian historian František Palacký. In 1848, Palacký declined an invitation to participate in the Frankfurt Parliament's attempt to create a Pan-German state. He gave two reasons: first, that he identified as a Czech, not a German; and second, that he believed Austria's survival to be a political necessity. Had Austria not existed, he wrote, it would have been necessary to invent it "in the interest of Europe, in the interest of humanity."[6] Palacký argued in his 1866 *The Idea of the Austrian State* that Austria's modern purpose was to defend the political equality of the many different national groups within its territory. In casting the Monarchy's purpose as political in nature, Palacký's justification bears comparison to two later ideologies, both with strong electoral power in Austria. Austro-Marxism envisioned the Monarchy as the basis for a future democratic workers' federation. By contrast, Christian Socialism aligned the Habsburg state with the mission of defeating both socialism and liberalism. Instead, it posited the supranational state as the protector of Catholic craftsmen and small tradesmen. What Austro-Slavism, Austro-Marxism, and Christian Socialism shared was the premise that Habsburg unity served a political end.

A third category of argument for Austria-Hungary's survival was cultural in character. This view was related to Palacký's Austro-Slavism, in so far as it presumed the intrinsic value of national diversity and Austria's power to protect it. This was the claim advanced by the monumental twenty-four-volume series spearheaded by Crown Prince Rudolf in the 1880s, The Austro-Hungarian Monarchy in Word and Image, which celebrated in turn each culture that made its home within the Habsburgs' lands. At times, cultural arguments in defense of Austria-Hungary took on spiritual overtones. The idea of Austria as a state built on and promoting Christian values, for instance, by no means withered away with the defeat of the Ottomans. After the outbreak of World War One, it was revived in a charismatic new form by the poet Hugo von Hofmannsthal, commissioned by the state to produce patriotic prose. Hofmannsthal reversed the relationship between Austria and Prussia as Lagarde had described it: now Prussia stood for the bodily virtues of efficiency and obedience, while Austria represented the ideals of piety and humanitarianism. On this view, Austria was the protector of the tradition of "spiritual universalism" that represented Europe's best hope for peace.[7]

Finally, a fourth argument for Austria-Hungary's existence was economic. It built on a tradition of northern European thought known as cameralism. Cameralism combined a theory of nature with a prescription for governance. On this view, nature was a divinely designed storehouse for the satisfaction of human needs. All wants could be satisfied through a proper understanding of nature's own "household." The goal of the cameralist state was to achieve a favorable balance of trade either by acclimating foreign crops to native soil, or by substituting native products for imports, such as beets for sugarcane. It was a program with special appeal to states like Austria with no overseas colonies. In this way, cameralism inspired a range of subsequent programs for economic development on the basis of the detailed study and expert husbanding of native natural resources.[8]

A modern economic justification for Habsburg unity took shape in the 1830s and 1840s, as British goods were flooding the markets of central Europe. The German Customs Union was formed in 1833 in order to protect central European agriculture and industry from British competition, but without the participation of Austria. In arguing for the expansion of this central European free-trade zone, Friedrich List cast it as a stage in the formation of a linguistically and culturally unified German state. In this respect, he echoed Fichte, who had argued in 1800 that in a "closed commercial state"—that is, one in which tariffs favored internal over external trade—"a higher degree of national honor and a sharply determined national character will develop very quickly."[9] That is, a commercially unified state would be a nationally *homogeneous* state.

Among Austrian industrialists, shippers, and merchants, there was considerable support for joining the *Zollverein*. But their vision of central European economic unity was coupled to the principle of national pluralism. In 1849, Vienna's new minister of trade, Carl Ludwig von Bruck, envisioned a central European commercial zone under Vienna's authority. Yet he insisted that this would remain a *multinational* state, and he explicitly denounced strategies of Germanization in southeastern Europe.[10] A former director of the Austrian Lloyd shipping concern in Trieste, von Bruck promoted dismantling internal barriers to trade and improving transport networks. According to this argument, the economic union of central Europe would stimulate agricultural and industrial development, and by that means cement the political unity of the Habsburg lands.

These four nineteenth-century versions of the Austrian Idea have not escaped the notice of historians. Rather, historians have judged them naive at best in the face of the geopolitical transformations leading up to World War One. At worst, these arguments have been seen as apologetics for German-Austrian and Hungarian oppression of the empire's minorities.

This may be as far as the history of ideas can take us. But what if we pose a slightly different question than the one formulated by Lagarde and Taylor? Instead of asking what idea this state rested on, what if we ask what resources Habsburg subjects possessed for thinking the idea of Austria. That is, what concepts, practices, representations, and material tools could they draw on to begin to situate themselves and their concerns in relation to this territory as a

FIGURE 6. *Allgemeine Charte der Österreichischen Monarchie zur Übersicht der neuesten Begraenzung und Eintheilung, mit den Nebenlaendern des österreichischen Hauses in Deutschland (General Map of the Austrian Monarchy Showing the Most Recent Borders and Divisions, Including the Neighboring Lands of the Austrian Dynasty in Germany)*, by K. J. Kipferling, 1803.

whole?[11] To reformulate the question in this way is to move beyond a history of ideas, toward a history of scaling.

Figure 6 shows the borders of the Habsburg lands as they stood in 1803, following the partitions of Poland and Austria's losses in the second coalition war against France. This image is noteworthy as one of the first maps ever to be published within the Habsburg lands depicting the entirety of their modern territories. As its German title emphasizes, it is an "overview." Until the last two decades of the eighteenth century, despite a thriving cartographic tradition, maps produced within the Habsburg lands had focused on local and regional views. The Habsburgs had relied on Dutch and then French cartographers for overviews of their territories. Accurate and complete military maps date back to the mid-eighteenth century, but they were state secrets, existing only in manuscript. Not until the middle of the nineteenth century did such maps circulate widely.

Following this cartographic boom, new textbooks, atlases, and even games trained young minds to appreciate the Monarchy as a territorial unit. Adalbert Stifter, for instance, described a geography lesson that employed maps sized in proportion to the area of the country represented. In this way, he affirmed, "an image of scalar relations is permanently imprinted on the active imagination of youth."[12] For lessons in Habsburg geography, educators could purchase a game that came with a jigsaw map and longitudinal and latitudinal measuring rods. Before play could begin, the instructions prescribed a study period of three months. Then children would be ready to invent journeys across the Habsburg lands, imagining their routes and calculating the distances traversed.[13] Maps like these formed a new set of tools for rethinking the idea of Austria, the relations among its parts, and its place in the world.

IMPERIAL SPACE

What nineteenth-century justifications of Austria-Hungary shared was a concern with the empire as a spatial unit. The economic argument associated with Minister von Bruck, for instance, was at heart about geography. Von Bruck stressed that trade would be enlivened by the "geographical relations" of the Habsburg lands: "The geographical relations between the Austrian crownlands and the Kingdom of Lombardy-Venice are such that all products that are imported and duty paid in one part of the Monarchy will, with very few exceptions, be consumed in the other part."[14] Analogously, arguments for the military necessity of Habsburg unity took a more broadly geographic perspective from the 1850s, incorporating new evidence about the medical consequences of travel, as we will see below. Likewise, cultural justifications of the multinational state were increasingly cast as maps or surveys, charting the distribution in space of factors like language use or architectural artifacts.[15]

What was emerging, then, was not a static "Idea," but a dynamic program for the empirical investigation of the Monarchy's resources and "geographical relations." To this end von Bruck quickly brought Vienna's Administrative Statistics Office under his direction at the ministry of trade, and he recruited men of science like the physicist Andreas von Baumgartner and the statistician Karl von Czoernig to work for it. In subsequent chapters, we will see how the ZAMG took up this agenda upon its founding in 1851, alongside the Imperial Geological Institute and other scientific bodies of the neo-absolutist era. The geosciences developed in imperial Austria in the second half of the nineteenth century as a means of fleshing out a new Austrian Idea by studying the concrete conditions of (economic) unity in (environmental) diversity.

This was a project with significant consequences for the development of the sciences. Gregor Mendel, for instance, the neglected author of the first mathematical law of biological inheritance, by no means confined his observations to his monastery garden. As Sander Gliboff has shown, Mendel's determination to quantify the organic world developed through his engagement with efforts to measure and map the natural conditions of plant and animal life throughout the empire—to systematize, in other words, Austria's celebrated diversity of flora and fauna.[16] "Far richer in geographical contrasts than most of the other European states," wrote Austria's premier geographer Albrecht Penck in 1906, "[Austria] beckons above all for the study of its own land. . . . The wealth of contrasts of all kinds makes Austria an observing field for geographers which is virtually unmatched in Europe."[17] The university in Vienna was supremely situated for such studies, sitting as it did at the confluence of hills and plains and in easy reach of the Alps. This feast of new phenomena nourished geography's natural-scientific turn in the late nineteenth century. "Without doubt there lies in the diversity of its land and people a good portion of Austria's charm," wrote the Vienna geographer Norbert Krebs. "And the way in which various natural and social phenomena interact strongly excites the interest of the educated person."[18]

Through these persistent reminders of Austria's natural as well as cultural diversity, Vienna's field scientists underlined the empire's status as a laboratory. That the Austro-Hungarian Dual Monarchy was an "experiment" has been a refrain equally common among the empire's inhabitants and its historians, its admirers and its critics. In 1897, for instance, the Austro-Marxist leader Victor Adler called the empire the "laboratory of world history"; the dissident Hungarian historian Oszkár Jászi used the same concept to describe the moribund empire in retrospect. The antiliberal satirist Karl Kraus agreed that the Habsburg capital Vienna was a laboratory, namely, an "experimental station for the end of the world." Even what arose in the empire's wake was hailed by the Czech leader Tomáš Masaryk as "a laboratory atop a graveyard."[19] The history of the field sciences in imperial Austria helps us appreciate the depth of this metaphor. This democratizing multinational empire was an experiment not only in a political sense. It was also a laboratory for the investigation of the nexus between nature and society, for probing the putative phenomenon of empire itself: a unity grounded in diversity, both human and environmental.

MONUMENTS OF NATURE AND CULTURE

In this respect, the natural sciences were developing in parallel to the human sciences. In linguistics, ethnography, architecture, and art history, and beyond, researchers were likewise setting out to document the Monarchy's cultural multiplicity.[20] Commonalities among these projects were overdetermined by their institutional histories, shared personnel, and mutual methodological influence. Von Czoernig, for instance, the director of the Vienna Administrative Statistical Office, oversaw projects on ethnography, art history, and climatology. Von Bruck also directed the commission charged with preserving Austria's art historical monuments, which later took responsibility for "natural monuments" as well. These projects shared a common grounding in the "positive" methods of natural history. Art historians, for instance, now insisted on the direct observation of original works of art, not copies. And they valued these objects not merely as things of beauty but as clues to an evolutionary process.[21]

More importantly, these projects shared a politics. They rested, first, on the principle that no cultural or linguistic tradition was too minor to merit scholarly attention. As Crown Prince Rudolf insisted, "none of the crown lands of Austria-Hungary should be regarded as unworthy of a loving, detailed depiction."[22] What the Czech statesman Masaryk called "small, detailed, mundane work" was thus as much a principle of scientific practice in the Habsburg lands as it was a tenet of the political philosophy of nationalism.[23]

What's more, these imperial surveys all foregrounded phenomena of mixing and exchange. In ethnography, for instance, the seminal three-volume survey by Karl von Czoernig charted a history of migrations across the Habsburg lands in order to demonstrate that the ethnic diversity of the Monarchy reached down to such minute dimensions that no division of the territory along national lines was conceivable.[24] Simultaneously, the Vienna School of Art History challenged the naïveté of many Romantic-nationalist celebrations of folk art by uncovering the historical hybridity of folk traditions.[25] Against nationalists who urged a return to "authentic" folk art, Alois Riegl insisted that folk art tended to be imitative of wider, cosmopolitan trends. The preservationist efforts of the midcentury Vienna School of Art History focused on the cosmopolitan dimensions of "Austrian" art, above all works of the Renaissance and Baroque eras. Thus their conclusions were prescriptive as well as descriptive: as von Eitelberger argued in 1870, "the progress of contemporary civilization rests precisely on the exchange of ideas among the most varied peoples."[26] Riegl later stressed the same point: "On the occasion when the unfamiliar meets the unfamiliar in a close and sustained relationship, the process of development is set in motion."[27] In this way, in an era when nationalism was reconceiving historical research as a quest for authenticity and indigeneity, a largely forgotten project emerged across the human and natural sciences to focus attention instead on the complexity of cultural flows.[28]

From the start, this research was intended for immediate public consumption. Indeed, as the art historian Rudolf von Eitelberger made clear in the 1850s, it was meant to teach the public a new way of seeing, a gaze that would transcend divides of class and nation. Thus "whole-state" (*gesammtstaatlich*) research in the human sciences resulted in a host of cultural productions, from atlases to exhibits to memorials, all of which conditioned a new way of regarding a landscape. Alois Riegl termed it a "view from a distance." It was a vision of the environment as an aesthetic composition, rather than as the setting for the daily struggle for survival. Riegl described the subjective effect as a "mood" or "ambience" (*Stimmung*). It resembled religious devotion, yet he traced it to the modern scientific view of a world interlinked by causal relations. It seemed to Riegl that the public of his day craved nothing so much as this experience of *Stimmung*. It offered "the reassuring conviction of the unshakeable rule of the law of causality."[29] It was a quest for redemption through art and the natural world.

For Riegl's successor as director of the monument's commission, Max Dvorak, this striving for a synthetic overview in the sciences was part of a broader transformation in the relationship between man and nature. Modern man had learned to take aesthetic pleasure in nature. His gaze took in "the diversity of natural phenomena . . . so that the full richness of the world down to the smallest wildflower and the most fleeting change in the atmosphere and quality of light has become a source of artistic sensations."[30] As we will see, this goal of appreciating nature's diversity at multiple scales resonated strongly with Habsburg scientists and cartographers.

In the 1890s, the monument-preservation movement also embraced the protection of "natural monuments," landscapes of particular scientific, aesthetic, or cultural value.[31] Between 1903 and 1906, the imperial education ministry, in conjunction with the crown land administrations, oversaw the production of an inventory of Austria's "natural monuments." Although proposed laws for the protection of natural monuments failed to win parliamentary support, the Zoological-Botanical Society continued to revise and expand the inventory of worthy sites in all the crown lands through the First World War.[32] Austria's nature-protection movement merits further research as a counterpoint to the nationalist framework within which nature protection was pursued elsewhere. Unusually for this era, Habsburg citizens were being urged to take responsibility for landscapes that were not part of their own national heritage.

Riegl's concept of *Stimmung* helped to explain how certain landscapes acquired a more-than-local value. Writing in the 1890s as director of the empire's monument-preservation commission, Riegl argued that the public had come to appreciate architectural remains due to what he termed their "age value." Like *Stimmung*, age value was linked to a view from afar, now in both the temporal and spatial sense. Just as the effect of *Stimmung* arose from an awareness of causal connections in space, age value corresponded to a vision of the cycle of universal birth and decay. Both *Stimmung* and age value indexed the subjective experience of recognizing that a particular object was causally related to a more general geographic and historical context. These concepts did important political work. Riegl argued that the monument commission should prioritize "age value" precisely because it stood above national differences. "While the feeling of pride of an Austrian, or of a Bohemian, a Styrian, a Carinthian, etc., or of a German, a Czech, a Pole, etc., regarding the monuments of the state or of a land or of a nationality always rests on isolation from others, whether foreigners or inhabitants of another crown land or members of another nationality, the feeling of age value is based on a sense of belonging to the entire world."[33] In order to affect a viewer, a monument did not need to speak to his or her personal or ethnic history. Anyone could experience the aesthetic pull of age value, simply by taking up the "gaze from afar," which Riegl associated with experiences of travel around the empire. Thus "the Bohemian searches to satisfy his deep longing for mood in something like a cathedral in Dalmatia, while a person from Styria finds it in Tyrolean wall painting, and the person from Silesia finds it in Salzburg's Italianate architecture."[34]

The twin preservationist projects, artistic and environmental, also faced a similar quandary. On what basis should the significance of a monument be assessed? Dvorak argued that art historians had often exaggerated the historical value of some objects at the expense of others, because their frames of reference were overly general. Reliance on nature's own scale of value would right this imbalance, he suggested. The gaze fit to judge the need for preservation was that of the "nature lover," who had "learned to appreciate the monument holistically [*in ihrer Gesamterscheinung*] as a part of a landscape and as an element of nature's beauty in the broadest sense." Judgments of historical value originated with "this love of nature, which is expressed when man bows in admiration before *what nature has created to be the measure of all things* [*was die Natur als Maß der Dinge geschaffen hat*]."[35] As a counterweight against universalist norms, the art historian would evaluate the aesthetic qualities of a monument in situ, as part of its milieu. He would let nature instruct him in the appropriate aesthetic response to a particular natural-cultural landscape.[36] Thus the imperial-royal art historian grounded his authority in his ability to judge the significance of monuments according to nature's own scale.

In short, the imperial-royal sciences were cultivating a new way of regarding the environment. Riegl's concepts of *ambience* and *age value* should thus be understood as the theorization of a gaze that was cultivated jointly by the natural and human sciences under the patronage of the Habsburg state in the wake of 1848. This was a spatially and temporally "distant" view of cultural landscapes, yet one attentive to local detail.

THE VIRTUES OF DIVERSITY

In the remainder of this chapter, we will begin to consider how this gaze constructed the natural diversity of the Habsburg Empire. That is, we will see how justifications of Habsburg power in the neo-absolutist era of the 1850s invoked Austria's great variety of physical environments in order to naturalize the phenomenon of the supranational state. Theories of the natural world informed arguments for transnational interdependence.

Von Bruck's claims for the economic benefits of Habsburg unity, for example, presumed a new form of knowledge about this state: a comprehensive overview of its material conditions. In 1850, this was more agenda than reality. The empirical investigation of the empire's natural and human resources was largely the work of the next half century. But the possibility of achieving such knowledge was the foundation on which von Bruck's argument rested. In line with Austrian critiques of German idealism, von Bruck cast political economy as a materialist and empiricist form of knowledge. In terms that foreshadow the subsequent *Methodenstreit* between marginalist and historicist economists, von Bruck quoted his close colleague Ernst von Schwarzer: "Freedom can be thought absolutely only in spiritual matters; however, matter is tied to a particular space as form. Economics is a practical and empirical science."[37] As an empirical science, economics had more in common with natural history than celestial mechanics. Von Bruck stressed the organic as opposed to mechanical character of the economic relations of central Europe, resting as they did on a "natural law of development."[38] The existing economic fragmentation of central Europe was "not natural," and the longer these "unnatural" conditions were allowed to persist, the more difficult it would be to effect the cure.[39] Von Bruck thus portrayed Austria as an "organism," whose strength rested on the healthy interdependence of its parts: on "this economic (continental and maritime) complementarity and mutual completion."[40]

In this vision, the heterogeneity of economic conditions in central Europe would catalyze trade, promoting economic and thus political unity. "For this reason the current of trade must flow freely from the north German ports to Triest, from the Mediterranean to the Danish Straits, from the Rhine to the lower Danube or in reverse."[41] As a publication of the ministry of trade put it in 1850: "Only by virtue of this vast territory, united even its diversity, bordering three seas, will it be possible to reconcile the economic differences and to develop the rich and varied treasures of its earth and soil and its advantageous locations into lively trade and industrial productivity."[42] Commercial freedom would ensure that "the continental and the maritime" would "complement and complete" each other. In von Bruck's imagery, the territory of the Monarchy was reimagined as a space of flows, a circulation of commodities, laborers, and capital.

This geographical vision did not stop at the Monarchy's borders. Rather than aiming for total autarky, von Bruck proposed to replace Josephinist prohibitions on imports with protective tariffs. This would serve to nurture nascent industrial and agricultural enterprises, while allowing central Europe to take its rightful place within international trade and thus international politics (*Weltpolitik*). Geography dictated in particular lively commerce between Austria and "Turkey," meaning both current and former Ottoman lands. Thanks to the Habsburg railway network and its Adriatic ports, the "powerful artery of European trade runs straight through Austria."[43]

Commentators at the time heard in this an echo of the old idea of Austria as the mediator between West and East. What was new, in their judgment, was to find "this belief so well framed and substantiated from the crucial materialist perspective." Economic geography was reinvigorating the old idea of Austria as "a cultural intermediary between the Orient and the Occident." Austria knew itself "to be the center and support of Europe"; it "repelled" some flows while "fulfilling" others. "The role of intermediary is natural to it, not only because it shelters and connects members of all the European nations in its embrace, *but also because of the diversity of its lands and waters*." The empire's waterways brought northern Europe, the Adriatic and Mediterranean, Poland and Russia all into Austria's "immediate economic sphere." "Thus Austria, a land of the most abundant diversity and economic plenty in its own right, represents likewise the central state that interconnects Europe's peoples and countries."[44] In this vision, the physical diversity of the Habsburg state, its environmental contrasts, were understood to create complementary markets and thus to catalyze trade. Commercial relations were reimagined as the metabolic system of the imperial organism. In the tradition of cameralism and of Linnaeus's equation between nature and nation, this justification of empire posited the human polity as a manifestation of nature's own economy.

In retrospect, however, the remarkable thing about von Bruck's proposal was his characterization of central Europe as uniform in climate. When he spoke of Austria's geographical variety, he meant its access to varied markets and routes, not a variety of regional climates. In that respect, the Habsburgs' dominions could not compare to the overseas empires of other European states, with their contrasts between "temperate" and "tropical" climes. And yet it was at this time that others were beginning to look more closely at conditions of climate and soil across this territory.

They discovered variations where von Bruck had seen uniformity. In the 1850s, the liberal Bohemian writer and statesman Ferdinand Stamm was looking into the role of climate and geology in Bohemia's agricultural and industrial development.[45] In an 1855 article for the liberal Viennese daily *Die Presse,* Stamm argued that Austria's climatic diversity was the key to its future economic self-sufficiency. Of the nine plants whose seeds could be counted on for sustenance, only three played the role of "daily bread" in any human society, and each of those was associated with one of the earth's climatic zones: "rice for the hot, wheat for the temperate, and rye for the north and the mountains." "You, lucky Austria, on your far-reaching land, grow the three most important plants and all the others mentioned!" Its varied regions could supplement each other's production in times of need. "There are no dry years and no wet years that could frighten . . . a land of Austria's breadth, stretching across two opposing climates." All that was needed was a suitable transportation network. "When this empire has drawn the veins of its rivers, its railway network and the finer mesh of its roads through its entire body . . . then its . . . inhabitants will be removed from the sphere of famine." In this sense, diversity was insurance against the variability of nature. Stamm appealed to a mental map of the territory's disparate yet interdependent environments. The Alps, for instance, served as a windbreak and as dry land during floods, while the melting snows on its peaks irrigated fields far and wide. "Austria never stands under one storm, never under one wind, never under the same blazing sky. . . . In the diversity of its climate and its soil conditions, Austria has its own compensation in the event of misfortune; it does not need other states." In short, "Austria is called a great power, because it has over 12,000 square miles; yet there are states with more land that are not great powers. Austria is indeed a great power, however, because its eagle stretches its wings into tepid and into warm air: diversity makes Austria great."[46] This argument would find tragic confirmation during the Great War, when the Monarchy's internal trade network collapsed and Vienna came close to starvation.[47] Until then, this economic interpretation of Austria's physical diversity would prove an important stimulus to the geological, botanical, medical, and climatological survey of its territory.

A MINIATURE "WORLD SYSTEM"

The decade after 1848 saw "Austria's re-creation," in the words of the statistician and naturalist Karl von Czoernig.[48] He was referring to the new rail lines, roads, canals, stone bridges, and telegraph wires that had bound the regions of the Monarchy more closely together. It was also in this period that a centralized banking network developed, while internal tariffs and restrictions on labor migration were lifted. Lively trade developed between the industrialized west and agricultural east, fueling growth in ways that began to even out regional economic disparities.[49] The economic historian Andrea Komlosy has recently argued that modernization theorists have been wrong to view Austria-Hungary as a periphery within the economic "world system." Rather, the Habsburg Monarchy was a miniature world system in its own right.[50] Komlosy insists that the heterogeneity of the Habsburg Monarchy was not its fatal flaw, but rather the source of its now well-documented economic dynamism. This was precisely how Habsburg subjects themselves perceived their situation.

Indeed, they were explicitly taught to see things this way, as we can see from later school texts. For instance, the 1910 *Österreichische Vaterlandskunde* for upper-grade elementary-school students, under the heading of "The Natural Endowment of the Monarchy," instructed that Austria-Hungary came closer than most other European states to the "ideal of a self-sufficient economic existence." Its capacity to meet its inhabitants' needs for agricultural and industrial products rested on its natural diversity. "The regions variously endowed by geography, climate, and soil also lend the most varied branches of economic production the most favorable conditions. Facilitated by natural avenues for trade, the exchange of its unique products has called forth an internal commerce that has also pressed powerfully outward in the last decade and successfully sought entry to international trade."[51]

Not everyone was so sanguine, however. As rail tracks lengthened and the Danube was regulated for travel by steamship, public debate turned to the question of how these improvements in transportation would affect the distribution of production. Would these changes invigorate the economic life of the empire as a whole, or would they magnify regional economic inequalities? Was the new vision of ecological/economic interdependence anything more than an excuse for perpetuating existing inequities? Following the lifting of internal tariffs in 1851, the price of grain rose in regions where it was grown for sale elsewhere in the Monarchy, just as many in western Austria had feared it would. Meanwhile, many Hungarians worried that they were entering a relationship of colonial dependency with respect to Austria.

Whether they were right is a question that continued to exercise economic historians long after the empire's fall. A consensus has emerged that the post-1851 economic relationship between Austria and Hungary worked to the benefit of both. Hungary seems to have been no worse off in this arrangement than the American South after the US Civil War.[52] Circa 1851, liberals argued against their critics that further improvements in agriculture and transportation would make the system work to the benefit of all. Achieving a more expansive market would make it possible to "balance out [*ausgleichen*] heterogeneous interests," such as those of the Monarchy's more agricultural east and more industrialized west. This contrast was rooted in contrasting natural conditions: the "virgin soil" of Hungary, on one hand, and the "harsh climate" of western Austria, on the other. What was needed, then, was a "rational" agriculture, one that matched each local "soil and climate" to an appropriate crop or manufacturing enterprise. Somewhat improbably, this anonymous author recommended turning Hungary's fields over to export crops like tobacco, wool, wine, and silk, while allowing Austria to become the "natural granary" of the German states of Bavaria, Saxony, and Silesia.[53] Other observers worried that the pursuit of autarky would lead Austria down the path of imperial Russia, where production tended to be specialized by region. They too emphasized that Austria displayed a historical tendency toward economic diversity that mirrored the heterogeneity of the natural world. Those regions that were most successful economically displayed "the most diverse industries alongside each other . . . just as the fields that yield the most abundant harvests are those on which the most diverse flowers spread their seeds."[54] Even among Italian nationalists in the spring of 1848, one could find leaders who clung to von Bruck's conception of central Europe as a zone bound together by its natural and human diversity. As the Italian version of the *Lloyd* proclaimed in the spring of 1848, the "natural conditions" of Trieste dictated "trade between the sea which sits in front of it and the continental lands to its back."[55] One could only conclude that division of this territory along national lines would be unnatural.

THE NATURALIZED ECONOMY

In the course of the nineteenth century, economic life seemed to be freeing itself from the control of nature. A host of technological improvements—including new means of fertilization, transport, refrigeration, and power generation and transmission—promised to liberate the geography of production from the distribution of natural resources. Thus Marx foresaw that capital, not land or labor, would be the decisive spatial factor in the future. With the rise of marginal economics, value became a subjective category, apparently independent of physical constraints. The historian Margaret Schabas argues that neoclassical economics "took man out of nature. The economy was seen to be the result of rational agency and, thus, no longer directly governed by natural forces."[56]

The "Austrian School" of economics is usually considered to have been at the vanguard of this subjectivist turn. Indeed, in Carl Menger's seminal *Principles of Economics* of 1871, we read that "land occupies no exceptional place among goods." A farmer renting a plot of land for a limited span of time need not concern himself with the land's history, with the origins of its fertility. "A buyer of a piece of land attempts to reckon the 'future' but never the 'past' of the land he is purchasing."[57] Nonetheless, Menger himself *did* reckon with the past. In his contribution to the *Kronprinzenwerk*, he narrated an environmental history of Bohemia, attributing the desiccation of its climate to population growth and associated drainage and deforestation.[58] In certain cases, then, Menger did take natural limits into account.[59]

More importantly, Menger's subjectivism never represented the mainstream of Habsburg economic thought. The misleading label of the "Austrian School" was affixed to him and his two major followers by his opponent Gustav Schmoller, for whom "Austrian" was a term of derision akin to "provincial." In fact, one-quarter of the membership of the historical economists' *Verein fur Sozialpolitik* was Austrian by the 1890s.[60] Perhaps a more representative example of late nineteenth-century economic thought in Austria was the textbook *Foundations of Political Economy*, which went through fourteen editions between 1893 and 1922. Its author was Eugen von Philippovich, professor of political economy at the University of Vienna from 1893 to 1917. Although he had studied with Menger, Philippovich insisted that even under industrial capitalism, the economy remained bound by natural conditions. He continued to employ the cameralistic concept of the *natürlicher Standort* or "natural location" of production, and to emphasize the economic significance of the environmental variability of the European continent:

> Yet agricultural production, founded on natural characteristics, that is, on the presence of renewable materials in the soil, demonstrates a dependence on Nature's influences, which cannot be equalized by the investment of capital and labor, which the land cannot elude due to its spatial fixity, and which eludes human calculation as to the quantitative and qualitative result of production. Insufficient investment of capital, inadequate labor, can certainly reduce the yield of a land favored by Nature, while generous capital investment and

> persistent work can increase the return; but they cannot cancel out the natural differences and eliminate completely the influence of Nature.[61]

This persistent tradition of natural historical reasoning in imperial Austria had at least three sources: explicit resistance to British economic liberalism; the lingering influence of cameralism, which had integrated political economy with agricultural science and forestry; and the prevalence of non-Darwinian evolutionary theories that attributed transformative, heritable effects to climatic conditions.[62] As the Ljubljana-based geographer Franz Heiderich put it in 1910, "Climate is one of the most powerful, indeed in many ways the decisive factor for organic life in general, as for the human economy in particular."[63]

That same year, Heiderich gave the opening speech at the fourth International Economics Workshop, a continuing education program where entrepreneurs and economists acquired knowledge of foreign lands and languages through on-site instruction. The 1910 workshop was held in Vienna, and Heiderich introduced it with a lecture titled "The Natural Conditions of Economic Life in Austria-Hungary," in which he began by insisting that economic life was "rooted" in the "geographical milieu." He was followed by Josef Stoiser, an expert on commercial geography at the University of Vienna, who gave a whole series of lectures on the various natural regions of the Monarchy.[64] Scholars like Heiderich and Stoiser had no use for a strictly subjective theory of value. Yet they were also bent on showing that classical economists had been wrong to assume that the fit between climate and mode of production was self-evident. On the contrary, it would now become the work of climatologists, geographers, and political economists to characterize each region's climate, and to deduce its place in the emerging "world economy."

IMPERIAL BODIES

What we might term the ecological justification for Habsburg unity was supported not only by research into the state's varied physical environments, but also by new understandings of the human body's relationship to its surroundings. Germ theories of disease became dominant in the late nineteenth century, but they never fully displaced medical attention to "airs, waters, and places." Throughout the nineteenth century, projects of colonization were understood to stand or fall according to the success with which settlers were "seasoned" to the new environment. Even as germ theories gained ground, colonial physicians conducted ever more detailed investigations into the environmental determinants of health and disease.[65] This was also true in imperial Austria.

Before the rise of germ theories in the 1870s, military and medical experts often blamed the poor health of soldiers in the imperial army—and thus Austria's military defeats of 1859 and 1866—on unhealthy environments.[66] They raised alarm, for instance, over conditions in Croatia, where soldiers were housed in overcrowded, poorly ventilated, and yet permeable dwellings that offered insufficient protection from the elements, including the fierce bora wind. Similarly, they recommended regulation of the Sava and Drava Rivers in order to remove the stagnant waters responsible for malaria.[67] More generally, they cited the experience of "climate variation" (*Klimawechsel*) as a threat to the Habsburg soldier's body. This claim found support in ancient medical wisdom holding that "variable" climates were less healthy than stable ones.[68] Environmental health remained a military concern even after the rise of bacteriology—and despite the army's decision in the 1880s to allow soldiers to be stationed in their home regions. Indeed, the first of fourteen volumes of *The Hygienic Conditions of the Locations of the Larger Garrisons of the Austro-Hungarian Monarchy* appeared in 1888.

Meanwhile, physicians began to reconsider the effects of "climate variation." In 1856 the Imperial-Royal Society of Physicians formed a Balneological Committee with the aim of producing a comprehensive survey of the Monarchy's health resorts or *Kurorte*. The committee appealed for assistance to "every friend of the fatherland, of natural science, and of ailing humanity." They solicited an exhaustive description of the climatological and medical properties of each locale—remarkably, even those whose curative properties were "questionable or even entirely insignificant."[69] Meanwhile, the newly founded ZAMG supplied health resorts throughout the empire with the instruments necessary to monitor local climatic conditions. "Climatotherapy" eventually came to be defined as "the study of the application of the varied climatic influences to therapeutic purposes," with an emphasis on variation.[70] This gave a strong boost to Austria's tourism industry, part of a development that Alison Frank Johnson has aptly termed the commodification of air.[71] At stake for Austria was the promotion of destinations ranging from the traditional spas of western Bohemia and Hungary, to the lakeside retreats of the Austrian Alps. By the end of the nineteenth century, this medical marketplace also included the so-called Austrian Riviera in Istria and Dalmatia and the high-altitude resorts of the Tatra Mountains and other regions of the Carpathians. The Habsburg economy depended increasingly on the appeal of spa and resort destinations that drew urban dwellers and the infirm to sunshine, snow, dry air, or ocean breezes. Climatic contrasts thus encouraged the economically vital circulation of people as well as goods.

Indeed, it was in this period that travelers through the Habsburg lands began to appreciate landscapes that had until then been described as wasteland. These included the craggy, arid karst lands and the grassy plains of Hungary. Adalbert Stifter described this shift in his story "Two Sisters," which revolves around characters who dedicate themselves to living off the hardscrabble highlands of South Tyrol. The narrator, accustomed to the fertile soil of his Bohemian home, had "no other concept of the beauty of a landscape. . . . Yet here I stood in a wasteland [*Oede*], where everything was wanting . . . and where nonetheless there appeared such a peaceful beauty that it seemed nature had laid a simple, sublime, epic poem before me."[72]

CONCLUSION

What emerged in the aftermath of 1848 was less a new Austrian Idea than a new way of looking at and experiencing the Austrian territory. In 1850, Minister von Bruck could still describe the climate of central Europe as relatively uniform. By the 1870s, the climatic diversity of Austria-Hungary was celebrated by men of science and statecraft alike. Diversity became visible because it became valued. Research across the natural and human sciences revealed minute variations of climate, soil, language, and culture, as well as resulting patterns of flow and exchange. The result was a new way to think the idea of Austria—not as a lofty abstraction, nor as a homogeneous national community, but rather as a physical space of circulation.

CHAPTER 3

The Imperial-Royal Scientist

"The Austrian Empire appears destined above all to become the most instructive school for meteorological and climatological relations. . . . Here an oceanic climate joins hands with a continental one. With such varied atmospheric conditions in coastal and interior regions, on alpine peaks and in valleys, there is nowhere better to study and grasp the interactions among them."[1] The year was 1852, the occasion a special meeting of the Vienna Academy of Sciences to mark the founding of the ZAMG, and the speaker was Karl Kreil, the institute's founding director (figure 7). Kreil had grown up in the hills of Upper Austria, where his father had risen from the peasantry to a low-level administrative position for the imperial education ministry. Young Kreil was fortunate to win free entry to a gymnasium where, unusually for the era, the natural sciences were emphasized. Subsequently, as an assistant at observatories in Milan and Göttingen, Kreil learned of the enterprise conceived by the famed explorer Humboldt and the astronomer Carl Friedrich Gauss: a network of precision instruments recording every variation of the atmosphere and geomagnetic field, at every moment, everywhere on the surface of the earth. It was Kreil's dream to bring this project to Austria, with its sprawling territory waiting to be charted with precision.[2] In Vienna's fledgling Academy of Sciences, Kreil found the support he needed.

The geographical image that Kreil evoked for the academy that day was a novel one. Until recently, there had been no accurate maps of the territory of the Habsburg Monarchy as a whole. Kreil himself had been the first scientist to conceive of—and then conduct—a geophysical survey of the entirety of the Habsburg realm. For a learned man to venture onto the Hungarian steppe,

FIGURE 7. Karl Kreil (1798–1862). Lithograph by L. Berka, 1849.

the Dalmatian karst, or even the high peaks of the Alps was a highly unusual enterprise. Now, with the founding of the Academy of Sciences, the ZAMG, the Imperial Geological Institute, and the Zoological-Botanical Society, a new generation of scientists would come to identify as experts on the nature of this space as a whole.

In this way, a new professional identity was born, that of the imperial-royal scientist. It has long been said that Habsburg civil servants were the keepers of the "Austrian Idea." This chapter asks why natural scientists invested in the idea of "unity in diversity" and with what consequences for Austrian history and the history of science.

ARBITERS OF LARGE AND SMALL

This book argues that the imperial-royal scientist built his public authority on a capacity to set local details in appropriate relation to a synthetic overview, to attend to minutiae without losing sight of the coherence of the whole.

It was this ability to judge rightly the significance of small things that gave the Habsburg field sciences their political resonance in the decades following the revolutions of 1848. In the next decade, the imperial-royal scientist attained mythic status in the fiction of Adalbert Stifter. Stifter is to Austria what Goethe is to Germany or Pushkin to Russia. A passionate amateur naturalist who had once aspired to an academic career in physics, Stifter strove to convey to readers of his novels and stories the nobility and aesthetic sensitivity of the man of science. He emphasized in particular the naturalist's capacity to assess phenomena according to a more enduring scale than personal inclination. In *Der Nachsommer* (Indian Summer), his most famous novel, the education of the protagonist can be measured by a shift in his perception of scale. The story centers on the friendship between a budding naturalist, said to have been modeled on the geologist and climatologist Friedrich Simony, and a wise older man, who bears more than a passing resemblance to the Vienna physicist Andreas von Baumgartner. From the older man, the younger learns such principles of art and science as how to paint *en plein air* and how to predict a storm. The key is attention to signs both large and small, both "coarse" and "delicate," both "in the immediate area in which one finds oneself" and across "a wider one."[3] Later in the novel, the narrator thanks his mentor for these lessons, which have succeeded in reversing his own sense of scale: "the large is small to me, and the small is large."[4]

Austria's new expert-lay scientific associations of the 1850s and 1860s were also developing a language with which to articulate this ethos. Consider, for instance, the words with which the botanist Eduard Fenzl opened a meeting of the Zoological-Botanical Society in Vienna in 1852, one year after the association's founding:

> What is needed therefore is the sober-minded, harmonious collaboration of all, as before; manly trust in one's own powers, free of faint-hearted self-doubt as of haughty overconfidence; the courage to appear in public and to engage in scientific debates; the adherence to the greater truth, which is formed by the appropriate combination of smaller ones into a larger whole, and which can only serve the substantive goals of the Society by means of the subordination of all selfish special purposes to the unique authority of science. This is the only lever that is needed, I tell you, to accomplish what the Fatherland is justified in expecting from our labors. It all lies in your hands, and will not escape them, gentlemen, as long as you maintain this one thing steadfastly: *an interest in the smallest thing that each of us produces, even if the subject lies far from your proclivities, training, and understanding.*

> *Our plenary meetings must remain the vital cement of this mutual fusion of interests.*[5]

Fenzl's words exemplify a way of speaking characteristic of the imperial-royal scientist. This rhetoric contrasted apparent and actual size, the naive and the expert view. In the wake of revolution, this was an explicitly political lesson. Narrow-minded attachments to petty pursuits would have to be set aside in the interest of collaboration. But what was "small" was not necessarily petty. What appeared small to one researcher might loom large for another, and might well prove essential to an understanding of nature as a whole. To serve Habsburg science one had to judge rightly the significance of "small" things.

This chapter considers how one became an imperial-royal scientist by learning to transcend the localness and particularity of one's origins. It is important to keep in mind that this perspective was not granted to all who served Habsburg science. In later chapters we will encounter remarkable individuals like the Bohemian botanist Emanuel Purkyně, the Galician meteorologist Max Margules, and the Anglo-German-Czech geographer Julie Moscheles—just a few of the creative minds who were sentenced to remain "provincial" naturalists, because they were marked by nationality, religion, or gender (or a combination thereof). Let us begin, then, with the particular places within the Habsburg Monarchy that offered a privileged vantage point for studying the earth as a whole.

PRODUCING THE LOCAL

Today's knowledge of the climatic history of central Europe rests in part on continuous instrumental climate records that begin in the middle of the eighteenth century at three locations within the Habsburg lands. One is Prague, where meteorological observations were initiated in 1752 at the baroque astronomical tower of the Jesuit college, the Clementinum. Another is the Benedictine abbey of Kremsmünster, near Linz in Upper Austria, where the monks kept meteorological records from 1763. Meteorological observations were also made in Milan from the 1760s, but the intellectual ties between Lombardy and the rest of the Monarchy were never close, and by 1859 it had been pried free of the dynasty's grasp. Prague and Kremsmünster are thus central to our story. Many of the leaders of climatological research in imperial Austria owed their careers to one or the other locale. Each center conceived of its research program as part of a distinctive local culture, tied to a specific natural and cultural environment, and yet each positioned itself as a window onto the atmosphere at large.

Kremsmünster

Bordering Bavaria to the west and Bohemia to the north, with the Danube Valley to its east and the peaks of Styria to its south, Upper Austria was a land of green hills and alpine lakes. Its wealth rested on the salt mines of the region around Salzburg, as well as areas of coal and iron mining. The high-altitude pastures lent themselves to traditional methods of farming, cattle raising, and fruit orchards. The region was also shaped by a violent religious history, as a major theater in the wars of the Counter-Reformation. By the eighteenth century, Protestantism had been defeated, and influential monastic orders had taken up residence. Their baroque monasteries still adorn the landscape of Upper Austria today.

The two men who contributed most to the institutionalization of climatology in the Habsburg Monarchy were both natives of Upper Austria: Karl Kreil and Julius Hann. What's more, both were graduates of the gymnasium run by the monks at Kremsmünster (see plate 1). Kreil completed his studies there in 1819 and Hann in 1860. By a further coincidence, both attended the school as scholarship students, coming from families of little means. Moreover, Kremsmünster also educated Adalbert Stifter, the man responsible for embedding the earth sciences in Austria's literary canon. Another pivotal figure for climatology, Josef Roman Lorenz von Liburnau (1825–1911), attended gymnasium in nearby Linz, where the teaching of the natural sciences reflected the abbey's influence.

A recent study concludes that the Benedictines "produced more Catholic Enlighteners than any other order" in the eighteenth century.[6] One reason for this was the willingness of Benedictine monks to engage with elements of Protestant thought. The writings of the early eighteenth-century rationalist philosopher Christian Wolff, for instance, were eagerly discussed at Kremsmünster. So it is not entirely surprising to find that the natural sciences at Kremsmünster were inspired by a "physico-theology" more often associated with Protestantism. The essence of physico-theology as it flourished among eighteenth-century writers in German was the effort to observe the wisdom of creation in nature's most mundane and often miniscule objects. At Kremsmünster, religious feeling motivated a highly disciplined regime of scientific observation.[7]

The abbey's astronomical observatory or "mathematical tower," as it was popularly known, was constructed between 1748 and 1758. Its collection of state-of-the-art scientific instruments was unrivaled in central Europe, thanks in large part to the efforts of Kreil's mentor, the astronomer Bonifazius

Schwarzenbrunner. It is said that Schwarzenbrunner was so single-mindedly driven to improve the instrument collection that he drove himself mad and died at the age of forty—but not before convincing the emperor to supply the finest clock, theodolite, and equatorial telescope.[8] In Hann's student days, the meteorological instruments included a thermometer, a barometer, a psychrometer and hygrometer for measuring humidity, and an instrument for measuring the ozone content of the atmosphere. Augustin Reslhuber, who was director of the observatory while Hann was a pupil, insisted that these costly instruments showed their true value because they were not left idle, but were put to excellent use by the students "for the glorification of the power of the Creator and the education and ennoblement of humanity."[9] Precision climatology was thus a spiritual endeavor.

The monks' dedication to the study of the heavens and atmosphere made a strong impression on students like Kreil and Hann. Both men were so attached to their teachers and school friends that they often spent holidays at Kremsmünster after graduating. The science of climate as it took shape in the 1840s and 1850s in Austria owed much to the abbey's naturalists. In works like Marian Koller's 1841 study of the course of temperature in Upper Austria, or Augustin Reslhuber's 1854 investigation of the water temperature of the springs around Kremsmünster, the monks conducted precise surveys of the limits within which climate varied. Such research helped to construct the "local," defining the specificity of the environment to which both Kreil and Hann felt such a strong attachment.[10]

At the same time, Koller imparted the farsighted lesson that the future of climatological research would require cooperation on a large scale, following the example set by Humboldt and Gauss for geomagnetism. Already in 1841, before communication by telegraph was even a practical possibility, he recognized the need for simultaneous observations across the globe. "One increasingly arrives at the inner conviction that all parts of our planet exist in perpetual atmospheric communication," he wrote, "and that the atmospheric conditions at one point on the earth are just one result of conditions on the rest of the earth." Koller envisioned an international project spearheaded by strong central states: "In this way, the whole becomes complex and the task difficult, far exceeding the physical and moral force of an individual. Its resolution can only be undertaken by means of united forces, by societies that are fortunate to rely on the protection and support of powerful authorities."[11] It would be another decade before Kreil could secure imperial support for such an undertaking. When he did, Kremsmünster and Linz became two of the network's essential nodes.

FIGURE 8. The Prague Clementinum, where regular instrumental meteorological measurements began in 1752.

Prague

As a local environment and as a window onto global phenomena, Kremsmünster was defined primarily in spiritual terms. Prague presents a contrast in this respect. The Bohemian capital, which nurtured several of the early leaders of climatology in imperial Austria—including Eduard Suess, Karl Fritsch, Emanuel Purkyně, and Friedrich Simony—had a more practically oriented scientific culture. To be sure, enlightened Catholicism flourished in the Bohemian lands as it did in Upper Austria. Karl Fritsch even claimed that he had been drawn to climatology since his school days because the Prague observatory was associated in his mind with the wisdom and grandeur of God.[12] Yet physico-theology was not as dominant an influence in Prague, and Bohemian intellectual life in this period tended toward realism, often in explicit defiance of northern German idealism. This orientation was expressed most famously in the universalist humanism of the former priest Bernard Bolzano, who inspired subsequent developments in central European analytical philosophy. More important for climatology was the rise in the late eighteenth century of a Bohemian patriotism centered on aspirations for economic development.

Bohemia's Patriotic Economic Society originated under the rule of Maria Theresa as a voluntary association dedicated to agricultural improvement and popular enlightenment. In a period when maritime European states were eyeing colonial territory overseas, proponents of development in the Bohemian lands focused their prospecting on the natural riches of their own territories. Just as they set out to catalog the distribution of useful plants and minerals, they also undertook climatological observations for the sake of agriculture and forestry. From 1796, the Patriotic Economic Society, in conjunction with the Jesuit observatory in Prague (figure 8), promoted systematic climatological and later phenological observations. By the end of the 1790s, records were also kept at six sites in Bohemia beyond Prague.[13] Karl Fritsch, for instance, began his scientific career as the coordinator of meteorological observations for the Patriotic Economic Society.

These studies of the local natural environment carried symbolic as well as practical value. The Bohemian patriots insisted that their land formed a natural unit, such that, as Palacký put it in 1849, "it is not possible to divide it without destroying it."[14] Given the significance of the natural environment and its resources to Bohemian patriotism, it may not be a coincidence that Palacký's son Jan became a biogeographer, concerned in particular with the relationship between climate, soil, and plant growth. Adopting his father's ambition of modernizing the Czech language, the younger Palacký even invented a Czech word for climatology (*vzduchosloví*, literally "aerography").[15] By the same token, one should not be shocked to find the beloved Czech novelist Božena Němcová cited as among "the pioneers of geobotanical studies of our lands."[16] Němcová penned a series of travel sketches at the request of the physiologist and patriotic activist J. E. Purkyně for his journal *Živa*, in which she highlighted the interdependence of local cultures and their natural environments.

As was the case at Kremsmünster, these scientific surveys served to define local Bohemian nature and culture, while also establishing Bohemia as a favorable vantage point onto the natural world at large. The Moravian poet Hieronymus Lorm (1821–1902) expressed these twin goals when he termed patriotic feeling (*Heimatsgefühl*) "the most powerful basis for delight in the natural world," yet added that what is most wonderful in nature is what can be observed everywhere.[17] In other words, local nature rose in value to the degree that it revealed something universal. In this vein, the botanical geographers Emanuel Purkyně and Ladislav Čelakovský sorted Bohemian plants not only for local practical purposes, but also to delineate a more general method for classifying plants in relation to altitude and climatic conditions. As we have seen, it was in this spirit of appreciating the more-than-local significance of particular details that Jan Evangelista Purkyně (Emanuel's father) introduced the inaugural issue of the very first Czech-language scientific journal in 1853, insisting that "in infinite nature nothing is insignificant, nor are the needs of man its only measure. The broad, indiscriminate [*bezohledný*] identification and comprehension of everything that presents itself to the senses—that is the task of the natural scientist."[18]

Thus the climatological research programs associated with Prague and Kremsmünster articulated their own reasons for attaching significance to the local and particular. What's more, each taught its own ecological lessons. Bohemia, with its gentle hills and dense forests, drew attention above all to the fine-grained dependence of vegetation on climate, while the Alpine foothills of Upper Austria inspired studies of the origins of mountain winds and squalls. Yet each of these local cultures endowed future imperial-royal scientists with a set of tools for investigating the small scale in relation to the large.

SCIENCE AND THE STATE

Still, it was a long journey from a training as a provincial naturalist to a career as an imperial-royal scientist. The professionalization of the sciences occurred in the Habsburg lands in roughly the same period as in western Europe and North America, but it followed a different course. Whereas British and American men of science could often look to industry or private educational institutions to support their careers, Austrian scientists were almost wholly dependent on the state. As university professors or gymnasium teachers or employees of a research institute, they were *Beamte* (civil servants), a class known for its identification with and loyalty to the dynasty. These scholars had a relatively small degree of professional autonomy and depended directly on the patronage of imperial ministries. Even when it came to academic hiring decisions, it was often the opinion of an official in Vienna (or, later, in Prague or Kraków) that carried the most weight.[19] Rarely did an imperial minister bother to read scholarly research. For these reasons, personal connections and a public reputation built on popular writings were essential to advancing in a scholarly career.

To appreciate these peculiar circumstances, consider the expression of startled admiration that the botanist Anton Kerner received in 1869 from Hermann Hoffmann, professor of botany at Gießen, in Hesse, upon having learned of Kerner's extensive research into phenology, the study of seasonal phenomena. Hoffmann expressed the greatest admiration for Kerner's "painstaking" data gathering and his "prudence" in their "successful interpretation." Yet Hoffmann had only been able to use Kerner's research "in a few places in my own work," because he had learned of it "only piece-meal through isolated excerpts."[20] Hoffmann published in specialist journals like the *Yearbook for Scientific Botany*, while Kerner's work appeared in the *Austrian Review*, a general periodical aimed at an Austrian audience. Hoffmann's letter suggests that the cost of attracting a nonexpert audience within Austria was the loss of a specialist audience beyond its borders.

For many Habsburg naturalists outside Vienna, the only path to a scientific profession was to take an active role in building the provincial academies, museums, and journals that could support such a career. As František Palacký put it, "I might complain that in Bohemia I alone have been burdened with work which in other countries is shared by governments, academies, and educational institutions. . . . I must be hod-carrier and master builder in one person."[21] Alternatively, the first step for the ambitious might be to move to Vienna. But life in Vienna, especially with a family, was extremely costly. In the early years of the ZAMG, Director Jelinek had to petition the education ministry to pay an adjunct like the young Julius Hann a salary that matched that of a Viennese middle-school teacher (who in turn earned even less than a teacher in the provinces).[22]

What would today be called the environmental sciences proved particularly vulnerable in this context, since they were expected to produce knowledge of immediate value to the state, for instance, for mining, forestry, and irrigation. Scientists were loath to risk losing a minister's favor by taking a controversial stance. Consider, for example, the lengths to which the naturalist and forestry expert Josef Wessely went to cloak a critique of the government in print. As editor of the *Vierteljahresschrift für Forstwesen* in the wake of the 1872 reorganization of the imperial forestry administration,[23] he was looking to make a veiled criticism of the forestry official Baron Julius Schröckinger von Neudenberg. So he changed the spelling of the adjective *schrecklich* (awful) to *schröcklich*. When the typesetter switched it to *schrecklich*, Wessely changed it back, adding a note to the copy editor to leave it that way, but to no avail. "And so my journal purged the only joke I'd made in the course of an entire year."[24] Wessely revealed this ruse in correspondence with Emanuel Purkyně, a younger naturalist in Bohemia, intending it as a cautionary tale. He urged Purkyně to express himself judiciously in print. As we will see in chapter 9, Purkyně did not heed Wessely's advice, and his career suffered the consequences.

THE IMPERIAL BODY OF KNOWLEDGE

How did the first generation of imperial-royal scientists win the authority to speak as experts on the territory of the Monarchy as a whole? Imperial-royal science was a corpus of knowledge embodied by its new practitioners, men whose power was supposed to rest in part on their firsthand experience of Austria-Hungary's physical and cultural contrasts. As Julius Hann wrote in his climatic overview of the empire for the *Kronprinzenwerk*, "Nature has made it easy for the inhabitant of Austria-Hungary to cultivate climatic research. If he has the Wanderlust and the means to satisfy it, he can—without crossing the border—allow climatic contrasts to operate on him directly such as no other European land offers within the same distances."[25] Such kinesthetic knowledge of the Monarchy's "physical contrasts," "climatic boundaries," and "transition zones" was essential to the embodied identity of the imperial-royal scientist, an identity gendered masculine and ethnically unmarked. It grounded the authority of their analysis of the relationship between parts and whole within the natural system of the empire.

Elements of the accounts of scientific exploration written by imperial-royal scientists will sound familiar to those versed in the rich historiography of the nineteenth-century field sciences.[26] Like naturalists working in western Europe, Habsburg investigators were often building on a long tradition of place-based, chorological research; and like scientific travelers in Europe's overseas colonies, they often had to struggle to make sense of unfamiliar landscapes and cultures. What was unique in the Habsburg case was the absence of a clear demarcation between the first mode of research and the second. The territorial continuity of the Habsburg lands made divisions between center and periphery, metropole and colony, far more ambiguous than in overseas empires. Like western European imperialists, Austrian elites often viewed themselves as bringing civilization to the "primitive" peoples on their peripheries, yet it was far more difficult for them to point to the place on the map where civilization ended and backwardness began. Indeed, the Dual Monarchy eschewed the term "colony," preferring to view newly acquired land (Galicia in 1772, Bukovina in 1774, Bosnia-Herzegovina in 1878) as "natural" extensions of its rule.[27] This pretension was in fact one ideological motivation for the state's support of the earth sciences.

What were the conditions under which scientists traveled through the Habsburg lands? To be sure, scientists traveled for purposes of pleasure or health as much as other members of the *Bürgertum* in the nineteenth century. Indeed, they sometimes managed to combine work, leisure, and health cure in one destination.[28] But field scientists also traveled in other capacities. One crucial to atmospheric science was the inspection of the stations of the imperial weather-observing network. As in astronomy, it was necessary to compare the instruments against each other at regular intervals, in order to ensure the uniformity of observations. Kreil was on the road for three months at a time in 1855, 1856, and 1857 to inspect the network's stations, of which there were only ninety at the time. Thanks to his lobbying to the ministry of education, the ZAMG received an additional 800 Gulden to fund these trips; but this was a battle that each of his successors would have to fight for themselves.[29] As Kreil insisted, it was only by personally checking each measuring instrument that he could be sure the measurements were fit to be compared.

Another form of travel characteristic of the field sciences was the expedition or survey, in which scientists would traverse a region on foot, recording its physical characteristics and sampling its minerals and plant life. In the context of overseas empires, the scientific surveyor seemed to stand for the impersonal, objective, and potentially ruthless authority of the colonial state.[30] In the Habsburg context, the surveyor was a far more benign and ambiguous figure, familiar to inhabitants across the empire since the days of Joseph II. The surveyor appears in Austrian literature of the nineteenth century as one who came by stories by virtue of having "gotten to know different regions and different people." As the surveyor-narrator of Stifter's story "Kalkstein" explains, "It's part of my profession that I interact with many people and make mental notes of them, and so I have acquired such a memory for people that I even recognize those I saw years before, even a single time."[31] This was the human side of survey work in the Habsburg Monarchy.

By the middle of the nineteenth century, university students and faculty were engaging in field expeditions as part of Vienna's new curriculum in physical geography. Since the eighteenth century, central Europe had fostered a tradition of excursions organized by natural-scientific voluntary associations, which brought together aristocratic and bourgeois members in patriotic celebration of the "local" environment, often the hills or lakes in the environs of a town.[32] But it was only in the middle of the nineteenth century, at Friedrich Simony's urging, that the territory of the Monarchy as a whole became the first and foremost subject of geological study. Until then, geological research had focused on mineralogy, while the biblical story of creation dominated the teaching of earth history. As Eduard Suess recalled of the earlier period, "Austria's incredible diversity encompasses the ancient mountain massif of Bohemia, the edge of the Russian plain, the far younger Alps and Carpathians, and the western periphery of the Aralo-Caspian depression. Knowledge of these great natural units remained entirely foreign to university instruction."[33] Suess, who became full professor of geology at Vienna in 1867, led students on outings primarily in the Vienna hills. Albrecht Penck, appointed professor of geography at Vienna in 1885, made excursions an integral part of the curriculum, and in 1896 the university received a subvention of 600 crowns from the education ministry to fund them. There was at least one each year, sometimes as many as three or four, each with the participation of up to two dozen students and professors.[34]

Usually the destination was within a few hours of the capital, but sometimes these tours ranged further afield. By the 1820s, the Habsburg lands had already entered the era of "modern tourism," thanks to the eighteenth-century state's energetic construction of roads and a system for their maintenance.[35] Nonetheless, in many parts of the Monarchy, travel remained arduous and perilous at midcentury. As one mining official later put it, "The difficulties of travel through large parts of the Monarchy placed geological field work at that time on par with the voyages of discovery to unknown continents."[36] Eduard Suess recalled that in the Alps in the 1850s, "one had to be quite content with an only partially watertight roof, a bundle of hay to sleep on, a bowl of milk and some black bread as a communal meal or with chicken-fat dumplings as a special delicacy."[37] Scientists often had the help of servants and horses, but in some cases they had to lug their own measuring instruments, specimens, and other equipment. This was an age when scientists were encouraged to observe and collect as widely as possible, so it was common for a researcher to harvest the specimens of several disciplines at once, carrying the equipment necessary to each. Scientists in the field would need some instruments that would function in extreme conditions, and others that, in the absence of detailed station data, would provide continuous baselines for comparison. For example, when Kreil undertook his geomagnetic survey of the Monarchy in the 1840s, he brought with him an instrument for measuring altitude and azimuth, two chronometers, three instruments for measuring declination and inclination, two astronomical telescopes, two portable barometers, three thermometers, two hypsometers for measuring altitude by means of the boiling point of water, as well as "many small tools and instruments, books, maps, etc."[38] Typically, the scientists would strike a base camp, then set out separately on day hikes. Since mountainous areas remained only roughly charted, scientists often played the role of land surveyor as well.

By the 1860s, field trips had become a core element of training in the earth sciences. Julius Hann gushed in his diary over one of his first such outings in May 1862, during his second year at the University of Vienna. It was a "turbid, hazy spring day" when he and his classmates set out for the hills on the Vienna outskirts with professors Peters, Sommaruga, and Moisvar. "Climbed around a lot in the meadows and quarries. Felt myself to be a foreign, tolerated guest in these beautiful surroundings. Listened with longing to discussions of excursions in the Alps. Envied the lucky ones who could all talk that way. What new pleasures life could offer through contact with such men. Winced with sadness when I looked back on my lonely, empty life."[39] As a provincial youth from modest circumstances, Hann saw himself as an outsider to Vienna and its academic world. The following year, Hann recorded a Sunday outing with Suess and several other distinguished naturalists in September. They discovered a marine fossil and met up with an officer of the Habsburg army who was engaged in a "scientifically correct" survey of the land. "Had a splendid time," Hann concluded. Looking back at this entry, he added: "I think happily of those first outings, as I climbed into our communal train car, where the participants, with the exception of Sueß and Arnstein, were entirely unknown to me. . . . Back then I had the highest, undiluted admiration for the great men of science, who seemed to me almost beings of another order." It was with "peculiar delight" that he overheard one of his companions saying that he had recently presided over a meeting of the geological section of the Congress of Natural Scientists in Karlsbad, where he had sparred with the notorious Otto Volger. "Among what scientific luminaries had fate dropped me? What names might be theirs? I had no one to enlighten me, since Sueß was sitting outside with the driver. So I could only listen devoutly to their conversation."[40]

Hann's accounts illustrate how field trips functioned to professionalize and socialize young naturalists. As Eduard Suess insisted, finding "joy out in nature" was something that could not be taught in schools. This capacity proved "that man is not yet fully urbanized," that he retains a remnant of the "savage life" (*Wildlingsleben*) of the ancient human past. "One may forgive the geologist, like the hunter, if he betrays something more of the savage."[41] Field excursions thus promoted camaraderie and helped to define a masculine scientific identity that partially transcended differences of class and nation. In some cases, these social bonds even developed into kinship relations. Among the most prominent examples: Neumayr was son-in-law to Suess, Hann was grandson-in-law to Kreil, Wegener was son-in-law to Köppen, Tietze was son-in-law to Hauer, and Wettstein was son-in-law to Kerner von Marilaun.

Nonetheless, a field excursion was not always a pleasant jaunt. Dangers of various kinds, real and imagined, lurked in the less studied regions of the Monarchy. Kreil's survey began in 1846 but was forced to break off in 1849 due to the revolutionary unrest in and around Hungary. Already the previous year, in the Banat, Kreil had suffered the indignity of being arrested as a spy; his loyal assistant Karl Fritsch was accused of making a map of military fortifications. To top it off, Kreil arrived home in Vienna sick with fever.[42] Climatologists also had to be on the lookout for thieves, especially given the expensive instruments they lugged with them. Minor attacks were not uncommon circa 1800, and possibly later as well.[43] One night on the border between Galicia and Hungary in 1870, three geologists witnessed the attempted theft of their coach driver's horse. In good humor, they reported, "Since there were three of us, the situation was less risky for us than for the horse." It was "almost a shame," they added, that they wouldn't be able to contribute a report on the arrest of a "pack of bandits" to the journal of the Imperial Geological Institute. Then, turning serious, they observed that the experience was "a warning to take the greatest caution for times when each of us will have to travel alone."[44] Reports of danger may have been exaggerated, however, given cultural associations between eastern Europe and North America's "wild West."[45] In 1858, for instance, Anton Kerner prepared for a scientific expedition to the Bihar Mountains between Hungary and Transylvania by requesting permission from the Hungarian authorities to carry weapons. He received authorization for "a double-barrel shotgun for hunting and a double-barrel pistol for personal protection."[46] In a published account of his research, Kerner recounted a terrifying encounter with Hungarian highwaymen or *betyar*. Only at the end of his account did he reveal that the whole incident had been a dream.[47] Real enough, however, was his sense of being cut off from contact with centers of civilization. Scientists in the field had to rely on mail coaches for sporadic communication with the institutes and ministries in Vienna.[48]

Beyond travel for leisure, health, the inspection of instruments, and survey work, Habsburg field scientists were occasionally on the road for yet another reason. Every successful scientist could expect to shift his residence multiple times during his career. As Jan Surman has shown, the Habsburg universities set in motion a characteristic circulation of scholars from provincial universities to the premier institutions in Vienna and Prague. Before liberalization in the 1860s, the education ministry had a twofold strategy for harmonizing the higher education system. It sought to bring in loyal, Catholic scholars from abroad, and to circulate scholars between universities, in an effort to create a common academic culture. It was common wisdom at the time that Habsburg universities were divided into "universities of entrance, promotion, and final station." Hence Theodor Mommsen's memorable quip: "sentenced to Chernivtsi, pardoned to Graz, promoted to Vienna."[49] After 1867, as influence over hiring decisions shifted from the education ministry in Vienna to the provincial administrations, this form of mobility diminished somewhat, and nationality became a bigger factor in academic hires. Still, scholars continued to move between Habsburg universities, and this movement continued to influence research trends. As Surman shows, Habsburg scholars themselves recognized the value of "intercultural mobility" to knowledge production, a recent theme in the work of historians of science and empire.[50]

Field scientists were no different in this respect. The example of Josef Roman Lorenz von Liburnau, president of the Austrian Meteorological Society from 1878 to 1899 and cofounder of Vienna's first university for agriculture, illustrates this well. Like Hann and Kreil, Lorenz was born in Linz, the eldest son of a local official. Lorenz first taught gymnasium in Salzburg, where he began to study the local moors. As a teacher in Fiume/Rijeka, where he managed to teach in Italian, he took up the study of the limestone formations known as karst. Not long after, he was recruited for the state-sponsored survey of the Adriatic coast. From there he was hired as a consultant to the agricultural ministry in Vienna, at which point he turned his research to the Danube. Each time he relocated, he redirected his research to a new environment.[51] These locales became part of his identity: when he was ennobled in 1878, he chose to call himself "von Liburnau," Liburnia being the ancient term for the area of the Adriatic coast where he had studied karst.

The research of Anton Kerner von Marilaun likewise bears the marks of a peripatetic career, which took him from Vienna, to Pest, to Innsbruck, and back to Vienna. As the climatologist Karl Fritsch wrote in his obituary for Kerner: "There is not the least doubt that his repeated change of residence, which brought Kerner into regions that differed starkly from the perspective of plant geography, thereby influenced him strongly. A man endowed with such a gift for sharp observation would have to notice that many species of plant that had previously been assumed to be homogeneous had a different appearance in the Hungarian plains as in the foothills of Transylvania as in the Alpine valleys of Tyrol."[52] As we will see in chapter 10, Kerner's sensitivity to this heterogeneity had surprising scientific consequences.

"CAR-WINDOW CLIMATOLOGY"

Experiences of travel were crucial to the emerging sense of the Monarchy as a geographic unit.[53] As travel across the territory was regularized, it became much easier to perceive coherence across the landscapes of Austria-Hungary. New forms of transportation, including the railway, mountaineering, and ballooning, proved essential to the work of climatology; they also produced a panoramic experience of empire.

The railway provided Julius Hann with the rhetorical framework he needed to write a synthetic description of the climate of Austria-Hungary for the overview volume of the *Kronprinzenwerk*, which appeared in 1887. Hann imagined a traveler making the half day's journey from a wintry Vienna to the warmth and sunshine of Fiume. He showed how rail travel could produce an experience of unity within diversity. As he pointed out, a tireless traveler could cover the entire east-west breadth of the empire, boarding in Czernowitz/Chernivtsi/Cernăuți and disembarking in Bregenz, without being likely to experience a temperature difference between start and finish of more than three degrees. Rail travel thus furnished the illusion of being able to survey the Habsburg climate at a glance.

Surprisingly, mountaineering played an analogous role.[54] As naturalists hiked their way up mountaintops and down into valleys, they frequently thought about the effects this trajectory would have on moving air. Heinz Ficker, an avid alpinist, invariably described changes in weather in terms of the modification of air masses as they traversed peaks and valleys. In this vein, during his research in Turkestan in 1913, Ficker often sketched the silhouette of a mountain in order to contemplate its meteorological effects. Likewise, he repeatedly contrasted landscapes that were "open" to air masses with those that were closed, going so far as to write of "invasions" of air into open valleys. Such images made it possible to envision the mountain chains of the Monarchy not as barriers, as they would appear on a two-dimensional map, but as features that structured a continuous, three-dimensional circulation.[55]

More dramatically, Ficker also became intimate with the Alps by hot-air balloon. The value of the aerial view was suggested by Crown Prince Rudolf's introduction to the *Kronprinzenwerk*. He proposed "a tour through wide open spaces, through multilingual nations, in the midst of constantly changing images," leading the reader along "the path of the mountains," as if on the wings of a bird. The reality of ballooning was not quite so graceful. As Ficker put it in his "Investigations of Föhn by Balloon," the research was "disturbed by many mishaps."[56] That much was to be expected when flying an inflated sheet of cloth directly into a windstorm. But manned balloons nonetheless became an important mode of climatological exploration circa 1900, and Heinz Ficker, Albert Defant, Wilhelm Trabert, and Wilhelm Schmidt were all accomplished balloonists.

One might expect that ballooning provided the ultimate panoramic view, but that was not in fact the aim. Rather, the balloonist sought an unmediated experience of atmospheric dynamics: to become a test particle in the wind field. And yet in reality his experience was mediated by the measuring instruments that he carried with him. Ficker cautioned that the readings of his instruments were "imprecise," since the balloon was buffeted like "a toy" by vertical air currents. More could be learned simply from the trajectory of the balloon, he concluded. As the British meteorologist and balloonist James

Glaisher described one flight: "Experienced a difficulty in reading the instruments. Lost myself, could not see to read the instruments."[57] It was precisely by "losing oneself" that the balloonist could begin to orient in new ways. Instead of the two-dimensional sense of space usually sufficient for everyday life on the earth's surface, the balloonist learned to find his bearings in the three dimensions of the atmosphere. Thus Ficker built up from his own kinesthetic experiences to a three-dimensional map of airflow over the Alps. In this way, he also learned to distinguish between air currents that were "only of an entirely local nature" and those with an influence on the "general flow."[58] The hot-air balloon thus became an essential tool of scaling.

This pursuit of climatology as extreme sport was by no means the norm at the time. By contrast, the Russian climatologist Alexander Voeikov was famous for traveling in luxury. As Julius Hann wrote to Wladimir Köppen from Vienna in 1886, "Voeikov is here, but I haven't seen him yet. He has a nice life, our friend! Little chance we'll ever get to combine science and pleasure like that!"[59] Voeikov's research relied primarily on station data rather than firsthand observations. When he finally traveled to Turkestan—in his view, a pivotal region for the climatic future of the Russian Empire—he rode in a private wagon of the Trans-Caspian Railway, supplemented by coaches and automobiles.[60] Other climatologists stressed the value of the climatological impressions of an observer in motion. Robert DeCourcy Ward—one of Hann's most fervent admirers—applauded what he termed "car-window climatology." Car-window climatology was "non-instrumental, unsystematic, irregular, 'haphazard' if you will." It thus complemented the highly regular, instrumental, numerical observations that were coming to define this science by the early twentieth century. "Travelers, even when passing rapidly through a country on the railroad, and, still better, when moving more slowly on horseback or on foot, usually have opportunities for making simple non-instrumental observations which will add greatly to the interest of their journey, and which, if the region is comparatively little known, may really be of considerable importance."[61]

ABSOLUTE AND LIVED SCALES OF MEASUREMENT

What could have been the value of firsthand impressions that were "unsystematic, irregular, 'haphazard'"? Consider that the second half of the nineteenth century saw the rapid expansion of instrumental weather-observing networks, but these networks remained incomplete. There was a flood of numerical data from permanent observatories, but it was compromised by geographical gaps that could only be filled in by the work of surveys. At the same time, late nineteenth-century scientists were conscious of living through a transitional era in the globalization of science: the closing of the heroic age of scientific exploration.

What did climatology stand to lose with this historical transition? "One would hardly be wrong to characterize the predominant method of climatology today," wrote the Heidelberg geographer Alfred Hettner in 1924, "by saying that it consists in the analysis of the quantitative measurements of meteorological stations." To view this as the *only* method, however, was to assume that all "precise" climatological measurements were instrumental and to be conducted at regular intervals over long periods of time. This was not the case. Hettner insisted that not all observations could be made with mechanical instruments, and station nets were not dense enough to capture local variations. Moreover, average values could not do justice to the "physiological" qualities of a local climate, such as the length of seasons or the characteristic winds.[62]

Hettner cited Austrian authorities like Hann and Supan on this point. In 1906 Hann had lamented that the weather journals kept by observers of the imperial network

> do not contain nearly everything that one would consider desirable from a climatographic perspective. That is, namely, such recordings as are made without instruments, as for instance the occurrence of the last damaging frost (*Reif*) of spring and the first of fall (which are not reported regularly enough to be able to analyze them); further the date of foliation, of the first buds, of the ripening of fruit of some widely distributed and familiar bushes, trees, and crops, in which the specifics of the local climatic idiosyncrasies, namely the influence of exposition, of a valley or cliff location, of the shadow of a mountain peak, etc., are expressed more easily, simply, and vividly than in instrumental readings.[63]

Heinz Ficker agreed wholeheartedly. As he put it in 1919 in his firsthand account of the climate of Turkestan: "on such a journey the eyes of the observer are the best instrument, which make possible precisely those determinations, which we seek in vain in the observation records of permanent stations."[64] The distribution of plants and animals was an important indicator of the spatial variation of climate. Local cultures also held clues. Thus an attentive climatologist would attend to the types of crops grown, the terms used for the winds, and the locations where people gathered to restore their health.

In short, climatologists were interested in variables that were not readily measurable with existing instruments. Instruments like thermometers and barometers had been developed precisely in order to isolate independent variables.[65] But living things respond to meteorological elements in combination.

Hence the interest in monitoring complex variables like evaporation, as well as other factors deemed essential to organic life, such as insolation, humidity, transpiration, and ozone content. Measuring these was often trickier than recording temperature or air pressure. Evaporation, a factor essential to agriculture, was particularly hard to measure, since it varies in complex ways with the other meteorological elements and depends on the degree of ventilation of the instrument. In cases like this, it was necessary to decide whether the variable in question was to be measured "as a practical task" or "as a meteorological and climatological element"—that is, whether a relative or absolute value was needed.[66]

The early twentieth century saw the introduction of devices designed to record factors vital to health such as ultraviolet radiation, ozone levels, and newly defined variables like "felt temperature." Many of these instruments required expert manipulation and were prone to errors. Measuring humidity with a hair hygrometer, for instance, relied on an organic rather than a mechanical response—the dependence of the width of human hair on atmospheric humidity. But not all hair behaves the same. Incorporating the body into the instrument introduced the same errors of personal difference that afflicted subjective, noninstrumental observations. Meanwhile, in the subfields of medical climatology and phenology (the study of seasonal phenomena among plants and animals), organic bodies were the only relevant instruments.

For all these reasons, Habsburg climatologists often prioritized observations that were mobile, full-bodied, and multisensory. This emphasis resonates strongly with Edmund Husserl's notion of "prescientific" experience. Indeed, Husserl's phenomenology is uniquely helpful for understanding the work of scaling in this place and time, since it took shape as a diagnosis of the state of the natural sciences in central Europe circa 1900. It was a quest for the "prescientific" experiences out of which science had developed historically and from which it derived its original human meaning. According to Husserl, the "prescientific" or "natural" world is known in large part through kinesthesis—the sensation of one's own bodily movements. Objects are "constituted" as observers move around them in space and time. The "natural" world is organized by a division between "near" and "far" as defined by one's own body—which Husserl, making the link to measurement explicit, called the "zero-point."[67] Husserl argued that conscious movement has the potential to break down this divide and replace it with an implicit scale permitting statements about *relative* proximity. Likewise, Habsburg climatologists argued that personal observations recorded in transit, often on the basis of sensory impressions, could correct and complement data from networks of permanent observing stations.

In phenomenological terms, their approach to scaling sought to reconcile the "lived" scale of firsthand measurements with the "absolute" scale of station data.

AUSTRIA AND THE GLOBALIZATION OF PHYSICAL GEOGRAPHY

Studying Austria-Hungary as a territorial unit could also inspire thinking on even larger scales. Before the advent of electronic calculators, analyzing large-scale geoscientific data was extremely laborious. Julius Hann, for instance, "lost all desire to take up a similarly wide subject again any time soon" after completing his study of the daily variation of temperature worldwide. What gave him the courage to undertake another global study a few years later was the experience of working through calculations of the same type for the Austrian half of the Monarchy. Through his "official post," the task fell to him to analyze the "very comprehensive" data of the Austrian meteorological network. "This brought it back into my head to carry out for Austria the investigation I had conceived earlier, in order subsequently to append some general comparisons for a larger portion of the earth's surface."[68] Two years later, Hann's *Handbook of Climatology* appeared, the most comprehensive survey and analysis to date of worldwide observations.

Indeed, the Habsburg territory proved good to think with. It freed the field sciences from the narrow horizons of the nation-state. This process produced results with global application, from the equations of dynamic climatology as worked out by Hann, Ficker, Margules, Defant, and Exner, to the climatology of prehistory as explained by Penck and Arthur Wagner, to Jovan Cvijić's (1865–1927) theory of karst formation. In this sense, the field of view of the imperial-royal scientist was crucial to the development of the understanding of the earth as a whole. The supranational structure of the Habsburg state shaped the logic of the field sciences, which strove at once for a view of the whole and the legibility of local differences. At the same time, these panoramic studies of the nature of Austria-Hungary contributed to a new understanding of the political phenomenon of the supranational state: a vision of the Dual Monarchy as "a structure formed not by chance but by necessity," in the words of Crown Prince Rudolf.[69]

Eduard Suess

Eduard Suess's tectonic theory of the formation of the large-scale features of the earth's surface took shape as he pieced together a mental map of Austria's geological landscapes. Born in London, Suess moved several times during his youth, living in Prague and then Vienna. Each new environment he encountered made its mark on him. A recuperative stay in the Bohemian spa town of Karlsbad, for instance, showed him a valley veined with granite, which was "so strikingly different in landscape as in geological structure from Prague and Vienna that I never tired of walking and observing closely." Once employed at the Royal-Imperial Mining Museum (Montanisches Museum), he traveled the Monarchy in the service of its mineral collection. He also took up mountaineering in the Alps, hoping it would help him to regain his strength. One bright September morning, after hiking for three and a half hours in darkness, he and a companion reached the peak of the Dachstein, a mountain made famous by Suess's senior colleague, Friedrich Simony. "The living map spread itself out wide below us," Suess recalled in his memoirs. By then, at the age of twenty, he had gotten to know several geologically different regions of the Monarchy: "the granite landscape of Karlsbad, the limestone and slate mountains near Prague, the tertiary landscape of Vienna, and one type of the Calcareous Alps." In Suess's mind, these juxtapositions posed a riddle about the history of the earth. "The contrast between the Bohemian Massif and the Alps was inexplicable to me; the elucidation of the riddle was from then on my life work."[70] Upon his return from a trip to France, he was struck again by the richness of the Habsburg territory for geological study, by "the diversity of our empire." A trip to Galicia planted in his mind the question of why the western Carpathians differed so markedly from the area around Kraków —"such a great contrast was hardly known anywhere else on earth at the time."[71] Already he was thinking on a global scale, wondering, for instance, about the distribution of plants and animals across the earth's surface and its implications for the history of the continents. He worked by differentiating between local and global phenomena, between evidence of processes restricted to one region and evidence of a truly planetary process.

For the questions that interested Suess, questions above all about changes in sea level, the Austrian Empire was ideally situated: "There the land is arrayed before us in unusual variety. Hardly anywhere in Europe are tectonic contrasts so plainly presented—contrasts between the Bohemian Massif and the Alps, between the portion of Russian table-land beneath the Galician plain and the Carpathians, the peculiar connection of Alps and Carpathians, the continuance of the Turkestan depression over the Aral Sea into the depression of the Danube and to Vienna, and much besides."[72] Such contrasts could only be explained with reference to a planetwide cause. In this way, he arrived at his theory of "eustatic" or global shifts in sea level.[73]

As he remembered it, this theory was born "on the plains of Eggenburg," an otherwise unremarkable patch of Lower Austria, which only took on significance in relation to the other landscapes stored in Suess's memory. "For the first time I was seized by the thought that such extended uniformity could not come about by an elevation of the land, but only by the sinking of the water level." Here lay the answer to his riddle. Subsequently, as he patiently sorted through the work of colleagues around the world, Suess came to see a planetwide pattern: stable continents on one hand, variable sea levels on the other. There was no need to suppose a force of mountain uplift, as so many of his contemporaries still did. His explanation of variations of the strandline was elegantly simple and universal: "The crust of the earth gives way and falls in; the sea follows it." In other words, the earth was collapsing in on itself like a desiccating apple. Suess attributed this bold hypothesis to his experience of the physical diversity of the Habsburg lands.

Albrecht Penck

Unlike Suess, Albrecht Penck arrived in Vienna in 1885, at the age of twenty-seven, fully formed as a geographer. Born in Leipzig, Penck had earned his doctorate at the University of Leipzig and his habilitation at Munich. To his former colleagues back in the German Empire, Penck's career appeared to be heading in a dangerous direction. They accused him of following the predilection of his mentor Suess for theorizing on a global scale. The historian of science Norman Henniges describes the approach to scaling common among German physical geographers at the time, which mitigated against large-scale interpretations and reinforced the hierarchical structure of the German geological survey.[74] By German standards, Penck was indulging in speculation. In Austria, on the other hand, he was celebrated for his ability to tease out the significance of local details within a global framework.

Like his colleagues in Vienna, and in disagreement with many in Germany, Penck came to see the Habsburg lands as a coherent unit: "Bohemia and Moravia, Hungary and the Alpine lands, form a complete whole and not simply the result of a clever politics of marriage. Vienna sits in the center of it as the nexus of important thoroughfares. . . . It is a center of crystallization of the highest quality." Under new political circumstances, however, Penck would later change his mind on this point. In an unpublished memoir written under the Third Reich, he explained why he had accepted a job at the University of Vienna in 1884 over an offer from Königsberg. He wrote that he been drawn to Vienna for the city, the university, and for his colleagues Hann and Suess—and in spite of his sense that Austria was "a state in the throes of decline."[75] Yet a memoir written under a dictatorship is a source to be read with care. Whatever Penck's private opinions on central European politics may have been during his twenty years in Vienna, he shared the opinion of his colleagues there that the territory of Austria-Hungary was an ideal field for modern "scientific" geography.

Penck's research benefited enormously from his time in Austria. He used the physical geography of the empire as lecture material, preparing his students for fieldwork. What's more, the eastern and Dinaric Alps were crucial sites for his studies of Europe's climate during the Ice Age. Small glaciers were, as he put it, the best "climatic measuring instruments." Observations of glaciers throughout the Alpine and southeastern lands of the Monarchy allowed him to determine snow lines for the Ice Age. These suggested that weather patterns had looked quite different than in the present, as had the distribution of vegetation. In 1906, as he prepared to leave Vienna for Berlin, he paid tribute to the empire he had served: "Far richer in geographical contrasts than most of the other European states, [Austria] beckons above all for the study of its own land. The wealth of contrasts of all kinds makes Austria an observing field for geographers virtually unmatched in Europe."[76]

Penck's experience of the physical diversity of the Habsburg lands mirrored that of Suess; not so his experience of the Monarchy's human diversity. Again treating his 1943 remarks with caution, consider this account of the process by which he had "oriented himself" "within the complicated physical composition of this state. I began to look around, traveled along the coast and back by way of Agram [Zagreb] and Budapest. There I became aware just how different were the levels of civilization in the individual regions that were welded together in the Dual Monarchy. Only the old Austria was German, all the others non-German, and I came to feel that the Germans were regarded everywhere with mistrust. I only got to know the western part of the Monarchy, but excursions with students indeed took me into the center of Hungary and to Bosnia-Herzegovina, as well as Dalmatia. I encountered Transylvania only later and Galicia and Bukovina not at all. That was a mistake." Avoiding Galicia and Bukovina was a mistake, presumably, from the perspective of 1943, when geographical knowledge of the Polish-Ukrainian borderlands was of strategic value to the Third Reich. What is striking about this passage is that the ethnic diversity so meticulously cataloged by Austrian geographers is reduced to the categories of "German" and "non-German." As Norman Henniges has argued, Penck's gaze was highly sensitive to geographic particularity, but dull when it came to human observations. Ethnographically, Penck saw in Austria only what he expected to see based on the crudest map of self and other.[77] He did not share the imperial-royal scientist's commitment to maintaining the legibility of human difference. Yet he did share a facility for thinking across scales, evident in his glaciological reasoning about the climate of prehistory. Even in the absence of any commitment to an Austrian Idea, Penck's science was nonetheless shaped by the structure of the supranational state.

Jovan Cvijić

Like Penck, Jovan Cvijić was not born into the Habsburg Monarchy, and yet its territory proved essential to his research. Cvijić arrived at the University of Vienna from his native Serbia in 1889, at the age of twenty-three, to study physical geography with Suess, Penck, and Hann. By 1892 he had completed a dissertation under their supervision titled "The Karst Phenomenon." The term *karst* was a German form of the Slavic word *kras*, meaning stony ground. It referred in particular to the landscape of what were known as the "karst lands" of the Dinaric Alps, a relatively barren landscape characterized by caves, fissures, and sinkholes—dry on the surface, but concealing rivers in its subterranean layers of porous limestone. In the rainy season, water would pool along the margins of karst fields, but it would soon disappear and leave the surface parched. According to Cvijić, the study of karst had progressed farther in Austria than anywhere else by the late nineteenth century, partly because scholars there had long taken a practical interest in the problem of supplying water to this thirsty land. The geological riddle of karst received renewed attention after the Habsburg army occupied Bosnia-Herzegovina in 1878. Irrigation and the construction of roads and railways demanded a better understanding of the karst environment, and Habsburg geologists were commissioned to study the area. It was in Austria, according to Cvijić, that the two dominant theories of karst formation had found their clearest articulations.[78] The more popular theory held that karst was the result of the collapse or fracturing of layers of rock. Alternatively, some argued that karst's peculiar form was due to water percolating through the rock and either eroding or corroding the limestone (that is, by a mechanical or a chemical process). By the time that Cvijić arrived in Vienna, karst was a hot topic.

Cvijić's approach to the subject was unprecedented in the breadth and detail of his comparisons. He meticulously classified karst from different regions, with the aim of elucidating the relationship between surface structure and subterranean hydrography. Particularly important were his observations of karst formations in the Habsburg lands of Carniola and Moravia, alongside the classic example of the Dinaric karst. Inspired by Suess's eustatic theory, Cvijić paid close attention to water levels, and he was able to show significant seasonal fluctuations. He concluded that the most important process in karst formation was the chemical dissolution of limestone.

It was Cvijić's 1893 publication that cemented "karst" as a universal term, beating out competitors like *le Causse*, used by Édouard-Alfred Martel to describe karst in France's Central Massif. Cvijić was hailed as the founder of the field of karstology, and karst was recognized as a phenomenon "distinct from the standard phenomena of fluvial geomorphology." Of course, there has always been the potential for confusion when using a regional name for a general phenomenon. But that is precisely the point of this example: to see how one local instance came to stand for a global phenomenon, and how Cvijić's knowledge of a few scattered fields of rock emerged as a global specialization.

Cvijić also surveyed the Balkans as an amateur ethnographer. By 1902 he had covered most of the peninsula. He had even studied parts of the Ottoman Empire, where he felt he was putting himself at risk, despite having more than one cavalry soldier to escort him. Among the phenomena he observed were the customs of the human inhabitants. As he put it, "A traveler conducting research who crosses a large region is involuntarily drawn to making anthropogeographic observations."[79] Cvijić distinguished four different types of inhabitants in the Balkans, dividing these further into "varieties" and "groups." He emphasized that the distribution of these populations reflected a history of migrations, driven in part by the difficulty of agriculture in the karst lands.[80] Despite the differences among these groups, many of which he attributed to their contrasting physical environments, Cvijić insisted that they were all part of a single "South Slavic" nation. Note that this position was by no means disloyal to the Habsburgs. Many Habsburg patriots thought similarly at the time and hoped that the Dual Monarchy would give the South Slavs a degree of autonomy parallel to that enjoyed by Hungary. Indeed, Cvijić's account of the South Slav nation might be described as a small-scale version of the Austrian Idea. It too rested on an ethnography of migration and exchange, and one that argued for unity amid diversity.

*

According to Eduard Suess, it was a supreme virtue of geological field research that it brought the scientist into contact with new cultures as well as new landscapes. In this way, Suess argued, earth science could provide an ethical education, a prophylactic against nationalist demagoguery.

> The scientist may be fortunate enough to gaze into the spiritual life of another nation. If he encounters there the same emotions, the identical sources of pain and joy, the same appreciation for what is noble and the same avoidance of what is base, then there awakens in him alongside patriotism a general love for humanity. It may be an abomination to the professional politician, but it germinates deep in every healthy human soul, and despite the politician's resistance, it germinates in his as well, or rather has the potential to do so.[81]

Suess's own political career bore out this thesis, as he consistently sought to promote causes that transcended national divides, such as the provision of clean drinking water, the regulation of the Danube, and the transfer of primary education from church to state. Since Suess died in 1914, he was mercifully spared the knowledge of the leading role that physical geographers would play in the warmongering and border disputes during and after the First World War. His formulation of the relationship between geology and universal humanism may have been naive as a general principle, but let us consider it instead as his interpretation of his own career. In his own experience, his ability to scale between local structures and global causes was intimately connected to his capacity for empathy. In this respect, Suess was carving out an identity for the scientist as Habsburg civil servant.[82]

CONCLUSION

In *The Climatology of Bohemia*, Karl Kreil described the study of climate as a negotiation between micro- and macroperspectives, between "the world on the large scale and on the small."[83] As we will see in chapter 6, Kreil's language mirrored Adalbert Stifter's often-cited defense of literary realism in the 1853 preface to *Bunte Steine*. There he justified his attention to ordinary people and their ordinary lives by analogy to geophysics, specifically the study of magnetic variations across the earth's surface. The relativity of scale emerged simultaneously as a methodological precept and an aesthetic principle in Austrian science and literature, its rhetorical power in one domain reinforcing its resonance in the other.

Others tested the political force of this trope. The Bohemian philosopher Josef Durdík, for instance, linked the Czech political project to natural science through the personality of the Prague astronomer Johannes Kepler. Kepler's calculations were "bitter work," but the goal, the "search for the whole," was ever present in his mind. He "explored the smallest details, but had always the liveliest thought for the whole of nature—that is also the mark of intellectual health."[84] More famously, the Czech nationalist Tomáš Masaryk, author of a youthful treatise on the classification of the natural sciences, labeled his political strategy "small, detailed, mundane work." He drew an explicit analogy to the significance of "small things" in the geosciences: "The world has stood and [still] stands on work. . . . The world is maintained only by work, and small-scale work at that, by constant work. As in geology, there are no catastrophes, nor have there ever been; what were once considered to be isolated, sudden catastrophes we now recognize to be the result of countless small and smaller influences."[85] Intentionally or otherwise, Masaryk was echoing the words of J. E. Purkyně upon launching the first Czech-language scientific journal, reminding his readers of the need to attend to small details.

More than mere rhetoric, this principle was meant to regulate the organization of scientific work in the multinational empire. It stood for a pluralist approach to the coordination of knowledge, without recourse to a single overarching system, nor to a hierarchy of explanations, nor even to a common language. And it kept alive the link between scientific knowledge and the quotidian, local purposes that had, historically, motivated its production.[86]

The work of scaling was political work in this sense: in Habsburg learned societies, naturalists negotiated the value of their varied local perspectives relative to the goal of a synthetic overview. Suess expressed this ethos in the preface to the monumental physical atlas of the Monarchy, *Das Bau und Bild Österreichs* (1903), which he coauthored with three distinguished colleagues, each responsible for a different region. He alerted readers at the start: "This is not a collective work. Just as each of the authors has collected his observations on his own journeys and has reached his conclusions independently, thus each has chosen to approach his account in his own way; and what is offered here is therefore not one representation, but rather four representations in one shared frame."[87] Thus the work of producing an overview of Austria-Hungary's natural foundations advanced like the piecing of a quilt, the seams attesting to the value of each contributor's local perspective. In Prussia, by contrast, the task of a wide geographic survey gave rise to a rigidly hierarchical structure for geological fieldwork, with the intention of eliminating individual subjectivity.[88] Not so in Austria. Austrian field scientists opted instead for what Marianne Klemun describes as a culture of consensus building, in which the synthesis depended on proper respect accorded to each contributor's regional viewpoint.[89]

In this way, the rhetoric of scaling translated the politics of science into the politics of empire and vice versa. Hann drew the parallel explicitly, likening the naturalist who neglects details to a haughty prince who ignores the petition of a poor woman.[90] But the prince could not afford to lose sight of the forest for the trees. Thus, in a 1906 essay on the globalization of atmospheric science, Hann expressed his satisfaction that meteorology had finally moved beyond the narrow-mindedness of *Kirchturmpolitik* or "church tower politics."[91] This was a term that had come into frequent use over the past decade as Habsburg-loyal statesmen denounced the rising politics of nationalism. On behalf of his scientific discipline, Hann was laying claim to the broad-mindedness of modern internationalism. The language of Habsburg climatology took shape in this political context, tacking between "detailed work" and the "synoptic" view.

CHAPTER 4

The Dual Task

Vienna's Central Institute for Meteorology and Geomagnetism defined itself against comparable institutions in other lands by virtue of its "dual task," as Kreil explained in the first volume of the institute's yearbook. It would be a "model institution," demonstrating the most modern, precise, and thorough methods of geoscientific observation; and, at the same time, it would be the central node of the empire's observational network, "which surveils everyone and, when necessary, instructs and assists." This language of duality succinctly expressed the opportunities and challenges of imperial-royal science. As Kreil elaborated, the "dual task" demanded that Habsburg scientists train their attention in two directions at once: one eye on the international scientific community, which sought results of global significance, and the other on the Habsburg populace and its varied needs.

> Our institute distinguishes itself significantly in this way from other institutions of the same kind, which either exist only in seclusion for themselves as observatories, or, without engaging in observation themselves, are concerned only with the analysis and publication of observations arriving from their stations. This dual purpose also makes necessary a corresponding division of our labor, which cannot be dedicated exclusively to one or the other, but must keep both equally in sight.[1]

This dual task demanded a dual vision, a gaze split into two fields of view. In this way, the concept of duality entered Habsburg science a decade and a half before the Monarchy itself became "dual." This chapter explores the meanings that Kreil attached to "duality" and its scientific and political significance.

Duality meant seeing double, from more than one standpoint at once. It is a principle contained in the very label "imperial-royal"—an invocation of the two identities of the monarch, as simultaneously emperor of the composite state of Austria and king of its individual crown lands.[2] In the sciences, duality signified the aspiration to embody simultaneously the virtues of a global *Wissenschaft* and a situated *Landeskunde*. Politically, duality delineated the unique responsibility of the imperial-royal scientist to speak both to an international audience and a national one. At a time when the profession of scientific "popularizer" was emerging in western Europe, the imperial-royal scientist was charged with bringing his research to the public himself. What motivated this distinctively public role was the fervent hope of the 1848 generation that the field sciences could motivate transnational cooperation both within and beyond the Habsburg borders. Kreil's concept of duality also tracks two strains of research: a "replicative" approach that treats the local atmosphere as a laboratory for the investigation of universal laws and a "chorological" approach that studies the local atmosphere as a unique piece in the jigsaw puzzle of the global circulation. The history of modern atmospheric science has depended on the complementarity of these approaches.

A METEOROLOGICAL *MITTELPUNKT*

Austria took the first steps toward an empire-wide weather-observing network in the 1840s, lagging behind the "global renewal of the earth sciences" by French, British, and Prussian scientists in the previous decade.[3] Austrian scientists were aware that they had been outpaced. They knew little about how the earth's climate and magnetic field varied spatially across the Habsburg lands. As Kreil wrote to Alexander von Humboldt in 1843, "The time will soon come in which the new science cannot rest content with the results of scattered European stations, but will study the magnetic force over every square mile."[4] Kreil began this undertaking with a geomagnetic survey of Bohemia, supported by the Bohemian Academy of Sciences. In 1844, he won imperial funding to extend his work across the Habsburg lands. Accompanied by his eager assistant Karl Fritsch, Kreil spent the next three summers on the road, devoting winters to calculation. The pair covered Upper Austria, Tyrol and Vorarlberg, Lombardy, Lower Austria, Styria, Illyria, the Adriatic coast, Venetia, Dalmatia, Moravia, Silesia, northern Hungary, Transylvania, and Galicia.[5] In order to coordinate measurements across such a vast space, they relied on the methods and instruments for which Kreil had won renown as director of the Prague observatory. They also enlisted collaborators who were suitably distributed across the territory, in Kremsmünster, Graz, Innsbruck, and Tarnow (Galicia). The ultimate goal was a permanent observing network spanning the entire Monarchy.

The founding of the Academy of Sciences in Vienna in 1847 lent new energy to this project. The idea for such an academy had been around since the days of Leibniz, but numerous initiatives had failed for lack of enthusiasm or because scholars disagreed on the form it should take. Meanwhile, provincial learned societies were founded in Bohemia in 1770 and in Hungary in 1825; Croatia (1866) and Galicia (1873) followed, and a new Czech Academy of Sciences and Arts formed in 1890. In the 1840s, as talk of constitutionalism and republicanism spread among central European intellectuals, the notion of an imperial academy found an unlikely proponent in the conservative chancellor, Count Metternich. He advised the emperor that the state should provide an outlet for new ideas rather than allowing them to fester.[6] Emperor Ferdinand gave the academy his approval and granted it the freedom to publish without interference from the imperial censor.

What distinguished the Vienna Academy of Sciences from other learned societies of its day was its explicitly supranational mandate. As historian Christine Ottner points out, the initial proposals referred to the academy as the "central node" (*Centralpunct*), indicating its intended role of assembling scholars from across the empire. From the start, however, opinions diverged on the proper relationship between center and peripheries. When the academy's first president drew up a tentative list of members, Metternich and his underlings criticized it for favoring residents of the imperial capital too heavily. They warned that the academy was in danger of becoming a local society for Vienna instead of a genuinely imperial institution. Archduke Johann, the inspiration behind the Joanneum in Graz, was among those who recognized what provincial scholars could contribute. The academy president acquiesced, and the initial list of forty full members included thirteen from outside Vienna, including representatives from Hungary, Bohemia, and Lombardy-Venetia. There remained disagreement over the language(s) in which discussion would be conducted, as well as complaints about the dues owed by provincial members, since it was unlikely that they would attend meetings regularly. The geologist Wilhelm Haidinger (1795–1871) even called on the government to provide scholars with railway tickets in order to travel to the meetings in Vienna in a manner "befitting their class."[7] The events of 1848, however, shifted opinion in favor of the centralizers, and meeting attendance by provincial scholars fell accordingly.

In climatology, the notion that Vienna was the Monarchy's "central node" was particularly resonant, but equally laden with ambiguities. Already in the spring of 1849, the Academy of Sciences had begun to plan an imperial institute for the study of meteorology and earth magnetism. In language typical of the reformist spirit of the 1840s, academy members lamented that Austria had fallen behind other states in this field of research, and "that it must go a step farther where possible in order to put right, in a sense, what has been neglected here for so long." The press likewise stressed the need to make up for lost time in "coming to know the general laws of atmospheric change in our climate."[8] Note the reference to "our climate." Of course, no one believed the climate of the empire to be uniform. Rather, the phrase evoked the political orientation of Kreil's project, its "whole-state" patriotism.

In a memo to the education ministry in July 1850, Kreil called for the creation of an observing network spanning the "entire monarchy." He argued that the study of meteorological and magnetic phenomena was of great importance due to their influence on human life and commerce, as well as for the investigation of the laws of nature. The Habsburg territory comprised so many different "climatic zones" that abundant observing stations would be necessary, and the ZAMG would require exemption from post and telegraph charges and funding to send scientists to inspect these stations periodically.[9] In public addresses, Kreil explained that the science of the atmosphere was a product of the "awakening of the spirit of research" in the eighteenth century, and that it depended on "association," since "the individual can only be productive in communication with others." By Kreil's count, at the time of writing there were ninety-four observing stations in the Austrian empire. It was, therefore, "without doubt high time to unite the dispersed and isolated efforts around one common purpose; to standardize and guide them; in short, to call to life an organic whole, an observing system."[10]

This patriotic language of organic unity and "communal goals" was keyed to the political circumstances of 1848. Climatology was exactly the kind of "outlet" that Metternich had in mind for the Academy of Sciences. It seemed comfortably apolitical in comparison with fashionable topics of historical and philological research. It would appease liberals with expectations of economic gain. And the cooperation and coordination it demanded would help to bind the peripheries more tightly to the imperial center.

The research of this network was to be broad in scope. At the time, "climate" designated not only atmospheric conditions but many phenomena of the organic world that stood in a relationship of interdependence with the atmosphere. Thus the factors to be observed included properties and phenomena that would be viewed today as part of meteorology, hydrology, atmospheric optics, and seismology, as well as "other unusual phenomena."[11] This list reflects the many practical applications that scientists foresaw for climatology. Kreil himself expected the greatest benefits for agriculture, public health, and shipping.[12] The network's observations were also used to assess the fertility of land in order to calculate property taxes, and as evidence in legal cases in which a defendant might claim innocence of a tort due to a meteorological "act of God."

The most innovative aspect of this program was the phenological observations—that is, records of periodic phenomena in the lives of plants and animals, such as the appearance of leaves, flowers, and fruits, the migrations of birds and fish, and the metamorphoses of insects. The origin of phenology as a science is usually traced to the Belgian astronomer Adolphe Quetelet in 1841. However, the Prague botanist Karl Fritsch had begun a phenological record back in 1834, before Quetelet. In recognition of this work, Fritsch was called to Vienna to become vice director of the new ZAMG.[13] He composed instructions for phenological observations far more detailed than those of Quetelet. The value of this program was immediately recognized by Austria's leading botanist, Franz Unger, who saw it as foundational to the burgeoning science of botanical geography. Like Kreil, Unger emphasized that "collaborative observations are most necessary." "Until now," Unger wrote, "individual naturalists have given the most contradictory answers to such questions. It is self-evident that here only comprehensive observations, undertaken on a large scale, by many people simultaneously, and through many generations, can achieve the goal."[14]

Phenology became an essential tool for visualizing the space of the empire and its history. Tracking the onset of the seasons introduced the perspective of migrating animals as they periodically traversed this space. It also evoked the point of view of human migrants, whether they traveled for seasonal employment—as many did in this era—or for seasonal stays at spas and resorts. When such observations were collected across generations, as Unger urged, they could also reveal temporal fluctuations of climate. Unger believed that human activities like agriculture could temporarily alter the climate, and he was evidently eager to collect evidence of such effects.[15] The result of these efforts was to add a dynamic, temporal element to the emerging map of climate across the Habsburg territory.

DEFINING DUALITY

As Kreil explained, the ZAMG would be a premier research institution, a "model" for observatories in other lands. In everything connected to meteorology and geomagnetism, it would not merely reproduce the highest state of scholarship but push beyond it. At the same time, the institute would be a clearinghouse for observational data from across the Monarchy, a "center of calculation," in Bruno Latour's sense. It would house the standard instruments against which all others would be calibrated, and it would instruct "friends of these disciplines" in their use.[16] However, unlike imperial institutions such as London's Met Office and Kew Gardens, the ZAMG also promised to serve directly the practical needs of the empire's provinces.

Imperial and Provincial

This promise was taken seriously by those outside the metropole. Consider, for instance, the assessment of the Moravian geographer Karl (Karel) Kořistka in 1861. Kořistka identified himself as a loyal Habsburg citizen. He praised the "whole-state" science of the past decade and the new institutions that were producing it, both imperial and provincial. He suggested, however, that the time had come to make these results useful to local communities. "Every friend of the fatherland who has attentively followed the intellectual efforts and undertakings of the last decade will confirm that these, steering clear of the space of political activity, have been directed, mostly with success, to the study of the natural composition, as well as the ethnographic, agricultural, and industrial conditions of the extensive lands of the Austrian state as a whole. . . . It seems to us that it is now time to synthesize the results of this work within the framework of individual provinces."[17] To illustrate how this knowledge could be applied, Kořistka edited a *Landeskunde* of Moravia and Silesia that included a chapter on the regional climate penned by Carl Jelinek, the director of the ZAMG at the time, and which relied on the data of its network.

The Moravian meteorologist and climatologist František Augustin also praised the ZAMG, but was more pointed in noting that it had emphasized synoptic-scale and basic research at the expense of small-scale, applied studies. "One cannot reproach the central institute in Vienna if it does not want to lose sight of the main goal of meteorology, which consists of the determination of meteorological laws and of research on the leading climatic factors, to which recently has also been added the conduct of synoptic studies." In doing so, however, it had neglected "ancillary interests such as detailed climatic research on individual provinces or the pursuit of meteorology with respect to field and forest management." Augustin argued that now, in 1885, it fell to the crown lands to right this balance between centralization and decentralization.[18] What exactly this entailed, however, was not yet clear. Two years later, Augustin would stand accused of plagiarism for having claimed a Czech translation of several pages of Hann's *Handbook of Climatology* as original work. The ensuing scandal raised the question of what provincial autonomy in the sciences would actually mean.[19] Augustin's accuser was a cartographer and young Czech politician. He prefaced his charge against Augustin with the phrase: "But the appearance deceives; truth prevails." This was an invocation of the Hussite motto associated with Tomáš Masaryk's campaign in the 1880s to expose the forgery behind what were claimed to be founding documents of the Czech nation—a motto that would later appear on the banner of the president of Czechoslovakia. This critic clearly hoped to see "Czech science" pursued in future with greater integrity and independence of mind.

In fact, the ZAMG worked hard to combine empire-wide research with local, practically oriented studies. In 1914, for instance, the institute was asked to weigh in on a proposal from officials in Bohemia for the establishment of stations for the measurement of evaporation, a critical factor for agriculture. The stated aim was to study the climate of Bohemia as closely as possible as an aid to the selection of crop varieties and fertilizers.[20] Director Trabert responded that the ZAMG had long overseen measurements of this kind in Bohemia and had put the results in the hands of locals, for instance, by furnishing vineyards with predictions of nighttime frosts. Trabert applauded the effort to build out the observing network in the "purely Czech part of Bohemia," which had remained "weak" despite the efforts of "Czech colleagues." But he objected that a Bohemian institution would "overstep its mandate" if it were to process the data itself.[21] In this way, the ZAMG would remain the center of calculation, even while tailoring research to local needs.

Private and Public

Kreil also described the ZAMG as "dual" in the sense that it was both a private and a public institution—part of a private academy, yet working for the public good. The implication was that the imperial-royal scientist could serve the state and still lay claim to a scholar's intellectual freedom. This was consistent with Kreil's own institutional affiliations. His position as director of the ZAMG had come with an appointment as full professor of physics at the University of Vienna, where faculty had fought for academic freedom in 1848 and won. Yet Kreil and his successors would struggle to maintain a balance between loyal civil service and intellectual autonomy.

Duality in this sense also implied that the ZAMG was not a typical nineteenth-century observatory, where astronomers or geophysicists would tuck themselves away and pursue research undisturbed by the public.[22] Seclusion was not an option. It was the responsibility of the institute's scientists to make regular treks across the Monarchy to inspect and calibrate measuring instruments throughout the network. It was also their duty to engage the public as scientific observers. Tellingly, Friedrich Simony chided the typical observatory for being more of a corner for scientists to retreat into than a window onto the world. He advised scientists to get out and travel. Meteorologists, he said, needed to develop a painter's eye for nature's variety, and this was not something they could acquire from the "narrow space of the observatory and from its instruments."[23] Far from sheltering scientists in a metropolitan tower, the ZAMG was to be their portal onto the empire at large.

Universal and Particular

Finally, the ZAMG's task was dual in the epistemological sense of balancing the quest for universal laws with attention to local peculiarities. "There is probably no scientific task on which local conditions have a greater influence than meteorological observations," Kreil wrote. Weather records reflected local geography and an "abundance of minor circumstances, to which one usually pays too little attention." In this category of "minor circumstances," he included the course of nearby mountain chains and rivers, the proximity of ocean or standing waters, the geological conditions and vegetative cover, and the situation of the measuring instruments. "Ignorance of these circumstances places everyone who wants to use such observations in danger of ascribing to the influence of the atmosphere what actually is a result of local conditions, and thus to draw false conclusions."[24] Atmospheric science therefore required working simultaneously at multiple scales.

*

In sum, the task of the imperial-royal atmospheric scientist was dual in three respects. First, he was meant to establish Austria's scientific reputation internationally, while also catering to the local and practical needs of each crown land. Second, he was both a scholar and a civil servant, autonomous in some respects but not in others. Thus he was responsible to a private society, the Academy of Sciences, and to the public, who took an eager interest in his science and felt entitled to its results. Finally, he was to pursue the universal laws of geophysics, but also the endless forms those phenomena might take within the borders of the Monarchy.

The ZAMG presented itself as dual in these ways when it approached the public to request their participation as volunteer observers. In 1869, Director Jelinek oversaw the publication of an official guide to meteorological observation. In keeping with Kreil's definition of duality, the guide explained that the observing network had two goals: to "determine the climatic conditions of a given area," on one hand, and to "generate the material for an investigation of the general laws of the atmosphere," on the other.[25] Accordingly, observers were instructed to locate their homes on "a good map," and to identify a window facing north or north-northwest. They were then prompted to describe how their local environments—whether flat or hilly, coastal or inland—modified the "general air currents."[26] As ambiguous as these instructions may have been, they effectively communicated the essential problem of disentangling locally contingent phenomena from those unfolding on a larger scale.

"PUBLIC FEELING"

The first foreign visitors to the ZAMG were surprised that such a modest-looking institution could be publishing such impressive results. Its quarters consisted of a tight space in a new building on Favoritenstrasse, already a busy part of town and growing ever busier. It was an improbably modest space, but chock-full of self-registering instruments of Kreil's own design. Kreil did his best to make the space work for precision geophysics. In order to sight the horizon, for instance, he first had to obtain permission to have a balcony built. Desperate as he was to make improvements, Kreil initially sought public support. But he soon learned to keep his head down, working "in silence and seclusion" on bustling Favoritenstrasse.[27]

To outsiders, it was a mystery why such an energetic and outspoken man as Kreil had fallen so quiet. Within a few years, the work of the institute suffered from lack of financial support, and Kreil gave up the annual publication of the network's observations.[28] Even to his own staff, his behavior appeared mysterious.[29] Had Kreil acquired a taste for seclusion? The evidence sooner suggests that the "darkness in which he hid himself" was a response to political circumstances. The controversy, since forgotten, went back to the early days of the Academy of Sciences. In the heady weeks of the summer of 1848, a group of scholars, led by the geologist Wilhelm Haidinger, had formed a rival association, the Society of Friends of the Natural Sciences, with a more populist and pluralist agenda.[30] It would address all branches of science and be "open to all."[31] Consistent with his pluricentric program, Haidinger simultaneously worked to develop parallel bodies elsewhere in the empire, including Moravia and Silesia, Pest, and Milan.[32] Although the Society of Friends of the Natural Sciences described itself respectfully as an outgrowth of the Academy of Sciences, it stood as a powerful critique of that institution. Its members lamented the exclusivity and narrow research interests of the academy.[33] Haidinger, who became director of the new Imperial Geological Institute in the fall of 1849, drew the ire of members of the new academy, and his institute fell under attack. Due to the academy's influence, the Geological Institute was threatened with budget cuts or even dissolution. When Austria's military loss in 1859 plunged the state into fiscal crisis, Haidinger's worst fears were realized. The Imperial Geological Institute was put under direct control of the Academy of Sciences, its budget was cut, and the state stopped paying rent on the institute's quarters at the Palais Lichtenstein.

What saved the Geological Institute was the beginning of the end of neo-absolutism, as the budget shortage forced the emperor to compromise with political liberals. Subsequent generations of Habsburg geologists reveled in the story of how their science had been rescued by the downfall of neo-absolutism. On the floor of the newly "strengthened" parliament, in the fall of 1860, the institute found its defenders: aristocrats who praised the practical value of geological research for mining and pointed to the international prestige of the Imperial Geological Institute. Count Andrássy, the future Hungarian prime minister, argued that science flourishes under conditions of competition, not monopoly, and the institute should therefore be allowed to remain independent of the Academy of Sciences.[34]

This was a victory for the principle of pluralism and a turning point for the politics of science in imperial Austria. With the liberal agenda gaining traction after 1859, Habsburg scientists could seek public engagement without fear of reprisal. For the field sciences, in particular, this was a transformative opportunity. The lay scientific networks and associations that developed across the Monarchy in the following decades proved essential to the fields of geography, geology, botany, zoology, ethnography, and last but not least, meteorology and climatology. Perhaps in celebration, Karl Fritsch delivered a pair of public lectures in Vienna on meteorology in the fall of 1861, "On Meteorological Observations" and "The Climate of Vienna."[35]

Kreil had led the ZAMG through its desert years, but he would not live to see the promised land. He passed away in 1862 at the age of sixty-four, without having published even the first volume of his projected climatological survey of Austria. At Kreil's death, "this institute found itself in a desperate situation" financially.[36] It was left to his successor, Carl Jelinek, to realize Kreil's vision for the Habsburg observing network. In 1865, the Austrian Meteorological

Society was founded with the goal of fostering public interest and participation. It published the premier journal of the discipline, the *Zeitschrift der österreichischen Gesellschaft für Meteorologie*, later the *Meteorologische Zeitschrift*. Jelinek's detailed "Instructions to Observers" was an international model and would remain in print for decades. Under Jelinek, the number of observing stations in the ZAMG's network grew from 118 to 238.[37]

By the time Hann took over the directorship from Jelinek in 1877, the public-spirited orientation of the ZAMG was firmly ingrained. And yet, internationally, it was increasingly evident that publicity was a double-edged sword for meteorological institutions. In Britain, as Katharine Anderson has shown, the Met Office's botched attempts at storm forecasting were drawing public outrage. To critics, it looked like a great deal of public money was being wasted to record data that was put to no good use. Asked by a member of parliament whether it was worth the expense of continuing to publish the Met Office's daily observations, the astronomer royal George Biddell Airy argued in 1877 that meteorological publications were a matter of "public interest."[38] Airy's invocation of "public interest" resonated in Vienna. Twenty-five years later, Hann was still repeating it for readers of the *Meteorologische Zeitschrift*. He pointed out that the director of the ZAMG was frequently asked by local authorities or even private individuals for meteorological data, to inform public services or economic calculations. He quoted Airy to the effect that "public feeling is an element not to be put out of question in matters of this kind."[39] What Hann called (in English) "public feeling" was a factor that could not be ignored in Vienna.

THE IMPERIAL NETWORK

With the founding of the ZAMG, meteorological observations could be made across the empire "according to a uniform scheme," "providing the basis for all investigations of the climatic conditions of the empire."[40] Yet this satisfaction masks some shortcomings. The Monarchy was not evenly covered by the new observing network (see figure 9). The station density was much higher in the western lands than in the east, higher too in the north than in the south. Allowing for differences in population density, Carinthia had the most stations per person, followed by the other Alpine lands and Bohemia; Hungary, Transylvania, Galicia, Bukovina, and Lombardy-Venetia had the fewest. In 1870 Galicia averaged one station per forty-eight square miles, while Tyrol and Vorarlberg counted one for every fourteen square miles.[41] Some claimed that the problem was a lack of interest in meteorology among non-Germans. The truth was hardly that simple.

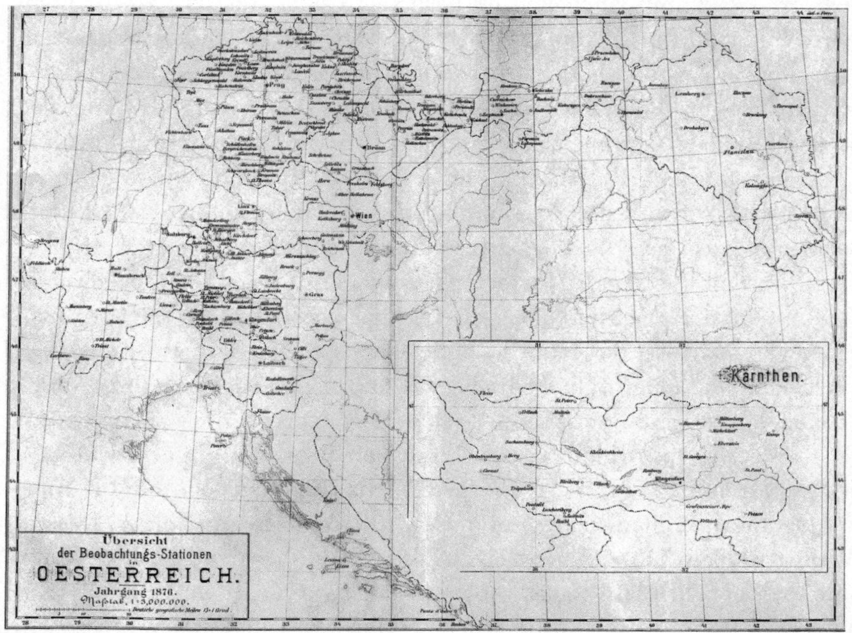

FIGURE 9. Distribution of meteorological observing stations in the Austrian half of the Habsburg Monarchy, 1876.

Since the new network was born amid revolution and war, it did not originally include Hungary, Croatia, and the Italian lands, where fighting continued through 1849.[42] In Hungary, meteorological observations had been recorded on and off since 1783 at the observatory in Buda and elsewhere. The geographer János/Johann Hunfalvy worked with the ZAMG to synthesize these, producing a climatographic overview of the Hungarian lands for the journal of the Austrian Meteorological Society in 1867.[43] Meteorological observations formed part of the charge of the new natural-scientific committee organized by the Hungarian Academy of Sciences in 1860. By 1863 there were eleven stations in Hungary, and by 1866 there were twenty-six.[44] A Hungarian Central Institute for Meteorology was founded in 1870, at which point there were 152 stations in operation in the Hungarian half of the Dual Monarchy.[45] Thereafter, Hungary took full responsibility for its own climatic research.

The Habsburgs' brief coastline on the Adriatic represented another gap in the network. Atmospheric observations along the coast were vital to Austria's shipping interests. Throughout the 1860s, Jelinek was in frequent contact with the imperial Hydrographic Bureau in Triest. In this region, however, the network faced unusual challenges, particularly after Austria's loss of Lombardy in 1859. Jelinek complained about the difficulty of enlisting unpaid collaborators in a region where Italian nationalism was rife.[46] He regretted that he, like Kreil before him, had no personal influence in the region, and he appealed to the imperial government for diplomatic assistance in the recruitment of observers. Interestingly, Jelinek justified this request not only with reference to knowledge of sea winds for shipping, but also, more fundamentally, "from a climatological perspective, because the conditions of the mainland differ from those of the islands." In fact, climatological research had been under way in Görz-Gradisca and along the Croatian coast since 1859, and in 1866 the government and the Academy of Sciences sponsored a cartographic and physiographic survey of the Adriatic, which encompassed both the periodic (climatological) and nonperiodic (meteorological) conditions of the coast.[47] To the ZAMG, the Adriatic coast was not merely a source of storm warnings; it was part of the larger picture of the climatic conditions of central Europe.

The most problematic region of all for the ZAMG was Galicia. Despite Vienna's attempts to seed new stations there, local societies took the research into their own hands. Galician geologists had set a precedent for doing so in 1869 by opting out of the Austrian geological survey. The Academy of Sciences in Kraków established its own commission for the production of a geological atlas of Galicia, although this was apparently based on surveys led by the Imperial Geological Institute in Vienna. In the 1880s a row erupted between Kraków and Vienna over the right to publish a geographical atlas of Galicia. Geographers in Kraków insisted that the Viennese survey had been carried out too quickly and the resulting map contained errors. They accused the Viennese of taking a "flagrant attitude" and "monopolizing scientific research."[48]

In meteorology, cooperation between Vienna and Kraków likewise got off to a rocky start. Instrumental weather measurements had been made in Lemberg/Lwów/Lviv since the 1820s.[49] Several local organizations later took up this work, among them the Balnealogical Commission, the Meteorological Section of the Physiographic Commission of the Academy of Sciences in Kraków, the Tatras Society, and the Bureau of Melioration (part of the Provincial Executive Board), established in 1857, 1865, 1877, and 1881, respectively. Not surprisingly, the stations were unevenly distributed, with their center of gravity in the wealthier western half of the crown land. Galician naturalists had hoped to get the network off the ground in the west, then use the Tatras Society to expand it eastward. Instructions for meteorological observations were first published in Polish in 1867, but were not identical to those issued by the ZAMG.[50] Some of the instruments used were supplied by Vienna, but others came from Kraków and were not calibrated against those of the ZAMG. From the available sources, it seems that many of the Galician stations sent their observations to the Meteorological Section in Kraków rather than to Vienna. These observations, meanwhile, were often of dubious quality. Despite Kreil's insistence on the exhaustive description of the observing site, most of the reports came with no description of the terrain. It may be that the quality of observations declined with time. When the ZAMG sent its inspector to Galicia in 1877, he found conditions at the seven stations he visited to be far from satisfactory. Thermometers were placed at different heights above the ground, precipitation was measured at different times of day, and some observers used local time while others used central European time.[51]

The root of the problem in Galicia may have been the ZAMG's insistence that observers work on a volunteer basis, without pay—a hard sell among the poor, largely rural population of Galicia. Indeed, in 1895, when the Hydrographical Bureau in L'viv took over the stations of the Kraków Academy of Sciences and began paying observers for their work, the supply of observations increased dramatically. These observers included teachers, state forest workers, and farmers, all of whom were undoubtedly happy to have the extra salary. Some even quit when they did not receive a raise. According to the only historical study of meteorology in Galicia, the Hydrographical Bureau was the sole agency that succeeded in collecting reliable climatological observations in Galicia.[52]

IMPOSING STANDARDS

Thus the Monarchy remained unevenly covered by the observing network, and where stations existed, they did not necessarily proceed uniformly. In fact, for years there was no uniform observing schedule—"because one cannot demand of voluntary observers that they keep to a schedule that, due to their other occupations, is inconvenient, often even impossible."[53] What then was the best way to calculate a daily average temperature for measurements taken in different places at different hours? The author of a 1901 study of temperature variability in Austria tested dozens of different formulas, an exercise that only served to emphasize the infinite number of possible combinations. None could be considered the "true" average.[54] How could this heap of numbers be transformed into a standardized set of variables with which to construct a continuous overview of the Monarchy's climate? Given the limits of standardization, the fundamental challenge was to devise a rational method for comparing data recorded by different observers at different stations in different environments at different times, and to represent the results without creating the illusion of greater certainty than was warranted. As Victor Conrad later put it, "the principal and fundamental aim of climatological methods is to make the climatological series comparable."[55]

A first class of obstacles involved the variable conditions under which measurements were made. Consider, for instance, the subjective character of observations of elements such as cloudiness and precipitation and the inconsistent participation of volunteer observers. Add to these disruptions to telegraphic communications, shoddy instruments, and the unreported substitution of one instrument for another. One commonly overlooked problem, for instance, was a leaky rain gauge. A small hole gradually grew, until someone thought to repair or exchange it—at which point, "the cycle began anew."[56] A researcher relying on a series of rainfall measurements from this station faced the problem of distinguishing between real variations in rainfall and those due to the varying leakiness of the container; real variations were just as likely as those due to the leak to exhibit a periodic character. The problem was one that climatologists summed up by calling such a measurement series "inhomogeneous"; a homogeneous series, by contrast, was one where variations were caused only by the meteorological elements under study. In principle, homogeneity could be judged via comparison to measurements from a nearby station, since real variations in rainfall could be expected to affect surrounding measurements as well.

Other problems arose from the fact that weather stations came into and fell out of use at different times. This could lead to skewed inferences: an average January temperature taken from a few years of particularly cold weather would make that location look misleadingly colder than the rest of the region. In order to compare a shorter series of observations with a longer one, it was therefore necessary to "reduce" the former to the time period of the latter. To that end, the station that had been operated consistently was termed the "normal" station and its period of operation was the "normal" period. "Reduction" meant a correction based on the difference between (or ratio of) the average of the meteorological element (e.g., temperature or rainfall) at the two stations over their coinciding periods. This difference was expected to be more constant than the element itself, assuming that the two stations were similarly affected by the same weather events.

This method of reduction became increasingly important to Habsburg climatology in later years, as scientists attempted to fill in gaps in their knowledge of the empire's peripheries. To take an extreme example, consider Victor Conrad's determination to complete his *Climatography of Bukovina* after the outbreak of the First World War. Until 1910, Bukovina had no meteorological station south or east of the capital, and the station density in Galicia and Bukovina was approximately one-fifth that of the rest of the Austrian half of the monarchy.[57] The provincial school ministry had complained that the forecasts published for eastern Galicia and Bukovina were useless for the latter, since the combination of rivers and mountains gave Bukovina an entirely different climate than Galicia. Conrad in turn blamed the problem on a shortage of suitable observers: the local population was "on too low a level with respect to intelligence and reliability."[58] In wartime, however, even these limited observations were beyond Conrad's reach, and he relied heavily on methods of reduction. Conrad's satisfaction with these approximation methods seems to have been tied to his assessment of Bukowina's climate as "monotone," although hardly benign. Its precipitation features, for instance, were "actually quite simple": the plains were dry, while rainfall increased with altitude, although no strict regularity could be derived from the scanty data. Instead, to demonstrate the region's tendency to extreme summer rainfall, Conrad relied on a personal anecdote: an afternoon downpour in 1912, when a "torrential river" flowed through the streets, carrying pieces of furniture, and "in which a few small dogs are said to have drowned."[59] Conrad evoked a peasant culture at the mercy of nature. By portraying such an "extreme" as "characteristic" of Bukovina's climate, Conrad reinforced his description of the region as "continental," a technical term implying extremes of heat and cold, but here also carrying the stigma of "eastern." Conrad quickly reached the conclusion that the "impact of the steppe climate" was so strong that "gradually even the great morphological differences of landscape and their influence on the climate are obliterated by it; the climate of Bukovina is monotone."[60] In this way, the climatologist circumvented the problem of scarce data. A land so closely related to the steppes was predictable even in its unpredictability.

Methods of reduction were also essential to the climatography of Serbia that Conrad produced during its wartime occupation by the Habsburg army. He described this work paternalistically as "a form of rescue effort," relying as it did on data the Serbians had failed to publish, produced by stations that had been "swept away by the war." In typical colonialist language, he insisted that "meteorological and climatological research is a cultural imperative [*Kulturförderung*]," and hoped that his efforts would serve Serbia's "reconstruction" "in the near future."[61] That the Serbian observations would require extensive statistical manipulation was self-evident, for they were riddled with "many printing errors and untrustworthy values."[62] Conrad's methods of reduction found an ultimate, if circular, justification in his conclusion that Serbia's climate was "easy enough to describe," "neat and clear," "an extraordinarily simple and easily predictable image."[63] Conrad's final judgment was that Serbia displayed "a central European climate with strong continental traits." Its stormy, dusty southeastern winds were a boon to agriculture, he admitted, but a bane to human health and comfort.[64]

A striking contrast to Conrad's simplifications can be found in the climatography of Bosnia-Herzegovina published in 1918. Austrians typically described Bosnia as a backward, mountainous land where the environment set rigid constraints on social development, and where forests were still "intact." Experts urged that the region's natural resources be managed directly by the state according to a "wise use" ethic. For this reason, the state placed a premium on the "rapid scientific conquest" of the colony.[65] The Habsburg military had begun to construct climatological stations in Bosnia shortly after Habsburg occupation in 1878. Hann had deemed these lands a "total terra incognita" for climatology, inhabited by a population whose attitude toward weather was "Oriental fatalism."[66] Of particular interest to the Habsburg authorities was data on rainfall in the karst region of western Bosnia and Herzegovina, which was judged to be inhospitable to agriculture and human health due to its excessive heat in summer and fierce bora winds in winter. The karst region attracted geological and hydrological studies geared to improving the land for agricultural use, a project for which climatic data was also essential.[67]

The 1918 *Climatography of Bosnia and Hercegovina* was published by the Institute for Balkan Studies in Sarajevo, which had been established by private means in 1904 and would be dismantled by the provisional Bosnian government in 1918 as an institution "that served Austro-Hungarian interests."[68] The study's author was the Prague physical geographer Julie Moscheles, a descendant of a cosmopolitan family who would describe herself in the wake of World War One as an Anglo-German "with no sense for national feuds."[69] Unlike Conrad in his studies of Bukovina and Serbia, and unlike Hann in his climatological sketch of Bosnia of 1883, Moscheles chose not to use reduction methods to compensate for significant gaps in the data. She found reduction appropriate to the "western and central European climatic region," but unsuited to Bosnia's highly variable blend of continental and subtropical traits. The climate of Bosnia-Herzegovina varied so greatly in time that a different "normal" station would be needed from year to year—and even for each meteorological element. Robert Donia has argued that the Habsburg authorities worked to subvert Serbian and Croatian nationalism in Bosnia by promoting a "multireligious Bosnian identity based on Bosnians' common loyalty to the territory of Bosnia-Herzegovina."[70] The genre of climatography would seem to be a valuable resource for fostering a territorial identity. In practice, however, the *Climatography of Bosnia and Hercegovina* suggested that even a territorial identity might be unstable: "In extreme cases our entire region falls first into one sphere of influence, then, shortly after, into another."[71]

Whatever their solution, Habsburg climatologists recognized that the method of reduction was not foolproof and that they often worked at its furthest limits of applicability. The more widely spaced the stations, the less secure this method would be. Hann summed this up in an empirical equation stating that the variation in the differences between series from two stations increases linearly with both the distance between the stations and their difference in altitude. In certain areas, however, this equation breaks down, and the variation in the differences in a given direction in space becomes *discontinuous*. Beyond that geographic divide, the normal station no longer serves as a useful reference point. The region is no longer "climatically coherent," to use a term that Victor Conrad introduced in the 1920s. Instead of throwing up their hands, however, imperial climatographers turned this incoherence into an analytical tool in its own right. Such lines of discontinuous variation served to identify "climatic divides."[72] In this way, statistical manipulations could bring local contrasts into focus, giving definition to a landscape viewed, of necessity, from a distance. Klein's *Climatography of Styria* likened this effect to scaling a peak in order to make the landscape below intelligible. Just as the view from above could reveal the geomorphology of the surface, so could statistics provide "a liberating overview."[73] Such was the power of statistics to resolve a disorderly natural world into a map of diversity.

A PRIVILEGED VIEWPOINT

Klein's analogy to a mountain outlook was not fortuitous. Mountain observatories can be understood as points of leverage in the effort to achieve a scientific, integrated vision of the Monarchy. Thus the short-lived Sonnwendstein station in the Semmering was promoted as the key to weather forecasting for the entire Mediterranean: with low-pressure centers arriving from the north, the Alps became "the weather wall or the weather hinge [*die Wetterseite resp. der Wetterwinkel*] of the Mediterranean region. What Iceland is for northern Europe, the chain of the Alps is for the Mediterranean." The southernmost Habsburg lands were said to require high-altitude observatories in the Alps as a "form of secondary defensive line." In this way, "potentially two thirds of all weather events in the south of Austria could be placed under surveillance [*unter Kontrolle*]."[74] These martial metaphors of "defense" and "surveillance" suggest that mountain observatories were associated with the fantasy of a vantage point from which to achieve a strategic overview of the empire.

At the same time, the view from above was popularly associated with a supranational, pluralistic perspective. Thus the Galician writer Leopold von Sacher-Masoch founded an "international review" of science and culture in 1881 under the title *Auf der Höhe* (At the Pinnacle). The journal's program was contained in its title, as the first issue explained:

> We will not rest content with excluding every narrow-mindedness, every hateful prejudice in political, national, religious, scientific, or literary matters; our goal is far more to make our review a neutral ground, on which no interests are presumed to be those of humanity in general, on which the major intellects of all nations and orientations can address each other openly and honestly, but always with respect.[75]

In this sense, the view from "the pinnacle" represented the pluralist ethos of imperial-royal science, as developed within the new state institutions of the 1840s and 1850s.

From the earliest days of European alpinism, the mountain had functioned as a site of science in two quite different ways.[76] One way was "replicative," in the sense introduced above: tall peaks furnished a physico-chemical laboratory, where experiments on inanimate matter or living bodies could be conducted under extremes of low pressure and intense radiation. This type of research sought universal laws, results that would be independent of any particular locale. A second mode of mountain research was "chorological": it focused on what was specific to a given mountain site, whether in its natural or human history. Such chorological knowledge might serve the needs of locals and/or contribute to projects of documenting the variety of natural conditions on a larger scale. As mutually exclusive as these two ways of approaching the mountain might seem, they proved mutually reinforcing in practice.

Although Austria's mountains and coastline were prized as windows onto the passage of storms, Habsburg scientists never saw mountain and coastal observatories merely as perches from which to watch weather pass. On one hand, in a replicative vein, they sought a basic physical understanding of how the physical existence of the mountain or coastline modified weather locally and beyond. But they were equally interested in coastal and mountain climates for their own sake, from a chorological perspective. As in the work of the Adriatic Commission, they investigated how these unique physical environments interacted with human and nonhuman life. From the replicative perspective, the coastline or mountain was a laboratory, a space for producing generalizable atmospheric effects under extreme conditions. From the chorological view, it was a field, a place where distinctive atmospheric conditions were collected like specimens, as clues to patterns of geographic variation.

Tellingly, the Austrians fell into conflict with American scientists on precisely this point. Americans conceived of mountain observatories far more narrowly as aids to forecasting. The high-altitude stations set up by the US Weather Service were not intended to contribute to basic research, nor to descriptive climatology. The Americans were skeptical of the replicative function of mountain observatories in Austria. Specifically, they disputed the mountain's status as a laboratory for the investigation of the physics of storms. On the basis of data from the Sonnblick, Hann claimed to have refuted William Ferrel's thermal theory of cyclones by showing that cyclones correlated with cool temperatures at high altitude. Ferrel retorted that the observatory had no means of determining air temperature nearby at the same height. A temperature reading at the Sonnblick might be below average for the season, yet still warmer than the surrounding air. More fundamentally, Henry A. Hazen and Henry Helm Clayton rejected the Sonnblick's claim to be a good approximation to the free air. Hazen charged that the Sonnblick was really part of an extended mountain range, "not an isolated peak," and for that reason its data alternately mimicked the local climatic effects of one or the other of its neighboring valleys (see figure 10).[77] Ferrel, Hazen, and Clayton rejected the Sonnblick as a site of replicative investigations.

This disagreement highlights the unusual status of mountain meteorological observatories in Austria-Hungary. Throughout the first half of the nineteenth century, Alpine tourism and scientific attention to mountains had grown hand in hand. The Austrian Alpenverein, the earliest mountaineering club in continental Europe, was founded in 1862 by natural scientists with the primary goal of opening up the mountains to scientific research. According to a survey of the "most important" mountain weather stations worldwide in 1900, seven out of thirty-two were in Cisleithania, all but one of which lay in the Alps; an eighth, in Habsburg-occupied Bosnia, had the much-touted distinction of being the only high-altitude observatory on the Balkan Peninsula. These observatories were buoyed by a unique set of political affiliations. First, mountain observatories were local projects of modernization: funded in part by regional mountaineering clubs and tourist associations, they helped to extend an infrastructure of transport and communication and to further open

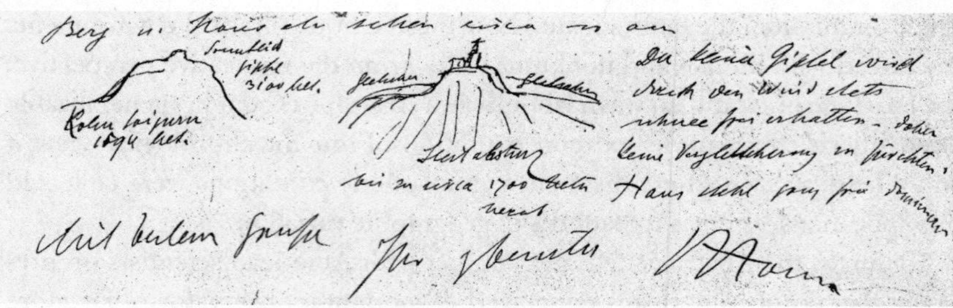

FIGURE 10. Sketch of the Sonnblick by Julius Hann shortly after the observatory's opening in 1886. In his disputes with American scientists, Hann argued that the Sonnblick was an "isolated peak" approximating the climate of the open air. He wrote to Köppen: "The small peak is kept consistently free of snow by the wind. So there is no danger of a glacier forming. The position of the house is entirely open and towering [*ganz frei dominant*]."

up the mountains to tourism. At the same time, the new high-altitude observatories were meant to help mitigate the impacts of modernization. By the 1880s, ethnographers realized that traditional ways of life in the mountainous regions of the Habsburg Monarchy were endangered. As putative "survivals" of an earlier stage of cultural evolution, mountain cultures were uniquely valuable to the burgeoning nationalist movements in central and eastern Europe that were intent on establishing their historical authenticity. This explains the fascination of both Polish and Ruthenian nationalists with the vanishing cultures of the Tatras.[78] In these respects, mountain observatories were provincial as well as imperial projects, helping to record and preserve elements of local culture and the local natural environment. Accordingly, the enormous costs of these stations was split between local voluntary associations and the imperial government.[79]

That is to say that mountain observatories had both an imperial and a local agenda: to connect isolated points to the empire's observing network and to record and preserve what was unique to their locales. Let's see how this dual program played out at three mountain observatories in three very different regions of the Monarchy: Carinthia, Bohemia, and Bosnia.

Sonnblick

At 3,105 meters the Sonnblick (see figure 11) was the highest year-round observatory in Europe at its opening in 1886. It perched like a doll's cabin atop the jagged peaks of the Hohe Tauern, aside massive glaciers, reinforced against winter storms known to pile snow thirty feet deep. Imagine, then, the surprise of visitors to find on this remote summit "the most modern lighting technology of the day." In an age when the streets of Vienna were still lit by oil lamps, the Sonnblick enjoyed electrical lighting, telephone service, and—the pièce de résistance—a mechanical lift to ferry scientists, tourists, and supplies from the valley to the mountain top. "It is indeed quite a wonderful effect, when the Edison lightbulbs suddenly twinkle against the night sky." To find a working telephone on a mountain covered in several meters of snow amazed even the scientific elite. The elevator that carried guests to the top struck an idiosyncratic balance between modern comfort and backwoods adventure. The machinist played up this incongruity by swaying the lift or stopping it dead at the most vertiginous moments, all in order to frighten the young women—whom the observatory's patron, Ignaz Rojacher, made a point of accompanying. Austrian scientists were fond of noting that the Sonnblick's communications technologies were unique in the high Alps: "Consider that in Oberpinzgau from Zell am See to Mittersill—that is, for a stretch of 30 kilometers—there is no telegraphic connection; and that a considerable part of this long and beautiful alpine valley, with its hamlets and ruggedly picturesque [*wildromantisch*] vales, must forego this modern means of communication." Thus the Sonnblick's claim to fame was not merely its technical accoutrements, but this particular juxtaposition of the "modern" and the *wildromantisch*.[80]

Indeed, the Sonnblick had an explicit modernizing agenda. Proto-industrialization had arrived early to the Hohe Tauern: its gold mines hit a modern high point in the late fifteenth to early seventeenth centuries, generating wealth that produced a bustling village and school near the Rauris gold mine, the future site of the Sonnblick. In the seventeenth century, however, the Little Ice Age began to transform the Rauris Valley. Mining suffered from the advance of the glaciers, although it did not cease entirely, and the village descended into poverty. The glaciers did not begin their retreat until the 1850s. This coincided with two crucial historical developments: the rise of alpinism as a leisure activity and the industrialization of the Austrian Alps. From the eighteenth century, the surrounding province of Carinthia slowly emerged as a center for iron production, although it was held back by the absence of coal. Facing the evidence that the mountain had been mined to excess, the mine's director, Rojacher, offered the highest point of the Rauris as the site of a meteorological station. According to another participant in the venture, it was Rojacher's bid to redirect the energies of the local population and to "elevate public spirit." The journal of the Sonnblick society presented these transformations as having successfully brought the Rauris Valley into the modern

FIGURE 11. The Sonnblick observatory in Carinthia, ca. 1915.

world. At the same time, the journal reported on archeological finds from the early modern heyday of the mining village, and published an ethnographic study of the survival of pagan forms of weather knowledge.[81] Thus the Sonnblick was celebrated at once for its role in modernizing the Rauris Valley, conserving its natural beauty, and salvaging its cultural traditions.

Donnersberg/Milešovka

Meanwhile, the turn of the twentieth century also saw the opening of mountain observatories in the Carpathians, the Ore Mountains (Erzgebirge/Krušné hory), and the Dinaric Alps. Their construction coincided with the monarchy's experiment with policies of decentralization, devolving authority onto provincial governments in the attempt to resolve nationalist standoffs. This was also the period in which the ZAMG experimented with the decentralization of its weather service, on the principle that more localized observations could better serve the needs of agriculture, trade, and tourism. Scientists at the ZAMG repeatedly endorsed proposals to give the individual crown lands greater control over the weather service; indeed these researchers were eager to be free of the responsibility for daily forecasting. In the end, however, only Tyrol was allowed to test an independent weather service. The ministry of education closed the debate on decentralization in 1913, ostensibly on financial grounds. The ZAMG also rejected demands to translate its forecasts into languages other than German.[82] All the while, however, provincial institutions were taking the matter into their own hands, setting up their own local meteorological networks and publishing their own results, some in Italian, Czech, or Polish.[83]

At the turn of the century, a group of German-speaking Bohemians active in the regional *Heimat* movement began to agitate for the construction of an observatory on the Donnersberg (also known as the Milleschauer or Milešovka) in the Ore Mountains, on the border between Bohemia and Saxony (see figure 12). The Donnersberg rose only 834 meters, a molehill Alpine standards. Yet its proponents claimed that, for meteorological purposes, a mountain was defined by isolation, not height.

Nineteenth-century descriptions of the Ore Mountains emphasized that they were surprisingly densely populated for such a raw, harsh land. The view from the Donnersberg revealed to an observer in the 1830s a "land blooming with craft production [*Gewerbefleiss*], covered in all directions with cities, market towns, and villages."[84] This was taken as evidence that Bohemians had successfully overcome constraints of soil and climate.[85] Yet industry still required protection from the hazards of weather. As the "Central Committee for the Construction of a Meteorological High-Altitude Station on the Donnersberg" wrote in its petition to the imperial government, Bohemia was an "orographically and meteorologically unique land," plagued by "weather damages and elementary catastrophes." Hence it needed forecasts based on local observations, and, therefore, a mountain observatory—at a cost of approximately 70,000 crowns. The plan's backers solicited designs from five architects. Tellingly, the winning proposal was for a miniature stone castle, replete with gables and a ten-meter-high tower with a panoramic balcony. The interior even followed local custom in placing a living room next to the kitchen on the south side. The observatory would be reachable in an hour or two on foot from the train station in Boreslau/Bořislav, making it relatively accessible by the standards of the era. In promotional materials for the observatory, its scientific director expressed the hope that it would "awaken in the many visitors a lively interest and contribute much to popular enlightenment in this field." For meteorology was a field where "ineradicable superstition and

FIGURE 12. The Donnersberg/Milešovka observatory in Bohemia, 1910 postcard.

archaic survivals, combined with the biases against any innovation, stand in the way of any kind of research like an impregnable bulwark."[86] The Donnersberg observatory, then, was to be a *Heimat*-tourism attraction and a venue of popular enlightenment.

Bjelašnica

A mountain observatory in Bosnia-Herzegovina was a modernizing project of yet another kind. As we have seen, the Habsburg state commissioned climatological studies of Bosnia in order to inform typical colonial projects of land improvement and river regulation. By 1894, the engineer Philipp Ballif had organized

FIGURE 13. The Bjelašnica observatory in Bosnia-Herzegovina, ca. 1904.

seventy-seven weather stations in Bosnia.[87] He and Hann advocated for a new observatory that would extend the ZAMG's network of high-altitude stations dramatically to the southeast. The imperial government opened the station at Bjelašnica in 1894, at a height of 1,067 meters and a cost of approximately 30,000 crowns.[88] Its data would be supplemented by the Romanian state meteorological network and Bulgaria's observatory in Sofia. Since Bjelašnica lay directly in the path of low-pressure troughs traveling from the Adriatic across Hungary to the Black Sea or across Austria to the Baltic, it was well positioned to telegraph storm warnings to Vienna and Budapest. Indeed, Bjelašnica was arguably more central than even the Sonnblick to central Europe's weather. The Bjelašnica observatory was thus a step toward the scientific conquest of the Balkans. Yet in the photographs that circulated in the scientific community (see figure 13), Bjelašnica lies shrouded in winter frost, accessible only through the kitchen window. These images reinforced the conclusion that Bosnia-Herzegovina was destined to remain a primitive land, a colony literally by nature.

CONCLUSION

These high-altitude observatories throw into relief the institutional "duality" of Habsburg climatology. The observatories on the Sonnblick, Donnersberg/

Milešovka, and Bjelašnica were "dual" in the sense of being both public and private institutions, with both imperial and local agendas. They reinforced the consciousness of imperial-royal climatologists that they were immersed in the medium they were studying, constrained to observe from one or another local perspective. This duality was essential to the character of climatology in the supranational state. After 1867, Vienna was only one of the Dual Monarchy's two capitals, while Prague, Kraków, and Zagreb were all eager to lay claim to equal status with Budapest. The "centrality" of the "Central Institute" was thus questionable. And in fact, after 1870, Hungary had a "Central Institute" of its own. In this sense, there could be no definitive perspective from which to gain an overview of the Monarchy. This was what the ZAMG learned when it sent scientists to Galicia to collect observations, only to find that many were being sent to Kraków instead. This duality—the center's awareness of being peripheral from a competing vantage point—determined the multiple scales on which climatological research unfolded.

* 2 *

The Scales of Empire

CHAPTER 5

The Face of the Empire

An unusual collection of objects went on display in the Vienna palace of the exiled Count Klemens von Metternich in the wake of the revolutions of 1848: fossilized creatures, mineral specimens, geological profiles, and panoramic paintings of the Alps. The exhibit was the work of the legendary naturalist Friedrich Simony (see figure 14). It was the event that launched his scientific career and, with it, the discipline of physical geography in imperial Austria.

Simony was born in Bohemia to an unwed mother and raised by her and his grandfather, a senior civil servant in Olmütz. After only two years at gymnasium, the fourteen-year-old boy was apprenticed to an uncle as an apothecary. By the late 1830s, Simony's pharmaceutical practice in Vienna was lucrative enough to allow him to pursue his love of natural history. He completed gymnasium through private tutoring, began attending lectures at the University of Vienna, and made his first long treks through the Alps. At this time, the winter climate of the high Alps was still a mystery to naturalists. Simony's notes on his early ascents are filled with recordings of temperature, atmospheric visibility, winds, and the forms of clouds. The peak that he called the most beautiful in all the Alps, the Dachstein, served him as an "aerial observatory," where he could hang his instruments and set about "to study occurrences in the kingdom of the clouds."[1] Among the occurrences that intrigued him were wintertime temperature inversions, when upper layers of the atmosphere become warmer than lower ones—a characteristic of mountain climates that held clues to atmospheric dynamics and the health effects of pollution. Over the next decade, Simony became renowned for his exploits as an alpinist, for his popular lectures on the natural history of the Alps, and for his exquisite landscape art.[2]

FIGURE 14. Portrait of Friedrich Simony (1813–96).

It was also in the course of the 1840s that Simony won the patronage of the most powerful man in central Europe, Count Metternich. Metternich's daughter Melanie saw Simony as a curiosity: "a very interesting man" who had given up his profession to study geography and to "spend his life on the summits of the Alps, where he remains even in winter." Count Metternich took a lively interest in the late Enlightenment project of a global physical geography and was in regular correspondence with naturalists like Humboldt and Leopold von Buch. In these last years before Metternich's fall from grace, Simony could often be seen coming and going in his lederhosen from his patron's palace on Vienna's stately Rennweg.[3]

It was within those walls that Simony managed to chart a new course for Habsburg science. By the late 1840s, he had expanded his already remarkable geological collection by another forty crates. He had also mapped the depths of Alpine lakes and produced a geological profile of the eastern Alps that measured six and a half meters.[4] At the Palais Metternich, these objects attracted the admiration of Count Leo von Hohenstein, the man whom the new emperor had entrusted with a top-to-bottom reform of the Habsburg universities. In a conversation that lasted three hours, the count displayed an insatiable curiosity about Simony's research. He was particularly intrigued by the hypothesis of the Ice Age, and even asked to take some of Simony's sketches with him for further study.[5] By the end of their meeting, the count had invited Simony to draw up a proposal for a university chair in this new field of study, "physical geography" (*Erdkunde*).

As we have seen, geography was one among several disciplines after 1848 that aimed to survey the vaunted multiplicity of the Habsburg lands and to present its findings to the public in the form of maps, atlases, panoramas, and museum displays. To this end, Habsburg geography was in search of new visual techniques. How could a representation of this territory produce an impression of overarching coherence while highlighting small-scale variability? How, for instance, could a single map do justice both to the topographical extremes of the Alps and the subtle gradations of the Hungarian flatlands? If the mathematical conundrum of squaring the circle was known in Leibniz's day as the *Problema austriacum*, then we might say that the technical challenges of visualizing "unity in diversity" became the "Austrian Problem" for the nineteenth century.

Certain solutions to this problem have become iconic. For instance, Karl von Czoernig's ethnographic map of Austria and Hungary applied stark color contrasts in order to make the visual argument that the ethnic diversity of the Monarchy reached down to such minute dimensions that no division of the territory along national lines was conceivable. Other responses have been forgotten, such as the Hauslab and Peucker color schemes for the representation of a broad but finely differentiated range of surface elevation, discussed below. In climatology, the Austrian Problem resulted in statistical and cartographic methods that made it possible to visualize locally confined, short-lived phenomena within a synthetic image of the climatic conditions of the Monarchy as a whole. By following these innovations, we will see how the eighteenth century's static, regional image of climate began to give way to a dynamic and multiscalar view.

MAPPING THE MONARCHY

It was not until the second half of the eighteenth century that maps of the Habsburg Monarchy as a whole began to be produced domestically. Until the eighteenth century, the Habsburgs had shown little interest in centralizing their power and standardizing their rule throughout their motley lands. Instead, they treated each as its own challenge, and the loyalty they inspired among their subjects took the form of personal fealty, not territorial patriotism. Symbolic of this style of rule were the multiple crowns that the dynasty owned—one for the Holy Roman Empire, one for Hungary, and one for Bohemia (plus lesser crowns corresponding to their smaller duchies and counties). The early modern Habsburg rulers had relied on Dutch and then French cartographers for overviews of their territories, while maps produced within the Habsburg lands focused on local and regional views.

After the Habsburg and Polish-Lithuanian armies freed Vienna from the Ottoman siege of 1683, Austria entered a new era of territorial expansion, decisively reconquering Hungary and acquiring land in northern Italy, the Balkan Peninsula, and Poland (though losing most of Silesia). These conquests provided an important stimulus to survey work and mapmaking.[6] Still, Austria's loss in the Seven Years' War (1756–63) was blamed in part on the lack of uniformly detailed, accurate maps of its territory, a necessity under the new conditions of mobile warfare.[7] Much of the detail of local topography was still unknown and unrepresented, and no attempt was made to represent the heights of hills and mountains precisely.

Soon after the close of the Seven Years' War, the army was commissioned to produce a new set of maps of all of Austria at 1:28,000, an unprecedented level of detail. This project, known as the Josephine Survey, was designed for military use. The maps remained state secrets, and only the highest-ranking military officers had access to them.[8] However, they also had civil uses. Joseph II had explicitly instructed his military cartographers in Transylvania to record economically relevant information, including soil quality, the distribution of manufacture, and natural resources, as Madalina Valeria Veres has shown. Joseph II used these manuscript maps to plan his travels to Hungary and to make decisions about economic policy.[9]

Behind this cartographic project stood the academic discipline of cameralism, which sought to achieve autarky in part by cataloging natural resources and investigating new processes of manufacture. While the Habsburg rulers of the sixteenth century had collected *naturalia* as a stimulus to wonder and a symbol of power, the cameralists of the eighteenth century recast the value of natural diversity in practical terms. Unlike the *naturalia* collections of the Renaissance, cameralist natural knowledge was explicitly territorial, concerned with the geographical distribution of resources. Cameralism thus helped to focus attention on the Habsburg lands as a single territorial unit. What's more, cameralism helped to set Austria on the path to becoming an integrated economic entity, with its own geography of centers and peripheries. Joseph II gradually eliminated internal tariffs, at the expense of some local producers, but to the benefit of what he now called "the totality that is the monarchy."[10]

Although ordinary Habsburg subjects never saw the new military maps of the Josephine era, they experienced the shift to a territorial conception of the Monarchy in other ways. As in Revolutionary France, centralization involved both the standardization of measurements and the rationalization of internal borders. Thus Maria Theresa introduced the Viennese post mile in 1756, which remained the standard measure of distance until the introduction of the metric system in 1871. In addition, the medieval tax system, based on an overlapping patchwork of feudal domains, was gradually remade as a uniform, continuous overlay of cadastral municipalities (*Kadastergemeinde*). The borders of these municipalities were drawn according to visible, natural landmarks, such as trees, hills, rivers, or boundary stones, in some cases diverging significantly from the human geography of a region. Of all the new internal divisions imposed by the centralizing state of the late eighteenth and early nineteenth centuries, these municipal borders remained the most palpable to Habsburg citizens up until the reforms of 1848 and 1867. They marked "on one hand the radius of self-governance and, on the other hand, the space in which each citizen was granted the right of residence." Surprisingly, however, this municipal grid was not inscribed on any map in the Josephine era; that project was not completed until 1861.[11] In short, spatial experiences of the Monarchy were changing, but in ways that were only beginning to be recorded visually.

From the 1780s Austria saw a boom in mapmaking, due to several factors, including the relaxation of censorship under Joseph II, which boosted the publishing industry; the success of Joseph's economic policies, which created a merchant class with money to spend on printed materials; and the founding of an engravers' academy in Vienna in 1768. Consider that between 1700 and 1779 only three hundred maps were published in the Austrian lands, while three hundred were produced between 1786 and 1790 alone. Among these maps were views of the newly acquired territories of Galicia and Bukovina, as well as of the Banat, Hungary's military border zone.[12]

The earliest published maps of the modern Monarchy as a whole reflected the eighteenth-century expansion of transport and communication.[13] Among these were representations of postal routes or *Postkarten*, in the original sense of the term. A 1782 postal map of the Habsburg lands (see figure 15) was festooned with an image of carriages driving down a winding road, which fades into the distance near the empire's southwest corner, as if inviting the viewer on a tour.[14] In a similar spirit, a 1785–86 map depicted a projected system of

FIGURE 15. *Post-Charte der Kaiserl. Königl. Erblanden*, by Georg Ignaz von Metzburg, detail, 1782.

canals that would connect Austria's major rivers, intended to "facilitate circulation" domestically and with foreign lands.[15] The plan itself turned out to be unworkable, because the engineer had little sense of the actual elevations in the Alps and Carpathians. Nonetheless, maps like these drew attention to the real and potential circulation of people, commodities, and information.

The late eighteenth century also saw the first cartographic representations of "the products of nature and industry," or what we would call economic maps of the Habsburg lands. As the *Wiener Zeitung* observed of the map in figure 16, "the main focus is on familiarizing a worthy public with the natural treasures and manufactured products of each province." As this image indicates, the effort to include all such information resulted in a map so overcrowded as to be virtually illegible.[16]

Two aspects of these eighteenth-century maps are particularly noteworthy. One is the omission of climatic information. Climate factored in these maps only in the form of notes about terrain that might or might not be travers-

FIGURE 16. Economic map of Carniola, ca. 1795.

able in a given season.[17] The kinds of climatic information that would have served agriculture or medical geography were not displayed on cartographic overviews, because they had not yet been centrally collected. Where climatic data was being recorded—as, for instance, by physicians working in the tradition of medical topography, or by agricultural improvement societies—it was a local project, without provision for standardized measurements across large regions. Second, it is evident that little attempt had been made as yet to impose order on the proliferation of cartographic symbols. A unifying aesthetic had yet to be invented. What these maps tell us, then, is that during the cartographic explosion of the late eighteenth century, a great deal of effort was expended to represent the geographical distribution and impressive variety of natural resources in the Habsburg lands. Yet there was little if any effort to develop an appropriate visual scheme for doing so.

WONDERS OF THE AUSTRIAN EMPIRE

The need for such a scheme was first felt after Napoleon's devastation of central Europe. Francis II, Holy Roman emperor, was stripped of his imperial crown. He was, however, allowed to don a new one. He became Francis I, "emperor of Austria," for the first time uniting the Habsburg hereditary lands under a single title.[18] In an effort to unite this population against Napoleon, Francis I attempted to inculcate a new whole-state "Austrian" identity. New publications sent out the call for contributions to a patriotic research project, or *Vaterlandskunde*, which would extol the many virtues of Austria's land and population. This literature was designed to counter not only defamations by the French, but also the more narrowly provincial forms of patriotism that Napoleon's occupation had enflamed. The new *Vaterlandskunde* developed what Werner Telesko has aptly called an "encyclopedic aesthetic."[19]

Among its early fruits were celebratory volumes such as *Marvels of the Lands and Peoples of the Austrian Empire* (1809) and *Natural Wonders of the Austrian Empire* (1811).[20] Both volumes were due to Franz Sartori, who was a writer and editor by profession and a naturalist by avocation. In his later years, while editor of the *Vaterländische Blätter*, Sartori undertook a survey of "Austrian" literature, by which he meant "not merely that of the Germans," but also "all the various peoples, whose language and education are advanced enough that they possess a literature." These included "the Bohemians and Moravians, the Slovaks and Poles, the Russniaks (in Galicia and Hungary), the Serbs (in Hungary, Slavonia, and Dalmatia), the Croatians, the Winds (in inner Austria and western Hungary), the National-Hungarians or Magyars, the Jews and Greeks." Others—"such as Wallachians, Armenians, Gypsies, Clementines, and Ottomans"—had not yet reached the point of having their "own" literature, but (Sartori implied) might one day be included in such a compilation. Sartori described this project as a continuation of the long Habsburg tradition of patronage for the sciences. He set before the reader "the intellectual output of a population of 32 million people, the centuries-long cultural history of the diverse Austrian nations . . . brought into full view in one grand image."[21] In this way, he suggested that the scholarly work of the cultural nationalist movements of the early nineteenth century could in fact ground a new Habsburg-loyal discourse.

In short, *Vaterlandskunde* developed hand in hand with provincial *Landeskunde*. It was an umbrella discipline, both natural-scientific and humanistic. Its encyclopedic style attempted to fold the proliferation of research on places and peoples into one aesthetically pleasing image of Habsburg "unity in diversity." It was both a "political-administrative" undertaking, a means of rendering the population and territory legible to government, and an aesthetic-ideological pursuit, cultivating an appreciation for the varied "wonders" of the Habsburg lands.

But Vaterlandskunde remained a marginal enterprise in Sartori's lifetime. What catapulted it to the fore were the revolutionary events of 1848–49. Facing the challenge of modern, popular nationalist movements, Habsburg-loyal intellectuals began to shape a supranational ideology for a new age. Liberals agreed that, politically, the state would need to strike a fine balance between centralization and decentralization, building a common legal and economic framework and a shared "Austrian" identity, but also allowing for local autonomy. To quote the liberal statesman Victor Franz Freiherr von Andrian-Werburg, the goal was the "preservation of provincial differences, without harm to the unity of the whole."[22]

VISUALIZING THE FATHERLAND

Central to the development of *Vaterlandskunde* as an aesthetic enterprise was the vision of Friedrich Simony. Having been invited by Minister von Thun to design a program of study for physical geography, Simony sent his proposal off to the education ministry in February 1851. Physical geography as a discipline at Habsburg universities would be responsible for a unified representation of the natural diversity of the Monarchy as a whole:

> The author is convinced above all of the need to present the subject consistently with as much possible reference to the rich evidence that the natural diversity of the Austrian lands itself offers. A vivid depiction in word and image of the most interesting and instructive natural phenomena of the different regions of the empire, derived from direct observation, will not fail to achieve the desired effect of awakening love and enthusiasm for the great, beautiful, united Fatherland. Travels to individual parts of the Monarchy seem to the author to be an essential part of the responsibilities of this professorship, so that the lecturer can gradually acquire the desired first-hand observations and simultaneously collect the material that will permit a fruitful presentation of the subject.[23]

In these lines we can trace many of the traits that would distinguish Habsburg physical geography in the decades to come. First, there is the patriotic value attached to it, keyed to the political circumstances of the aftermath of 1848. We hear, too, echoes of the program of *Vaterlandskunde* that had taken shape earlier, in the wake of the Napoleonic Wars, with its panoramic eye for the natural marvels of the Habsburg lands. At the same time, we find the outline of a research program, one that stresses firsthand knowledge of the "individual parts" of the Monarchy—knowledge that can only be acquired through scientific travel and the patronage of natural historical collections. Finally, Simony refers to the need for "vivid" (*lebendig*) representations of natural phenomena "in word and image" (*durch Wort und Bild*). These keywords invoked a pedagogical principle that was rapidly gaining ground at the time. In Austria it was associated with a wave of interest among Bohemian patriots in the writings of the seventeenth-century heretic Jan Ámos Komenský, or Comenius, who had argued for the pedagogical value of visual aids. Simony elaborated that these "vivid" representations would take the form of "panoramas and profiles, characteristic landscapes and images of individual natural historical points of interest or objects, then too graphical representations of various kinds," all of which would need to be assembled and perfected for use as accompaniments to scientific lectures.[24] Simony was firm about the requirements for such visual aids: they would need to be produced on a grand scale to be suitable for public presentations, on wall-sized sheets of paper, and should be kept in their own archives to protect them. In order to allow himself time to produce such an ambitious set of illustrations, Simony requested that his lectures be kept to a minimum. Remarkably, Thun agreed. Simony would be permitted to teach only in the winter semester and to take leave from teaching whenever he might need to.[25]

Ironically, Simony never took advantage of this freedom to explore the territory of the Monarchy beyond the Alps, Bohemia, and Carniola. Not even the Carpathians tempted him. Although he helped to articulate a vision of imperial science for the postrevolutionary era, he was a product of an earlier generation, and his own scientific interests remained focused on the natural history of the Alps.

Nonetheless, he developed exquisite solutions to the Austrian Problem. Most of the meteorological maps and diagrams with which Simony illustrated his lectures were too large to be published and have since been lost. But many of his meticulously observed landscape paintings and drawings have survived (see figure 17 and plate 2). These demonstrate his innovative methods for capturing fine-grained detail within compositions that are, nonetheless, clearly legible as foreground, middle ground, and background. The only thing more remarkable than the level of detail in Simony's Alpine panoramas is how he managed to maintain the sense of depth. Albrecht Penck, Simony's successor as chair of physical geography at Vienna, noted that Simony achieved this effect in part by choosing his standpoint appropriately, in part by sketching the foreground in bolder strokes, the background in finer lines. He found that he could make the background retreat by filling it with so much detail that it grew

FIGURE 17. *Markt Aussee*, Styria, undated drawing by Friedrich Simony. Note the numbers that identify the finely drawn distant peaks.

darker.[26] Photography could not achieve the same effect, and so Simony often added details to the backgrounds of his landscape photographs.

Simony's techniques, while keyed to his scientific goals, also reflected the rise of *plein air* painting in Austria. The art historian Thomas Hellmuth suggests that a defining feature of Austrian Biedermeier art was its attention to natural detail, combined with the effect of a unified, all-at-once aesthetic impression. Hellmuth points, for instance, to the country scenes painted by Ferdinand Georg Waldmüller in the 1830s and 1840s, including those painted in Simony's favorite corner of the Alps.[27] A particular ambition of these artists was to capture the nuanced, fleeting effects created by changes in the atmosphere.[28] Simony addressed the relationship between landscape painting and meteorology in a series of popular lectures. He aimed to convince his audience that the atmosphere deserved the painter's closest attention. "How charmless, how dull land and sea would appear . . . if the whole endless diversity of meteorological processes did not bring drama and variety to the scenery." Atmospheric changes could transform a landscape in dramatic ways. For instance, visibility increased sharply on the coldest of winter days, as he knew from the stretch of seventy-two hours he had spent on a three-thousand-meter-high Alpine peak in February 1847, when "Alpine peaks twenty and more miles away displayed clear outlines and even individual details of their forms."[29] In the course of these lectures, Simony explained such technical concepts as the difference between absolute and relative humidity, the thermodynamics behind rising air currents, and the classification of clouds. He was, in effect, teaching his audience to see the atmosphere not as mere backdrop but rather as the scene of dramatic confrontations between warmer and colder, wetter and drier air currents. Even the most fleeting features of a landscape, the clouds, depended on the "interaction of the temperature, humidity, and movements of the layers of the atmosphere, on the position of the sun, the composition of the earth's surface, and the vegetative cover of the land."[30]

The novelist Adalbert Stifter brought Simony's aesthetic lessons to an even wider public. The protagonist of Stifter's *Nachsommer*, evidently modeled on Simony, learns that he was wrong to eliminate the optical effects of the atmosphere from his landscapes. His mentor teaches him to see what he had previously tried to ignore:

> Thanks to my friend's critique, it suddenly dawned on me that I had to observe and acquaint myself with something that until then had always appeared to me as insignificant. Objects take on another appearance due to air, light, mist, clouds, and nearby objects; this I had to fathom, and I had to take the causes as the subject of my research, as I had previously taken the things that immediately strike the eye. In this way it would be possible to succeed in representing objects that swim in a medium and in an environment of other bodies.[31]

Simony and Stifter thus taught a wide audience to see the atmosphere as a connecting medium and as a dynamic element of the landscape. These were essential contributions to the visual repertoire of Austrian *Vaterlandskunde*.

THE GEOLOGICAL EYE

Geology was the first field to confront the Austrian Problem head on. The Royal-Imperial Mining Museum (Montanistisches Museum), founded in Vienna in 1814, consolidated minerals and gems that had previously been arranged haphazardly in princely *Wunderkammer*, reorganizing them along modern geographic lines. This demanded knowledge of the spatial distribution of minerals across the Habsburg lands. What was needed was a standardized, uniform "geognostic" map—one depicting areas defined by their principal type of rock—for the Habsburg territory as a whole.

The production of the first such map was directed by Wilhelm Haidinger and completed in 1845. He constructed it on the basis of existing regional maps, supplemented by his own field research. His account of this process is revealing: "Every more detailed map displays differences in the composition of rocks that are locally of great importance, but which disappear at lower magnification, assimilated to one or another neighboring formation."[32] For the geologist to whom such local variations mattered, the result was long, needless labor in the field to recover the lost details. The job of the geological mapmaker was thus to judge the significance of local details for the representation of larger geological structures. In this sense, Haidinger's project can be seen as a forerunner to the global geological synthesis achieved a generation later by Eduard Suess—who, in 1850, was still a university student working part time for Haidinger at the imperial mineral collection. Thus Haidinger faced the problem of choosing an appropriate scale for his imperial overview: "The highly non-uniform composition of the Monarchy, the important mountain chains on one side, the great Hungarian plain on the other, all seemed to demand a different treatment. Variation on one hand, uniformity on the other, demanded a different scale for the same overview." Indeed, the complexity of the geological strata in Bohemia, South Tyrol, and mining regions would not permit of too coarse a scale. And yet the map could not be produced in too many leaves, for it was intended to "reach a wide public by means of a low price"—not least in order to stimulate further research.[33] Haidinger eventually decided on 1:840,000, which was the scale of the road map published by the Imperial-Royal Military Geographical Institute.

Haidinger presented the 1845 map as a work in progress, a baseline for further investigation. He fully hoped that members of the public would apply their local knowledge to the task and supply him with corrections. Indeed, he had presented his early drafts of the map to colleagues at regional scientific societies throughout the Monarchy, in order to obtain critiques and revisions. As he made clear, a geological overview could only emerge from the closest cooperation between provincial societies and a central institute. The future of Habsburg science hinged, in his eyes, on centralizing the production of knowledge partially but not completely. He therefore invited "all friends of the geological knowledge of our land" to submit additions and corrections.[34]

One of the first major undertakings of the Imperial Geological Institute upon its founding in 1849 was to replace Haidinger's 1845 map with a truly uniform cartographic overview of the geology of the Habsburg lands. Thus by 1863 the geology of "the entire territory of Austria-Hungary, from Lombardy [*sic*] to Bukovina, from Dalmatia to the Elbe gap" had been mapped at a scale

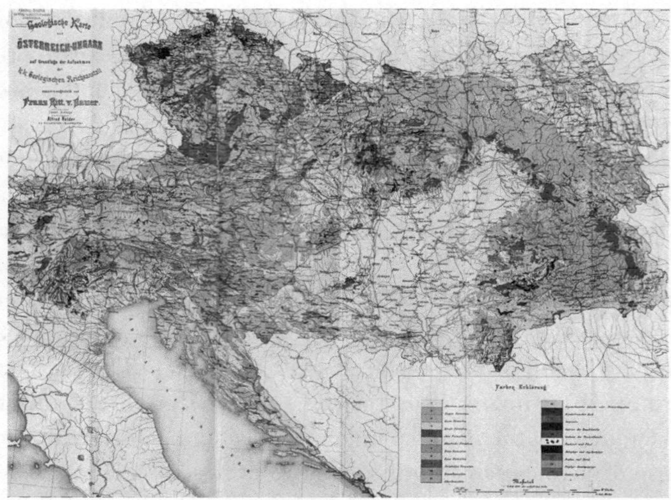

FIGURE 18. "Geological Map of Austria-Hungary, on the Basis of the Survey of the Royal-Imperial Geological Institute," by Franz von Hauer, 1867.

of 1:144.000.[35] The importance of this achievement was that it enlarged the field of study beyond an individual crown land. Already in the course of his survey work in the 1840s, Haidinger had begun to think like an imperial-royal scientist, to see continuities that could not be grasped within a narrower field of view: "One could deal with one crownland after the other. But the nature of mountains makes such a division along artificial borders untenable." Haidinger insisted that the greater part of Austria was dominated by one "mountain system," including both the Alps and Carpathians, the unity of which must be made visible. Here was a novel, coherent, and majestic vision of the physical structure of the Monarchy.[36]

CLIMATIC CARTOGRAPHY

Many of the obstacles to producing a geological overview of Austria had counterparts in the mapping of climate. But the challenge was compounded because the climatic map was an entirely new genre. Until the nineteenth century, climate zones had been drawn based simply on latitude. In 1817 Alexander von Humboldt introduced *isotherms* as a means of representing the empirical distribution of climate across the earth. Isotherms are lines connecting points of equal temperature, a graphic method of interpolating temperature between locations at which it has been measured. Nineteenth-century scientists appre-

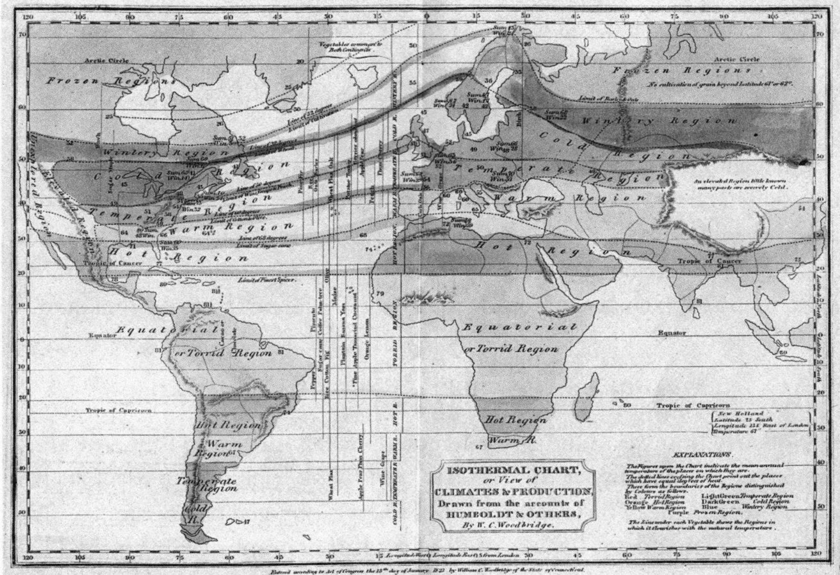

FIGURE 19. First map of global isotherms, ca. 1823.

ciated how much isotherms revealed at a glance. In the form of closed circles, they indicate a region of elevated temperature; where the lines bunch together, temperature changes rapidly over a short distance.[37] Humboldt, however, was working with very limited data: fifty-eight stations for the entire surface of the earth, of which only two lay in Asia and Africa (see figure 19).

Isothermal maps became more realistic later in the nineteenth century, as observing networks grew denser and more extensive, and as new methods for analyzing and representing their data were introduced. In 1848 Heinrich Dove published a world map of temperature distribution with data from nine hundred stations, and in 1864 a new one drawing on two thousand stations worldwide. Maps of climate also began to represent factors other than temperature.[38] Alphonse de Candolle and Wladimir Köppen introduced influential climate-classification systems based on the distribution of vegetation, while Albrecht Penck devised one based on humidity. As more data became available, competing schemes emerged, each suited to different applications, from plant geography to geology to human biogeography. Yet classifications like these—based on individual meteorological factors or on the effects of climate on living things—were occasionally criticized for attending to climate's effects rather than its "essence."[39] The Habsburg physical geographer Alexander

Supan attempted to get at that essence. He distinguished "continental" from "oceanic" regions according to the average annual range of temperature (the difference between the temperatures of the warmest and coldest months). Nonetheless, Supan acknowledged that no classification scheme could ever be absolute.[40]

What all these schemes had in common was the absence of weather. They represented climate as a static variable, a long-term average, in which phenomena on the time scale of weather—clouds, storms, and gales—were invisible. How could the dynamic nature of climate be made manifest? How could a single map represent a temporally and spatially fluctuating phenomenon?[41]

AN ATLAS OF THE HABSBURG WORLD

By 1898, when Austria-Hungary celebrated the fiftieth year of Emperor Franz Josef's reign, Habsburg field scientists could look back on the previous half century with pride. The efflorescence of topography, geology, seismology, hydrography, climatology, botany, zoology, and ethnography in this period had succeeded in "placing applied geography on a scientific footing."[42] The result of all this activity was an overload of information and the need for new approaches to the Austrian Problem.

Between 1882 and 1887, Eduard Hölzel's printing house in Vienna brought out the first thematic atlas of Austria-Hungary. It was a pioneering work, applying innovative methods for visualizing physical geographic and demographic ("statistical") data.[43] Even after the life of the Monarchy, this atlas would continue to be cited as a crucial source of demographic information on the region—as "still the best cartographic introduction to Austrian problems."[44] The atlas opened with a series of maps of the distribution of temperature and rainfall, presenting these as the most fundamental scheme for the division of the Monarchy into natural regions. Two classifications were proposed: one, based on temperature (see plate 3), identified twenty-two "natural climatic regions"; another, based on rainfall, distinguished seven such zones. From climatology the atlas moved on to hydrography, geology, and vegetation, and from there to various demographic and ethnographic maps. The later maps refer back to the earlier ones in order to emphasize the interdependence of environmental and human factors. For instance, Josef Chavanne's "Forest Map" referred the reader back to the maps of elevation, geology, rainfall, and temperature, and also ahead to the maps of economic production, in order to grasp the variety of factors determining the distribution of forest cover. The atlas facilitated the appreciation of these connections by imposing a uniform design. Of twenty-five maps in total, nineteen were drawn at 1:2,5000,000, sixteen at 1:5,000,000. Related if not identical color schemes were used throughout. German toponyms were used where available, and the keys and explanations were exclusively in German. Amusingly, one double-page spread presented four uniformly scaled and colored maps of the Monarchy, illustrating the distribution of towns, hail fall, illiterates, and pigs. As the text accompanying the map of elevations explained, "Austria-Hungary in particular is a country in which one can clearly follow the interactions between the physical conditions of the land and the physical-cultural conditions of the populace; in both ways, the complex structure of the land offers a diversity of conditions of development and formation."[45] According to this logic, the representation of fine gradations of physical-geographic features might also allow for a more nuanced appreciation of cultural variation.

But how could a small-scale map (that is, one representing a large area of the earth on a small sheet of paper) encapsulate all the relevant local detail? How, for instance, was one to represent elevation on a map that encompassed the towering heights of the Alps and the flat expanse of the Hungarian plain? One might choose a scheme to capture the dramatic contrasts between peaks and valleys, on one hand, or one sensitive to small variations in altitude, on the other. But could both goals be achieved at once? Here was a classic instance of the Austrian Problem, and the 1887 atlas exhibited some of the more ingenious solutions to it.

At the time, there were several methods in use for representing elevation. Hatching—shading by means of fine lines—was the oldest, and it had become a speedier process thanks to the substitution of lithography for copper engraving around 1800. Contour lines had also been in use since the late eighteenth century. A slightly newer technique was the profile, or horizontal cross-section, developed for physical geography by Alexander von Humboldt.[46] Austrian cartographers chose instead to illustrate elevations by means of color. They were responding in part to popular demand for more accurate maps of the Alps. Using a range of colors could give an immediate visual impression of elevation, particularly for coarser-grained overviews. The goal was "plasticity," meaning a visual effect of three-dimensionality. The first such scheme to be widely used was developed by Franz von Hauslab (1798–1883), a military cartographer in Vienna. According to Hauslab's rules, colors should be layered from lighter to darker with increasing altitude, with each tint clearly distinguished from the previous one in the series. This system was put into practice by cartographers in Vienna and Prague, including Friedrich Simony, and their maps circulated widely in school atlases and wall maps of the Habsburg lands. By contrast, elevation maps in Wilhelmine Germany tended to use a simple two-color scheme, green for lowlands, brown for mountains.[47] The 1887 atlas of Austria-Hungary employed Hauslab's color scheme for elevation, deviating only in the choice of pale blue for the highest peaks, the realm of eternal frost.

It was not simply the application of color that distinguished the Austrian experiments in mapping elevation; it was the depth of theoretical engagement with the problem of representing vertical scale. A competing system was developed in the 1890s by Karl Peucker, a geographer from Prussian Silesia who worked for a Vienna publishing house. Drawing on research on the physics, psychology, and physiology of color perception, Peucker invented a color scheme that, he argued, gave a more lifelike impression of elevation. While Hauslab had advised darker colors for higher areas, Peucker called for "richer" colors, according to his own definition of richness. He also gave cartographers a more precise vocabulary by distinguishing between two priorities: 1) *Meßbarkeit* or measurability—that is, the degree to which relative magnitudes on the map corresponded to the relative magnitudes of the objects they represented; and 2) *Anschaulichkeit* or visualizability, the degree to which relative magnitudes on the map *strike the eye* as standing in the same proportion to each other as the real-world objects. Peucker's system was highly influential within and beyond Austria-Hungary. In 1913 it was adopted for use by the International Map of the World, translating a solution to visualizing the supranational state into a key to representing the globe.[48]

The Austrian Problem was not confined to elevation maps. The 1887 atlas of Austria-Hungary showed how it cut clear across what we would call physical and human geography. The political stakes were highest for the map of language use. Produced by Franz le Monnier and based on statistics from the 1880 census, this map went furthest in the application of color to visualize local variation. Von Czoernig's 1855 ethnographic map had only represented the dominant language for each municipality. Le Monnier observed that it had thereby neglected "smaller minorities." His solution was to apply multiple colors to a linguistically mixed municipality: colored dots represented a minority language spoken by 10–29 percent of the population, while stripes stood for a language spoken by 30–50 percent. This representational choice had clear political implications: Germans now appeared to be "not only the numerically strongest nation, but also the most widely distributed."[49] Le Monnier accentuated this effect by using pink not only to represent German speakers but also to draw political borders. His map, however, was an outlier, the only one to suggest that Habsburg unity derived from German culture.

Other maps in the atlas instead carried the message that unity arose from conditions of mutual dependence deriving from environmental circumstances. Consider the map of river basins. Read in conjunction with the rest of the atlas, this map illustrated the role of physical factors like geomorphology and climate in shaping the waterways of Austria-Hungary "in their character as factors in commerce and transport." The accompanying text underlined the impressive variety of conditions of elevation, geology, temperature, and vegetation along the Danube. "To the degree to which these elements either in their extremes or in their various degrees *enter into a local relationship of mutual dependence*—to that degree the hydrographic effect will also vary." The map thus displayed the factors influencing hydrography "both *individually and in their connections and mutual dependence*." Its goal, then, was to represent local variations, both natural and human, as interdependent parts within a greater whole. It achieves this effect by depicting more than half of the Monarchy as lying within the Danube basin, uniformly shaded blue. Smaller regions defined by the Danube's tributaries are indicated only by text, so as not to break up the unifying effect of that blue expanse. In this way, the map subsumed the diversity of hydrographic conditions along the Danube to the unifying concept—and uniform image—of "the largest river basin in Europe."[50]

"BUT ENOUGH OF ISOTHERMS"

The 1887 atlas of Austria-Hungary included seven climatic maps: three for temperature and four for precipitation. All display a particularly thorny instance of the Austrian Problem: how to depict overall climatic patterns by means of data coming from stations influenced by very different local conditions. This was a defining challenge of scientific internationalization, which Habsburg scientists addressed within the borders of their own state.

Indeed, the job of drawing isotherms brought the Austrian problem to the fore. In mountain ranges, small-scale climatic variations thwart the tactics of large-scale climatology. In particular, the change in temperature with height is on the order of one thousand times greater than the horizontal variation in temperature. Saussure made the first measurements of the decrease of temperature with height in the Alps in the eighteenth century, but had not been able to discover a universal law. Julius Hann had applied thermodynamics to this question in the 1860s, reasoning that rising air expands as it ascends into regions of lower pressure and loses heat in the process. But this decrease in temperature (known as the lapse rate) did not appear to be uniform. When Josef Chavanne constructed the first map of isotherms in Austria-Hungary in 1871, he concluded that "the law of the temperature decrease with height has by no means been determined."[51] Subsequently, mountain observatories, kites, and balloons indicated that the lapse rate varies from one region to another and from one season to another. "But enough of isotherms," one Austrian climatologist exclaimed in 1909. "The task of the climatographer is more narrowly circumscribed and is supposed to describe real conditions."[52]

For the 1887 atlas of Austria-Hungary, Chavanne therefore abandoned the method of reduction—which "can furnish only tenuous values"—and with it the drawing of isotherms. Instead, he used colored closed curves, modeled on the Hauslab color scheme for elevation, to show the average temperature distribution in different seasons, without reduction to sea level (see plate 3). Alexander Supan had set a precedent for this in 1880, when he was working with data in part from stations on high plateaus. He had sought a standard correction in order to reduce the measurements to sea level. Depending on season, however, he found that this correction would have to be positive in some cases and negative in others. He therefore chose to use uncorrected values, which "allow the law of the horizontal distribution of the annual fluctuation of temperature to emerge in the clearest way."[53] What a map like this revealed was that the more important geographical influence on temperature was not absolute height but the relative height of mountains with respect to valleys. At the time, some isothermal maps had begun to display the influence of land and sea, bending warmer contours inland, for instance.[54] But few attempts had been made to visualize the influence of orography. As Chavanne observed, the effect of elevation on temperature depends less on absolute height above sea level than on the relative height of mountains and valleys—hence his omission of contour lines. The lowest average temperatures were found in valleys where mountains blocked the influx of warm air from the south and west. In other words, Chavanne's statistical and cartographic methods began to make visible climate dynamics—specifically, how the form of the earth's surface modifies atmospheric motions.

The goal of visualizing climate dynamics is evident in another of Chavanne's maps for the atlas, the distribution of days with storms. Earlier maps of storm distribution had tabulated stations reporting the same number of storms, then divided Europe rather crudely into regions of summer storms, winter storms, and, in the east, a region without winter storms.[55] Chavanne's map was constructed along entirely different lines, since he intended it to display the dynamics of storm generation, not merely their frequency. It reveals local contrasts that were obscured by the older maps. By keeping local variations in focus, the map makes visible the dynamics of storm generation, not merely their frequency. Namely, it reveals the influence of topography on storms; Vienna, for example, displayed a greater storm frequency than Wiener Neustadt, because circular valleys open to the south were more prone to storms than their surroundings.

Remarkably, Hann achieved a similarly dynamic effect in his *Atlas of Meteorology* (1887) by means of an entirely different method. Hann constructed his map of global isotherms (see figure 20) by omitting data from high-altitude stations and reducing all temperatures to sea level, employing a lapse rate of 0.5 degrees per one hundred meters. He justified these choices in detail in the text accompanying the maps. It might appear to be a disadvantage that elevations with a more gradual lapse rate would appear warmer than their surroundings, while those with a steeper lapse rate would appear cooler. But this was not a problem in Hann's eyes. According to Hann's definition, isotherms were meant to represent the temperature of the lowest layer of the atmosphere—"since valleys are the most commonly inhabited part of the earth's surface."[56] In other words, isotherms were supposed to represent climate in its significance for human life. Yet Hann also wanted his isotherms to illustrate the *dynamic* effects of mountains on temperature. As he explained, mountains could potentially have a warming effect by shielding against cold winds; or they might have a cooling influence, either by preventing the outflow of air that had undergone radiation cooling, or by shielding against warmer air currents from the sea. The point was to be able to *see* these orographic influences on the map, to visualize how mountains diverted air currents and redistributed heat. To do so, Hann argued, required a uniform lapse rate. What was visible on Hann's map, then, was the *real* influence of mountains on the distribution of temperature, by which he meant their dynamic effect. So, for instance, on the map of January isotherms one could see an extreme cold island in Siberia that was characteristic of the valleys during anticyclonic conditions, with a high-pressure center above. The Russian climatologist Alexander Voeikov objected that this cold island was an illusion, since the air in the mountains was significantly warmer than in the valleys when anticyclonic conditions prevailed. Hann disagreed, seeing the cold island as a genuine feature of the climate and insisting that valley temperatures were those most relevant to human life. "These are in reality relatively cooler or warmer parts of the earth's surface, and it seems to me positively advantageous that maps of isotherms express these conditions."[57]

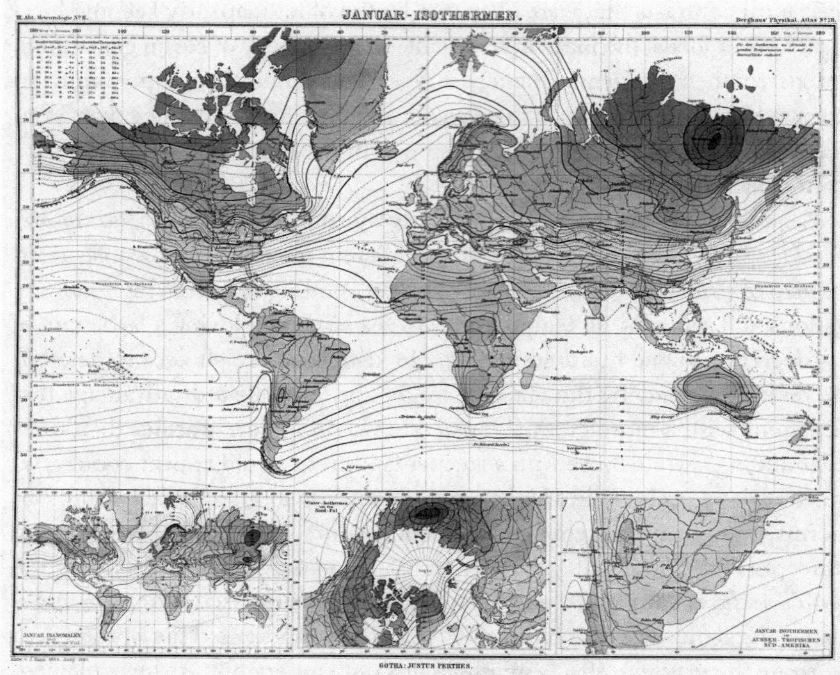

FIGURE 20. "January Isotherms," by Julius Hann. In *Atlas der Meteorologie* (1887). Compare to the 1823 map (figure 19); note how Hann's isotherms bend according to local conditions. Note, too, the cold island over Siberia, explained in the text.

CONCLUSION

When constructing a climate map from station data, a basic question is how to evaluate the significance of measured differences of temperature and pressure between neighboring locations. Do these reflect actual variation, or are they artifacts of differences in the measurement process? So, for instance, in his study of the daily course of temperature in Austria, Josef Valentin insisted that elsewhere "in Europe it is not easy to find meteorological stations situated in such diverse conditions as in Austria."[58] Valentin noted that Heinrich Wild's study, *Temperature Conditions in the Russian Empire*, had omitted data from the Habsburg stations in Vienna, Prague, Salzburg, and Kraków for this reason. Valentin did not dispute Wild's judgment, but he believed that "these locally influenced temperatures are indeed not without value." He argued further that the Austrian data could not be treated with the methods used for Russia. Although corrections based simply on latitude and longitude might have sufficed for the "more uniform orographic conditions of the Russian Empire," it was "obviously not to be considered for Austria, with its highly varied surface forms."[59] In this way, the influence of topography on daily temperature variations could become an object of study in its own right.

In short, the goal of doing justice to Austria's physical diversity lent significance to measurements that might otherwise have been dismissed as anomalous. Habsburg scientists saw the Austrian climate through the lens of the Austrian Idea. They believed this territory to be uniquely diverse, so they produced statistical and cartographical methods that were adequate to visualizing that diversity. Thus Hann gave credence to divergent measurements from nearby stations in cases where he was able to find a dynamical explanation for those differences in terms of local geography.

Look again at Hann's map of global isotherms. How is it to be read historically? From one perspective, it is a stage in the development of our current digital models of climate change. It displays empirical data essential to modeling the earth's energy budget, and it was in fact put to this purpose by a researcher in Prague during the First World War.[60] Hann, however, intended it not as data to be fed into a calculating machine, but as a form of visual communication. This map was meant to be legible at multiple scales, conveying both global patterns and local peculiarities. Indeed, for Hann, the laborious work of standardizing measurements from across the world served the higher purpose of being able to visualize the *dynamics* of climate at multiple scales—as determined by planetary-scale forces and by the local forms of the earth's surface. Strategies of visual representation are thus part of the story of the shift to a new understanding of climate in the late nineteenth century: from climate as a fixed feature of portions of the earth's surface to climate as a dynamic system of energy transport across scales.

CHAPTER 6

The Invention of Climatography

In 1901, the Austrian Academy of Sciences celebrated the fiftieth anniversary of the ZAMG. The festivities took place in typical imperial-royal fashion in the ornate auditorium of the Academy of Sciences.[1] Archduke Rainer, the academy's honorary curator, delivered the opening remarks and then turned the floor over to Wilhelm von Hartel, the liberal minister of education. It was Hartel's privilege to announce the ambitious new project of the ZAMG. Its fifty years of observational data would "soon appear in a monumental work, which will give a detailed picture of the climate of the so varied parts of our kingdom, for the benefit of all."[2] This chapter is concerned with the literary ambition behind this project—the dream of doing justice to these "varied parts" while producing a coherent description of the kingdom as a whole.

Seventeen volumes of this "monumental work" were planned in all, for each of which the ZAMG received a subvention of 2,000 crowns from the ministry of education, roughly the annual income of a domestic servant in one of the major cities of the empire.[3] Over the next eighteen years—the final ones of the Habsburg Monarchy—the imperial Academy of Sciences oversaw the publication of the first nine volumes. In 1918, the education ministry declared that it would continue the publication of the remaining volumes, with the enduring goal of producing a unified conception of the defunct monarchy.[4] Two lingering volumes emerged from the Academy of Sciences of the Austrian Republic in 1927 and 1930—ironically, in step with Robert Musil's satirical tribute to the imperial past in the first volume of *The Man without Qualities* (1930). Despite its initial pretensions, then, the series ultimately could not claim comprehensiveness. One volume, titled *The Austrian Coastal Land*, only got as

TABLE 1. Volumes in the Klimatographie von Österreich Series, 1904–1919

Province	Author	Year
Niederösterreich	J. v. Hann	1904
Küstenland (Triest)	E. Mazelle	1908
Steiermark	R. Klein	1909
Tirol und Vorarlberg	H. v. Ficker	1909
Salzburg	A. Fessler	1912
Kärnten	V. Conrad	1913
Bukowina	V. Conrad	1917
Mähren und Schlesien	H. Schindler	1918
Oberösterreich	P. T. Schwarz	1919
Related Climatographical Studies of Habsburg Provinces after 1919		
Küstenland	E. Biel	1927
Wien	A. Wagner	1930

far as Triest. During the war, the volume on Carniola was abandoned due to insufficient funding; Bukovina was published in the absence of data lying in enemy hands; and Moravia together with Silesia was squeezed out in 1918, but Bohemia never appeared, apparently because Prague refused to return the materials to Vienna after 1918. Galicia, insufficiently furnished with observatories, was also absent.[5] Clearly, the coverage was spotty and skewed toward the empire's Alpine lands, reflecting the greater density of the observing network in that region, scientific interest in the Alps, and the professed difficulty of finding suitable local authors in the "periphery." (See table 1.) Most glaring, however, was the absence of the projected final volume, an overview "of the climatic conditions, peculiarities, and contrasts, as well as the weather patterns of all of Austria."[6] Such a volume would have had to analyze data from roughly four hundred observing stations and approximately 241,000 square miles, stretching from the dry plains of Bukovina to the snowcapped peaks of Vorarlberg to the islands off the Adriatic coast. It would have been the most ambitious project of empirical climatology to date.[7]

Climate, as Julius Hann liked to observe, is a statistical abstraction. What relationship did it bear to weather, the object of our direct experience? How could static average values be made to reveal the dynamic processes that gave rise to them? These quandaries also constituted a literary challenge, a problem of genre. Genres—such as atlases, travel narratives, or nature writing—allow new information to be sorted and digested. They provide frameworks that generally conform to expectations and yet are flexible enough to be adapted to new purposes. They are generative of meaning, producing an effect that is "deeper and more forceful than that of the explicit content of the text."[8]

The term "climatography" entered the English language in 1813 and was used in German by the 1830s (OED). It was not yet an established genre, however, because long-term, regional weather data did not become available until the middle of the nineteenth century. No observing network before 1800 survived for more than two decades.[9] Climatographies were the work above all of the continental empires of Austria, Russia, India, and the United States. Imperial climatographies responded to the injunction to define the internal borders of an empire's "natural regions," as a means of rationalizing and integrating the economic relations of its parts. As we will see, the process of inferring boundaries where variation was continuous required attention to phenomena on multiple scales.[10] Climatography as a genre thus held the potential to make visible local-global interactions and patterns of change.

What follows, then, is the story of the invention of an environmental genre. The Climatography of Austria took shape at a time when geographers were hunting for new literary strategies. The American geomorphologist William Morris Davis argued in 1904 that modern geography could no longer be written as if the earth were only "the home of man" and not "the home of life" in the widest sense. He proposed the term "ontography" for "the organic half of geography." Ontography would be the record of the spatial distribution of physiological responses to the environment; placed in temporal sequence, this would become ontology, a record of "the sequence of organic responses to a changing earth."[11] Climatography was another solution to this literary challenge—the challenge, in Rob Nixon's terms, of how to "plot and give figurative shape" to environmental changes "dispersed across space and time."[12]

This chapter begins with the formulation of the Austrian Problem in literary terms in the 1840s and 1850s, by two men with shared backgrounds and interests: the novelist Adalbert Stifter and the earth scientist Karl Kreil. Kreil was seven years ahead of Stifter at the gymnasium at Kremsmünter, where both men conceived the ambition of a career in physics. Kreil would go on to found the ZAMG; Stifter, frustrated in his scientific aspirations, would found a literary tradition that gave center stage to the natural world. We will see how Kreil and Stifter each honed a written style that lavished attention on "small things"—from the behavior of insects, to the morphology of mosses, to the fluctuations of air pressure—but did so in ways that elucidated their significance within an imperial and a cosmic order.[13] From there we will explore the development of the genre of climatography as a solution to the Habsburg problem of scale. To this end, it mobilized four principal points of view: that of a central observatory, recording atmospheric phenomena as they pass into and out of its field of vision; that of a central observer looking down from above at the effects of atmospheric phenomena across the surface of this land; that of coordinated observers throughout the territory, collectively turning their eyes to the heavens; and that of an individual surveyor moving across this territory, registering the atmosphere both mechanically and sensuously.

THE LEGACY OF COSMOGRAPHY

But first we need to consider what climatography was not. Cosmography, in both its ancient and Renaissance forms, was a genre that blended description of the physical and human characteristics of an environment. Recent scholarship has characterized Renaissance cosmography as a profoundly humanistic and imaginative genre with a utopian impulse. Like its classical antecedents, it was part natural history, part ethnography, and often took a narrative form. By the end of the sixteenth century, cosmography effectively "dissolved": the disciplines of mathematical cartography, astronomical navigation, hydrography, and geodesy broke off from those of descriptive geography, ethnography, and natural history—the former falling to the mathematical "cosmographer," the latter to the "chronicler."[14] Thereafter, the mathematical and descriptive modes of environmental representation moved ever farther apart. By the late nineteenth century, the human sciences were attempting to define their methods against those of the physical sciences. Climatography disrupted this trend and with it the hardening distinction between explanation and understanding, facts and values. It sought to represent—in word and image—the *meanings* of meteorological measurements of a region for its human inhabitants.

In this respect, the invention of climatography bears comparison to Alexander von Humboldt's revival of cosmography. Published in German in 1845, Humboldt's *Cosmos: Sketch of a Physical Description of the Universe* came to be seen as a model of popular science writing. Its goal was to reveal the interdependence of all parts of the natural world. Particularly suggestive for naturalists in the Habsburg world would have been Humboldt's insistence on nature's "unity in diversity":

> Nature considered rationally, that is to say, submitted to the process of thought, is a unity in diversity of phenomena; a harmony, blending together all created things, however dissimilar in form and attributes; one great whole . . . animated

> by the breath of life. The most important result of a rational inquiry into nature is, therefore, to establish the unity and harmony of this stupendous mass of force and matter.[15]

Instructive too for Habsburg scientists was Humboldt's pursuit of a literary style that would animate rather than merely reproduce the results of empirical measurements: "Descriptions of nature ought not to be deprived of the breath of life. Indeed, the mere listing of general results produces just as wearying an effect as the accumulation of too many observed details."[16] Humboldt chose, for instance, to present the highly empirical and theoretically contested field of terrestrial magnetism not on the basis of empirical measurements, but by means of verbal pictures of the hypothetical causes of magnetic variation in unseen realms within the earth and in the upper atmosphere. In these ways, Humboldt's "physical description of the universe" set a pattern that Habsburg science could easily have followed.[17]

But this is not what happened. To be sure, imperial-royal scientists took inspiration from *Cosmos*, just as Kreil's initial geoscientific survey of the empire was inspired by Humboldt's example. Julius Hann read all four volumes of *Cosmos* as a gymnasium student in 1859, in the very spring of Humboldt's death, and they brought him "great pleasure," as unfamiliar as "this manner of serious, profound entertainment" seemed to him then. Upon beginning the book's fourth volume, he again relished "that warm breeze that clothed all the earnest scientific descriptions with a light and charming scent."[18] Nonetheless, imperial climatography would be a genre utterly distinct from Humboldtian cosmography.

This was already clear from Hann's rejection, in his 1883 *Handbook of Climatology*, of the idealism implicit in Humboldt's definition of climate in terms of human perceptions. To Humboldt, nature's unity was imposed by "the eye of the reflecting observer." By contrast, climatography was about climate as the shared reality of all living things. To the degree that its perspective was anthropocentric, it was so for the sake of practical aims, not Romantic ideals. Imperial-royal scientists seconded Humboldt's position that many "great" things in the universe appeared "great" only at the scale on which humans happen to exist. But they never proposed to leave aside human interests altogether. Climatography, on the contrary, was a genre dedicated to practical goals, appealing to a broad readership whose livelihoods depended on anticipating atmospheric conditions.

Climatography also broke with cosmography in its commitment to—some might say obsession with—the representation of local detail. As critics at the time pointed out, Humboldt tended to neglect local variability in the interest of formulating universal generalizations: conclusions about volcanoes in general, for instance, rather than about the variety of volcanic phenomena. Likewise, he presented his botanical map of Mount Chimborazo as a universal correlation of plant life with altitude, ignoring the distinct forms of mountain vegetation in different regions of the world. Climatographers instead believed that local variations mattered, even within the framework of an imperial, and ultimately planetary, overview.

Climatography also broke new ground in its resolve to make quantitative measurements intelligible to the public, rather than minimizing their presence in the text. Humboldt acknowledged the pleasure that naturalists took in exact measurement, but concluded that the overemphasis on quantitative analysis made the science of his day appear "wasted" (*verödet*) in the eyes of the public.[19] Climatographers, on the other hand, did not doubt that they could teach readers to extract meaning from numerical measurements.

A final, more subtle way in which Austrian climatography differed from Humboldt's cosmography was in the significance it attached to the use of the German language. Humboldt, who had previously published primarily in French, suggested that *Cosmos* could only have been written in his native language. Writing in the years of nationalist agitation leading up to the Frankfurt Parliament, Humboldt took a Romantic, patriotic, even mystical view of language: "The word is, therefore, more than the mere sign and form, and its mysterious influence still reveals itself most strikingly where it springs among free-minded communities, and attains its growth upon native soils."[20] By contrast, the choice to publish the *Climatography of Austria* in German was a matter of expediency: German was the multilingual empire's lingua franca. To claim that German or any other language gave privileged access to the workings of the universe would have been anathema to the ideology of imperial-royal science. Where Humboldt assumed an ideal affinity between language and thought, Habsburg scientists retained numbers and tables in recognition of the limitations and opacities of words.

In short, Austrian climatography stood in uneasy proximity to Humboldtian cosmography. Both carried the message of unity in diversity, and both attempted precarious feats of scaling. Yet, in keeping with the rejection of German idealism by many nineteenth-century Habsburg thinkers, climatography held fast to the irreducible complexity of the local and particular, and to a quantitative precision that was meant to serve a wide variety of practical endeavors.

THE VIEW FROM SAINT STEPHEN'S

In 1844, a year before the publication of the first volume of Humboldt's *Cosmos*, a book appeared in Pest with the title *Vienna and the Viennese in Pictures Drawn from Life*. Its editor was the thirty-nine-year-old Adalbert Stifter, who contributed twelve of the literary sketches, all in a breezy, irreverent, and cosmopolitan tone uncharacteristic of his later writings. Like Stifter, the other authors were not natives of Vienna, and the book was intended for those who did not know the city well. In his preface, Stifter assured readers that the volume they held was no dry compilation of statistics. It was, instead, an array of images "as in a kaleidoscope," which would allow the reader gradually to "paint an image of the life and work of this imperial residence for himself."[21] Austria's human and natural diversity was becoming a motif of patriotic tributes of the 1830s and 1840s, and this "kaleidoscope" presented Vienna as a microcosm of this multiplicity. *Vienna and the Viennese* was an early experiment in achieving the literary effect of a multiscalar view.

The book's opening sketch, "The View from St. Stephen's Cathedral," literally constructed an overview of the city. From this towering vantage point, Stifter pointed out the battlefields of Aspern and Wagram, sites of an Austrian victory and subsequent defeat by Napoleon's army. Vienna was thus the heart of an empire that "sits in the parliament of peoples and helps to determine the fate of planet earth." He identified the "mighty road" that had been built by Karl VI in the early eighteenth century over the Semmering, through the karst, all the way to "*our* port Triest," and which "connects *us* to the entire South,"[22] and another street, leading out from the city toward dusty, yellow fields, the road "to Hungary and the Orient." Stifter's use of the first-person plural placed the emphasis on unity. He juxtaposed aerial and ground-level perspectives, the close-up and the distant. From above, Vienna was "an enormous plain onto which this sea of houses spills"; from below, it was a dense stream of "endless" streets. For the departing traveler, meanwhile, it was merely that "tiny speck, Vienna." Stifter's perspective might be termed statistical, as individuals recede and the aggregate comes into focus. And yet he relied on a prestatistical episteme of signs and resemblances. Going about their daily business, the Viennese did not realize that they were "blithe and lovely letters, with which the muse writes the terrible drama of world history." The techniques introduced here for scaling between the individual and the social aggregate recall older models, such as Mandeville's descriptions of the hive in *The Fable of the Bees*. They also anticipate more modern approaches, such as Tolstoy's scenes of Napoleonic warfare in *War and Peace*.

"The View from St. Stephen's" is, above all, a view of circulation. Stifter proposed that the city's inhabitants were "the heartbeat of a great monarchy. . . . The blood, the simple red balm, flows happily through all the veins of the entire body, and never suspects that it built this marvel of a body itself." This arterial metaphor anticipated post-1848 interpretations of Habsburg unity as an effect of commercial "circulation." From above, Vienna revealed itself as the site of a confluence of streams of peoples and things, a nexus of commerce and cultural exchange: "Things and peoples mixed in ever-denser traffic, the rarified nervous fluid of international bonds." The individual "neither intends to promote the public interest, nor knows how much he is promoting it." Just as commerce figures here as a source of unity, so is money a tool of commensuration. It is the "one thing that swallows all the others up," the invention that makes it possible to carry a "duchy in a pocket."

Throughout, Stifter contrasted the narrow perspective of the city dweller down below with the wide view of the observer above. In this way, he juxtaposed divergent temporal as well as spatial scales. For the individual is also blind to patterns that unfold on the time scale of multiple generations. "This society builds unceasingly and diligently on a structure that it does not recognize, according to a plan that it does not know." Stifter's essay was designed to enlighten readers with respect to this plan. It was a lesson in scaling between the world historical and the personal, juxtaposing the history of the Habsburg Empire and the world at large with the stories of Vienna's inhabitants, each with its own particular joys and sorrows.[23] In short, the essay modeled shifts between different magnitudes of perception and different degrees of emotional distance.

In this respect, Stifter's contributions to *Vienna and the Viennese* demonstrated techniques of scaling appropriate to subjects of a supranational state. Climatology was not incidental to this program. In "Viennese Weather," Stifter began by noting that every great city or small village has its own distinctive weather. Although he denied that "the subject is ridiculous," the essay veers between science and satire. On the serious side, Stifter expressed concern about smoke pollution and pointed out the atmospheric effect now known as an urban heat island (a phenomenon that had been documented by Luke Howard in London).[24] On the other hand, he relentlessly mocked the scientific climatology of his day. In particular, he lampooned the pedantry of naturalists who distinguished ever-finer climatic regimes, identifying not only *Unterschiede* (contrasts) but *Unter-Unterschiede* (subordinate contrasts), such as "a peculiar suburb-weather or even authentic square- or alley-climates." He caricatured the weather "connoisseurs" or "collectors" who took a perverse pleasure in meteorological events that sent their neighbors running for cover. Stifter poked fun at their precious measuring instruments and at a fictive institute for "urban meteorology," with departments such as the "humidity committee," "the bureau of rainbows," and "the eclipse senate." In this spirit, he played contrasts of scale for comic effect. Thus he claimed that a city as large as the Sahara would be as dry as that desert. And he proposed that Vienna was a climatic microcosm of the Monarchy as a whole: "Everyone knows that the Alps have a rawer climate to their northern slopes, towards Switzerland, than in their southern ones, towards Italy—and are whole rows of houses not Alps of this sort? Which of us doesn't know that the south face of the Archduke Karl's palace has a mild Italian climate?"[25] Here is an image of Vienna as a replica of the Monarchy in miniature, a scale model of its natural and human diversity. Yet this is scaling in the mode of farce.

"Viennese Weather" was thus Stifter's revenge against professional science, following the rejection of his applications for academic posts. The satirical tone also reflected his discomfort with urban environments. Cities, he wrote, "spoil" man's "only form of nourishment that he can get for free, unadulterated, and in unlimited quantities"—that is, air. Modern man was trapped between leaving the city to breathe fresh air and remaining in it in order to earn enough to afford all his other "forms of nourishment." And yet Stifter equivocated, refusing to take seriously the hypothesis of anthropogenic climate change as it stood at the time. The essay worked too hard at humor to convey genuine concern about the environmental impact of urbanization.

In these ways, "Viennese Weather" forms a striking contrast with one of Stifter's better-known journalistic pieces from the same period, an account of the solar eclipse visible from Vienna on 8 July 1842. There Stifter's theme was the contrast between abstract knowledge and lived experience. Being able to calculate the trajectories of heavenly bodies was no preparation for the moment when broad daylight morphed into night. As in "The View from St. Stephen's" and "Viennese Weather," "The Solar Eclipse" sought to relate an experience shared simultaneously by all residents of the imperial capital. "From the attic windows of the houses all around heads were peering, from roof ridges figures stood and stared, all towards the same place in the sky . . . and how many thousands of eyes in the surrounding mountains might be looking towards the sun at this moment." The eclipse evokes at once the human place in the cosmos and the place of Vienna in the empire. Here too Stifter compares the narrow scope of daily urban life with the vast dimensions of the empire; from where he stands, he can see all the way to Hungary. Here, however, the emphasis falls on the cosmic scale, and the aesthetic is that of the sublime. Like "Viennese Weather," the essay emphasizes the limits of modern science. "In mathematics," he complains, "this space [the heavens] is nothing more than large." He tells us that God cares nothing for calculations. In this way, Stifter underlines the arbitrariness of human systems of measurement, which he contrasts with a God-given sense of "true measure" and with the "language of nature."[26] Even so, the structure of Stifter's essay suggests the power of astronomy to *prepare* humanity for such a revelatory experience of the cosmos's true scale, precisely by means of its observing instruments and its calculations. In the end, "The Solar Eclipse" suggests that the science of astronomy, widely disseminated and coupled to firsthand experience, may have the capacity to right man's sense of proportion.

Stifter had effectively thrown down a gauntlet to naturalists. As the satire of "Viennese Weather" suggested, the climatology of the 1840s was still mere place description. It was unable to convey the significance of the local in relation to the imperial, let alone the cosmic. In retrospect, it is not hard to identify what was missing. Without attention to the relationship between the particular and the general, between local winds and the planetary circulation, climatic description produced little more than a jumbled heap of "local contrasts." This was the literary challenge to which dynamic climatology would rise.

AN INTELLECTUAL MICROSCOPE

> It was once remarked against me that I only represent the little things, and that my people are always ordinary people. . . . The flow of the air, the rippling of the water, the growth of the grain, the waves of the sea, the greening of the earth, the gleaming of the sky, the twinkling of the stars I consider great. . . . If a man were to observe the needle of a compass, whose tip always points north, every day at the same time for many years, and to record its variations in a book, an ignorant person would surely dismiss this first step as insignificant and as a waste of time; but how awe-inspiring would this insignificant thing become and how fascinating this waste of time, once we discover, that these observations have been made across the whole surface of the earth, and that, from the tables thus produced, many small variations of the magnetic needle often occur simultaneously and in the same degree at all points on the earth; thus that a magnetic storm passes across the whole earth, that the whole earth simultaneously feels a magnetic shiver.
>
> Adalbert Stifter, 1853[27]

In 1849, the playwright Friedrich Hebbel poked fun at writers who made much of "beetles" and "buttercups." Such details could only interest a writer who had no insight into his characters' inner lives. Stifter retaliated four years later in the preface to his story collection *Many-Colored Stones*. His response is often quoted by literary historians as a paradigmatic statement of realism, but it deserves equal attention from historians of science. Stifter, after all, had set out to become a physicist before launching his career as a novelist. Had he not been turned down for teaching posts in Prague, Linz, and Vienna, he might never have written his beloved stories of rural communities and the natural world.[28] Stifter justified his attention to "little things" by analogy to geophysics. In nature, as in human life, Stifter argued, often the little things are most significant, once they are recognized as instances of a more general pattern, perceptible to observers everywhere. He offered the example of the study of magnetic variations across the earth's surface: small effects that, when taken together, reveal a higher law.

It was in these same years that Karl Kreil spearheaded the project of measuring magnetic and meteorological variations across the surface of the Habsburg lands. Kreil and Stifter can be said to have been pursuing different answers to the same question: the search for a less arbitrary scale of significance than narrow personal interest. In 1838 Kreil had moved to the observatory in Prague, where he began to collect the data that would form the basis for his *Climatology of Bohemia*. This was to be the first volume of a climatology of the entirety of the Austrian lands. According to his biographer, it became "the major work of his life," and he labored on it until his death. He saw it as a departure from his earlier publications on astronomy and geomagnetism, which were "strictly scientific, abstract works." Instead, the *Climatology* would draw on Kreil's own observations and would aim to be of practical use, particularly in agricultural communities like that of his childhood in Upper Austria. The challenge would be to give it "vitality and freshness." Indeed, Johann Franz Encke, head of the Berlin observatory, wrote Kreil to say that he hoped the book would be more than a "mere sea of numbers," that it would work through its data and arrive at some attractive results.[29] This was precisely Kreil's intention. Like Stifter, he was trying to perfect a new mode of writing that would represent the true significance of the small scale in relation to the large.

In the version published posthumously in 1865, Kreil initially defined the goals of climatology (as opposed to meteorology) as follows:

> Climatology represents the facts that have been uncovered by the observer as clearly and convincingly as the observations and the means of obtaining them allow; it writes the history or natural history of the atmosphere in the script of numbers, in order to identify therein the influence [of the atmosphere] on the earth and its form, and vice versa. For the eye practiced in this script,

> its tables are the true image of phenomena in the atmosphere. . . . By juxtaposing these tables and translating the numbers into spoken language, often into graphic representations as well, its task is completed, and it can leave it to the meteorologist, to the zoologist and the geologist, the botanist and the physician, the architect and the farmer, to draw from it what they need for their ends.[30]

According to Kreil, climatology offered description, while meteorology provided explanation, but this was not a strict division. Just as a historian sometimes ventured to speculate on the cause of human actions, so might the climatologist draw on meteorology to suggest how climatic conditions could be explained. Kreil used "climatology" to refer both to the discipline and the genre, much as "cosmography" had been used in both senses in the Renaissance. He did not employ the term "climatography," but his definition of climatology as a mode of representation squares with subsequent definitions of "climatography": it was concerned with the human meaning of meteorological measurements.

Climatography was invented to mediate between mathematical measurements of climatic factors and the subjective human experiences with which they might be associated, such as fluctuations in health, hunger, and prosperity. As Kreil explained, the physician wanted to know the distribution and variability of temperature, the quantity and distribution of rain, and the direction and strength of wind; the engineer and architect wanted to know the height of floodwaters, the strength of storms and the direction from which they came; the farmer asked after the extremes of temperature and the length of the seasons, the quantity of rain, and the frequency of hail.[31] As Wladimir Köppen wrote in his *Klimakunde* (1906), climate science "offers the farmer, the industrialist, the physician, the basis on which to judge the influence of the usual course of these phenomena at a given place on the growth of plants, on industrial processes, on diseases, etc."[32] Indeed, one finds references to climatographies in handbooks for travelers, in works on scientific agriculture, in medical guides, in advertisements for spas, even in military strategy. In this sense, climatography was defined by its uses. It was explicitly addressed to a wide variety of ordinary people who needed knowledge of natural conditions in relation to human welfare: the length of seasons, the likelihood of frost, the freezing of waterways. Among the intended users of a climatography were the many people "on the move" in nineteenth-century empires: farmers and colonists, merchants and shippers, medical doctors and patients, military captains, tourists and explorers. As H. F. Blanford explained in his 1889 *Practical*

Guide to the Climates and Weather of India, he was writing not primarily for "meteorologists and physicists," but for members of "the general public to whom the weather and the climates of India and of its seas are practical and not scientific objects of interest." He therefore aimed for "clear and concise language," free of "all technical forms of expression."[33] In other words, the implied users of a climatography were those imperial subjects who had the freedom and mobility to be able to put environmental information to use. In the Habsburg lands—where passports were not required for domestic migration after 1857, and where the Constitution of 1867 officially granted freedom of movement—this was a wide public indeed.[34] This is the most decisive difference between the early modern genre of cosmography, which remained a state secret, and the modern genre of climatography. That is to say, the historical relationship between these two genres tracks in part the emergence of an audience for whom environmental measurements *could* acquire meaning.

Thus it fell to the author of a climatography to model the *interpretation* of measurements. In doing so, Kreil's *Climatology* used the very same language that Stifter had used against Hebbel. "Everywhere there is a macrocosm and a microcosm," he insisted, "a world on the large scale and on the small—the latter just as important, often more important than the former."[35] Both authors were instructing their readers in the interpretation of nature on the small scale. And both were defending themselves against those like Hebbel or Encke who might accuse them of attending too closely to meaningless particulars.

In this vein, Kreil noted that naturalists were often too quick to posit large-scale natural laws, taking "pleasure in a lovely discovery, before the small-scale phenomena, the quotidian processes, have been properly appreciated." Echoing Stifter, he continued, "Thus there can and should be much included in climatology that would seem insignificant from the present perspective of the practical fields that it bears on, for instance the state and variations of air pressure, which is of no direct significance to any of these fields, but which in the mechanics of the atmosphere must be seen as one of the most powerful controls." Like Stifter, Kreil drew analogies to other domains of nature. In the animal kingdom, for instance, the "great monsters of prehistory" had vanished, while "the smallest animals continue to live and work spryly. For thousands of years they were neglected, and only now have we recognized their importance. The same is true in the physical realm." Despite the rise of uniformitarian geology, Kreil noted, naturalists were still drawn first to "phenomena on a grand scale." Likewise, in atmospheric physics, scientists had rushed to draw conclusions about the laws governing weather on the largest scale, before attending to the many small influences to which their measuring instruments were sensitive. (Kreil may have been referring here to studies by Bohemian plant scientists demonstrating the variability of temperature near the ground—see chapter 9. His standpoint also prefigures the subfield of "fluctuation phenomena," pioneered by Austrian physicists in subsequent decades.[36]) Although Kreil had been drawn into geophysics by Humboldt's example, he was resisting Humboldt's haste to synthesize. He insisted that one must "observe the quotidian processes with an intellectual microscope and disregard nothing that does not fit into the regular course. The smallest deviation, that has intervened perhaps a thousand times already, only to be passed over as observational error or so-called random variability, can, if properly confirmed, become a lamp that illuminates a formerly obscure chamber of science."[37] Playing with the inversion of scale, Kreil likened the narrow view through a microscope to the wider field of a beam of light. This was, in part, an attempt to forestall criticism of his *Climatology* for its limited geographic scale. "Many will believe that it would have been more advantageous to have immediately extended climatological consideration to the entire territory of the Austrian Monarchy, instead of dividing it into different parts." However, he insisted that "our Monarchy" was built up of such radically different physical parts that it was most efficient to treat it province by province. Bohemia, having the longest continuous period of observation, plus a natural orographic border, was a logical starting point for this project. What Kreil left unsaid was that, in the early 1860s, the Monarchy's physical diversity was still a rhetorical trope awaiting empirical demonstration. This would require new literary techniques alongside the visual tools of cartography and landscape painting.

Kreil's *Climatology of Bohemia* represents the fusion of two traditions of climatological study: the physico-theology of Kremsmünster, where he had been a gymnasium student, and the *vlastivěda* or patriotic science of Bohemia, where he had become a scientific leader in the provincial capital. From physico-theology, Kreil drew a religiously inspired appreciation of the role of small-scale nature in a divine cosmic plan. From patriotic Bohemian science, he took the conviction that the detailed work of local nature study would bring practical economic benefits. Out of these two unique cultural traditions, Kreil was weaving a new approach to planetary science.

And yet, despite Kreil's commitment to producing something more than a "sea of numbers," readers of the incomplete book that he left behind upon his death in 1862 were indeed at risk of drowning in measurements. The text functioned mainly to guide the reader's eyes through the numerical tables. Climatography as a literary genre had yet to be invented.

REALISM AND THE RHETORIC OF SCALING

While Kreil was trying to build a language for climatology, Stifter was honing a literary language that shared many of the same aims. Even at the time, readers recognized how closely Stifter's style of observation mimicked natural history. Friedrich Simony, for instance, attributed to Stifter the capacity to observe both as a painter and a naturalist.[38] More importantly, the relativity of scale functioned as an aesthetic principle in Stifter's fiction, much as it functioned as a methodological principle in the geoscientific research that radiated out of Kremsmünster.

In a pivotal early scene of *Der Nachsommer*, for instance, Risach, the wise naturalist, predicts that the afternoon's threatening clouds wil not bring rain, even though his measuring instruments suggest otherwise: the barometer is falling, and the hygrometer indicates maximal humidity. Here Risach insists on a first shift in scale; these indicators refer only to "the small space in which one happens to find oneself; one must also consider a larger one."[39] Thus Risach draws the narrator's attention to the sky and to the folk knowledge necessary to interpret its signs, like the shapes and movements of clouds. "Quite rightly," he remarks, "science often relies on knowledge gained from long experience." Then Risach proposes a second shift in scale, observing that "all the signs that we have spoken of so far are quite coarse . . . and are usually only recognized by us by means of spatial variations—which cannot even be observed by us if they do not attain a certain magnitude." Scientific instruments translate signs that are too large for our senses, such as the motions of the atmosphere, into ones we can read. But such instruments pale in comparison to the "finer contrivances" whose workings remain a mystery to us. These contrivances are nerves: not human nerves, which tend to be overtaxed, but the nerves of animals, especially the insects and spiders. With patient, regular observation of these little creatures' habits and "domestic arrangements," one can learn to use them as reliable weather indicators. All this would be patent, Risach concludes, if only people could adjust their sense of proportion: "Many men who are accustomed to view themselves and their pursuits as the center of the world consider these things to be small. But it is otherwise for God. Something is not large just because we can lie our ruler along it many times, and something is not small because we have no ruler to measure it." By the novel's end, the narrator has fully assimilated this lesson.

Stifter enacted such reversals and inversions of scale in many of his stories. Amitav Ghosh has argued that a defining feature of modern fiction is the constancy and passivity of the nonhuman world.[40] Stifter's fiction is exceptional in this respect, persistently redirecting attention from the human characters to the nonhuman backdrop. The narrator of "The Bachelors" ("Der Hagestolz"), for instance, interrupts the characters' conversation to spotlight the world around them: "While they spoke of what they believed to be great, around them transpires only what they likewise believed to be small: the bushes continue to grow, the fertile earth continues to sprout and begins to play with spring's first little creatures as if with jewels."[41] This passage even suggests that nature has its own scale of values, independent of human norms, and thus its own definition of jewels. Indeed, one might say that all of Stifter's fiction questions the priority usually given to plot over description. In a similar ploy, the narrator of "Two Sisters" challenges the reader to juxtapose the temporality of human life with that of nonhuman nature: "If, with your feelings and thoughts, you remain outside of the present moment, and are not dragged along by it, then everything restless, covetous, passionate hastens past. . . . If you observe nature then . . . what hustle and bustle here, what persistence there!"[42]

Elsewhere Stifter presented the reader with a new lens through which to view the world: a telescope. Halfway through "The View from St. Stephen's," for instance, the narrator instructs the reader to "take the telescope," and asks, "What do you see?" Here, as in "The Solar Eclipse," and the opening of "The High Forest" ("Der Hochwald"), Stifter leads the reader on a tour of a landscape, shifting between a synoptic gaze and the spyglass's wandering, magnified view. In other stories, the characters use a telescope to collapse distances on earth (as in "The High Forest") or between earth and sky ("The Condor").[43] In yet other instances, Stifter's exercises in scaling entail imagining the world through nonhuman eyes. In "The High Forest," the figure of the lake as eye—a "Naturauge"—helps to place the human scale of the story into a much longer temporal frame. Viewed through this more-than-human lens, the human tragedy symbolized by architectural ruins is dwarfed by the ecological tragedy of the destruction of the forest. In other cases, the human eye is trained to see anew. In "The Kiss of Sentze," instruction in scaling is provided by "small things" themselves. Mosses, in particular, teach people to appreciate nature's wondrousness (*Verwunderlichkeit*). The narrator learns to see a whole world of variation in a sample of moss: "I saw in the collection a greater number of mosses than I would have thought possible. I saw affinities, connections, and transitions. In the pressed leaves I saw the bounty of forms and was astounded by their delicacy and distinctiveness." As one of the most ancient phyla of plants, and one known for its slow growth, mosses evoke not only a contraction of the spatial scale of observation but also an expansion of the temporal. Against the backdrop of the failures of 1848, these smallest living things teach the lesson that "only things in nature are entirely true [*nur die Naturdinge sind ganz wahr*]."[44] Here Stifter's inversion of plot and description turns the historical drama into mere background. But it is a telling background, in that the political turmoil is the incentive to learn to see in this new way. Like Kreil's scientific methods, Stifter's literary techniques of scaling were a response to the emerging ideal of supranationalism.

To Rainer Maria Rilke, the Prague-born poet of the fin de siècle, such shifts in scale seemed to be the driving force behind Stifter's entire oeuvre. It seemed possible that Stifter's "inner calling became inevitable the moment that he, one unforgettable day, first attempted to draw an exceedingly remote point in the landscape into view through a telescope and then, with vision utterly staggered, experienced a flight from rooms, from clouds, from objects, an amazement of such abundance that, in these seconds, his open, astonished mind grasped world."[45] Rilke took from Stifter both the language of scaling and the motivation to make a reader see the world according to a scale other than that of human desires. As he described this insight in a letter to a painter friend,

> most people hold things in their hands to do something stupid with them (as, for example, tickling each other with peacock feathers), instead of looking carefully at each thing and asking each about the beauty it possesses. So it comes to pass that most people don't know at all how beautiful the world is and how much splendor is revealed in the smallest things, in some flower, a stone, the bark of a tree, or a birch leaf. . . . *The small is as little small as the big is big.* There is a great and eternal beauty throughout the world, and it is scattered justly over the small things and the big; for in the important and essential there is no injustice on the whole earth.[46]

The lesson, as Rilke put it elsewhere, was to "learn from the things" and "surrender to earth's intelligence."[47]

THE DYNAMIC ELEMENT IN CLIMATOGRAPHY

The literary strategies of climatography deserve to be considered alongside Stifter's techniques of scaling. Hann's 1904 *Climatography of Lower Austria* was intended as the template for all subsequent publications in this series and was hailed as a "model" for "monographs on the climate of other countries."[48] In the introduction, Hann expressed his literary aims thus:

> My work can and will not aspire to anything more than to furnish the necessary, appropriately constructed numerical framework [*Zahlenskelett*] for such a true climate description as I envision as an ideal, but which cannot even be attempted here. A true climate description would need to exhibit a piece of living nature, the total effect of all interacting meteorological elements, which is literally what we mean by climate, pointing to the connections to the local factors that determine the natural surface vegetation, as well as, especially, to the agricultural and industrial circumstances of the region, to the location of human settlements and their way of life, in so far as these depend on the average conditions of the atmosphere and their variation, as they influence man in his sites of residence.[49]

Even by the standards of German prose, this last sentence is striking for its length and convolutedness. By drawing it out and inserting so many active subjects, Hann underlined his definition of climate as a dynamic, evolving system with many moving parts. He strove to make climatography more than a "numerical framework," more than a "sea of numbers."

The *Climatography of Lower Austria* is a text shot through with movement. It follows the winds as they blow across the surface of the earth, transformed by the mountains and valleys they encounter. It shows how the winds—some of local provenance, some sweeping in from afar—are, in turn, the bearers of weather. It surveys climatic regions that are only more or less enclosed, each standing in communication with other regions—thanks, once again, to the persistent work of the winds. This dynamism is the key to climatography's effect of scaling.

That is, climatography conveys the significance of the local and particular by revealing each part as a component of a dynamic whole. Each volume of the series interprets the peculiarities of a region in relation to their participation in larger-scale patterns, whether in real time or as monthly or annual averages. Each distinguishes between strictly local winds, like those arising in enclosed valleys, which are "of minimal climatological interest," and air currents that traverse larger distances. These larger-scale trajectories are often described in relation to itineraries a human traveler might follow. Thus, for instance, the average annual temperature in Tyrol can be seen to fall as one follows the river Inn downstream. Similarly, several of the texts employ phenological data to track the arrival of spring across a region. Some imagine following the blooming of a particular flower or the ripening of a given fruit. More anthropomorphically, winds are said to follow "roads" (*Straßen*) through the mountains. Climate is described not only in relation to the flow of air, but also with regard to the circulation of people. A region's status as a *Sommerfrische* (summer resort), a *Wintersportplatz* (winter recreation area), or a *klimatische Kurort* (climate therapy resort) is taken as initial evidence of its climatic qualities. Reference is made to the cooler summer temperatures that draw visitors to a *Sommerfrische*, to the plentiful snow and mild winters of a *Wintersportplatz*, and to the abundant sunshine of a *klimatische Kurort*. Notably, when the circulation of people does not follow the circulation of air, this discrepancy earns mention. This was the case for the western Alpine towns of Langen and Sankt Anton, which displayed a difference of monthly average temperature of up to one degree Celsius. These are "two locations, separated only by the Arlberg, which are connected to each other—yet sure enough, not climatically—by the 10-km long Arlbergtunnel."[50]

Climatography is surprising in this respect. We would expect it to be a genre of place, like Humboldtian cosmography. We might expect it to resemble much of what is classified as "nature writing" in the United States, with its privileging of the local and its almost Heideggerian fascination with "rootedness."[51] As an imperial genre, however, climatography—like early modern cosmography—was not about place in this sense at all. It was instead about circulation: about the actual and potential mobility of air, goods, and people through an empire. The science with which it was in dialogue was not a physics of static forces and stable equilibrium; it was, rather, the emerging science of dynamic climatology, focused on the motion of air masses and their modification by the earth's surface.

Stifter too was fascinated by the challenge of representing motion, both in literature and the visual arts. Indeed, he believed that the aesthetic effect of art turned on the reader's or viewer's experience of movement, whether this derived from the artist's evocation of motion or the motion of the viewer's own gaze. As Risach reflects in *Der Nachsommer*, "Movement stimulates, stillness satisfies, and thus arises the spiritual closure that we call beauty."[52] Consider, in this light, Stifter's series of paintings and sketches entitled "Movement," which occupied him for several years in the late 1850s and early 1860s. They sought to capture the sheer essence of motion, as made visible, for instance, in the forms of clouds or falling water. One such study (see figure 21) centers, paradoxically, on a large stone, immobile in a shallow creek. Motion exists in the painting only in the subtle pattern of the flow of water around the rock. It is the viewer who transforms this into an image of motion, as the composition leads the gaze from foreground to background and back again in a clockwise circulation.

Climatography adopted a related technique. It acquired its dynamism only in part from the depiction of movement in nature. Equally important was the

FIGURE 21. *Die Bewegung I*, by Adalbert Stifter, ca. 1858–62.

implied movement of the observer. In this sense, climatography's dynamic rhetoric was performative. As we will see, it not only described but promoted the mobility of imperial subjects.

WANDERLUST

Climatography produced an effect of "lifelike" description not simply by virtue of its textual qualities, but by serving as a model for relating measurements to experience. Indeed, the genre depends on an implicit narrative: that of the imperial naturalist, learning to situate his own local knowledge within the continental framework of the empire he serves.

In 1887 Crown Prince Rudolf introduced the "overview" volume of *Austria-Hungary in Word and Picture* with an invitation to the reader "to a tour through wide, wide open spaces, through multilingual nations, in the midst of constantly changing images." When Julius Hann took up the challenge of writing a "climatic overview" of the empire for this volume, he likewise framed it as a journey. He imagined a traveler making the mere half day's journey from a wintry Vienna, "with its monotonous snow cover, the dusky cloud-covered sky and uncomfortably frosty temperatures," to Fiume and its "mild air full of sunshine and painterly light." From Fiume, one could—if only the tracks existed—continue down the Dalmatian coast and into the full bloom of spring, or, moving a short distance west to Ljubljana and into Carinthia, one could rejoin winter in the "Austrian Siberia." This was a new spin on an old trope, for Hann introduced into this picture of endless variety a human factor: a traveler conscientiously exposing himself to the "climatic contrasts" and letting them "operate on him directly."[53] He didn't fail to refer to the empire's famed *Kurorte*, such as Arco and Riva, or to point out that there were locations in the Austrian Alps that could rival Davos for cold-weather treatments. In this essay, however, it was not the destination but the journey that mattered. Rail travel set the proper pace for assimilating the empire's variety. It is a conceit that puts one in mind of the travels of Count Morstin in Joseph Roth's "The Bust of the Emperor": "As he traveled around the center of his multitudinous fatherland, what he responded to most were certain specific and unmistakable manifestations that recurred, in their unvarying and still colorful fashion, at every railway station."[54] Nineteenth-century rail travel, the agent of the standardization of time, suggested that diversity could be digested deliberately and methodically.

While Roth's Count Morstin tracked the recurring motifs of imperial insignias and coffeehouses, Hann's traveler followed the buds and blossoms of spring. As Hann explained, tracing the appearance of seasonal vegetation from one location to another gave a more vivid (*anschaulich*) idea of climatic contrasts than did the readings of a thermometer. The journey metaphor thus became an exercise in phenology, the study of seasonal natural phenomena and their geographic distribution. To follow spring on its course through the empire was to search for regularity in the midst of change—in Roth's words, "the familiar in the variable"—a quest that found visual expression in two engravings of spring vegetation. The first (figure 22), in Silesia, showed a peasant hut with animals and a mother and child against a vast, flat, and still barren expanse. The second (figure 23), on the island Lacroma near Ragusa, appears to be the ruins of a palace, overgrown with ferns and dappled with tulips. In the context of Hann's chapter, these images are a variation on the theme of the coexistence of primitive and decadent, depicting not the *Gleichzeitigkeit des Ungleichzeitigens* but near simultaneity: the two-month delay between spring's arrival in the Mediterranean and in Silesia. Following spring's course through the empire from the southern coastline to the low-lying western lands, up the mountains, and into the east took two and a half months in all. The metaphor Hann repeatedly fell back on to describe the progress of spring was not an awakening (perhaps too redolent of nationalism) but a "conquest." One after another, spring brought the crown lands "under its rule [*Herrschaft*]"—

FIGURE 22. *Spring on the March: Silesia.*

except, that is, on the tallest peaks, where "winter has made its permanent home." The imperial metaphor persisted into Hann's account of the empire's characteristic air currents: "By far the largest part of Austria stands throughout the whole year under the sway [*Herrschaft*] of the Atlantic air currents." Galicia and Bukovina, however, "stand wholly open to the cold-invasions out of Russia from north-east and east, while the other crown lands (except Silesia) are more withdrawn from these in part due to the mountains, in part due to their western position."[55] In Hann's account, wind, weather, and seasons swept triumphantly through the empire—not unlike the crown prince in his aerial geography of his father's lands.

This image vividly illustrated one of the central theses of the Climatography of Austria: the empire was as much a climatic as a geopolitical transition zone (*Übergangsgebiet*) between the temperate, maritime west and the severe, continental east. As such, its climatic contrasts could be understood as *continuous* transitions. The language of continuity was essential to the Climatography. In Styria, for instance, the climate revealed "an uninterrupted series of transitions. The central European intensifies into the eastern European on one side and is tempered on the other side into the coastal climate. But one searches in vain for an abrupt change; it does not take place in bounds, nowhere does one find the one or the other type sharply pronounced, because everywhere the influence of the mountains resonates like a soft, mellow over-

FIGURE 23. *Spring on the March: Lacroma.*

tone."[56] Here the implicit imagery of atmospheric waves lent itself to a more typical Habsburg metaphor: unity as musical harmony.

This imagery recurred elsewhere in the travel literature of the late Habsburg Monarchy, for instance, in Friedrich Umlauft's *Tours of the Austro-Hungarian Monarchy* (1879, second edition 1883). This was a lavishly illustrated volume commissioned by the education ministry in Vienna, meant as leisure reading for a popular audience and to serve the needs of instructors. The preface announced the goal of "stimulating interest in the less familiar regions" of the Monarchy and "awakening the *Wanderlust* of a wide circle [of readers]." Umlauft set out to integrate cultural and environmental description, such that "first folk culture, then climatic or geognostic circumstances receive more detailed consideration." Like Hann, he emphasized "the contrasts of landscape" and "climatic contrasts" that a tourist of Austria-Hungary encountered. In the

Alps, it was "as if you had covered hundreds of miles in the course of one day. It is no wonder that just these phenomena alone can call forth a full range of spiritual arousals and sensations."[57] Circulation, continuity, and the embodied experience of difference were the motifs that climatology contributed to the literary representation of Austria-Hungary.

THE DISCOVERY OF DIVERSITY

To compose a climatography was to reveal borders. It fell to the author to identify, on the basis of qualitative and quantitative evidence, the "natural borders" of the region under study, as well as the internal borders between one "climatic zone" and another. In some cases, the transition from one zone to another was gradual; in other cases, sharp. A historian considering this process might well wonder: did climatography discover divisions that existed in nature, or did it construct new categories of difference? It is a defining feature of climatography that it made this epistemological concern explicit.

Tasked with composing the climatography of Tyrol, Heinrich von Ficker faced a dual challenge: the absence of even a single mountain observatory to supply data on higher altitudes, combined with the presence of what he saw as the two "completely separate climatic regions" of North and South Tyrol. North Tyrol was an example of a "central European" climate region, modified by a mountain range that lent it relatively warm temperatures, ample sunshine, and the notorious foehn winds. South Tyrol, however, was an "Alpine variant" on a Mediterranean climate. For practical purposes, Ficker acknowledged the necessity of providing an overview of the province as a whole—it was, after all, a "political unit." But how to compare such different regions, especially given the holes in the data? On one hand, Ficker called on a colleague in zoology to provide information on the climatic dependence of Tyrol's flora and fauna, since these allowed "the climatic borders to be far more sharply defined than with purely meteorological data." Indeed, the zoologist noted the presence of many "Mediterranean" species in South Tyrol. Yet he contributed little more than a laundry list of species, nothing approaching a systematic "overview." By contrast, Ficker himself achieved his most striking overview by calculating—separately for North and South Tyrol—the average temperature at the region's average height. This revealed a north-south temperature gradient three times greater than the global standard. Here was a striking illustration of Ficker's thesis that North and South Tyrol were split by a climatic divide. Yet Ficker was able to restore continuity to his picture of Tyrol by identifying "a transitional region" in the Eisack Valley, near the famous Brenner Pass, with moderately cooler summer temperatures and moderately greater temperature variability than farther south. He concluded by paying homage to the empire's diversity:

> For the climatographer Tyrol is one of the most interesting regions of the monarchy. The climatic contrasts are stark and yet in their distinctive features mostly a blessing for the province. To assess these differences, to represent them in numbers, is an attractive task for the climatologist. Yet the contrast imprints itself in memory far more strongly when a spring journey of a few hours across the Brenner reveals to the senses the image of an abrupt change of climate like no other. The climatologist's burden is lightened in this way, since his numbers mingle effortlessly [*knüpfen ungezwungen*] with the common ideas that every educated person almost intuitively associates with the concepts "North Tyrol" and "South Tyrol." Stark contrasts make the climatologist's work easier, particularly in regions that belong, thanks to their scenic beauty, to the most famous of the continent.

The image of crossing the Brenner—the alpinist's gaze—suggests continuity rather than mere contrast. As Ficker explained, "The great contrasts between north and south Tyrol are blurred from great heights."[58]

Yet the question arises: was such a border real or perceived, an element of the natural world or a statistical artifact? The Climatography provided no conclusive answer to this question. Its authors often paused to discuss the choice between one statistical method and another. In the analogy proposed by the *Climatography of Styria*, the choice of a statistical measure was like the choice of a height from which to view a landscape. With the adoption of new methods, "insignificant differences swell into round numbers; and these, after further partition, differentiate themselves again from each other. And in places where he could find no contrast, he flips the averages around, divides them according to season, graphs the annual curve and finally finds to his satisfaction a difference—there, where everything had earlier looked the same to him, where before he had discovered only small deviations."[59] Far from attempting to elide the ambiguous nature of "climatic borders," the authors of the *Climatography of Austria* made the rhetorical choice to accentuate their indeterminacy.

CONCLUSION

Unlike environmental genres such as the chronicle or parish register, climatography did not explicitly record change over time. It defined climate as a statistical description of weather over a period of a few decades, and it treated the resulting distillation as existing outside of historical time. It depended further on the possibility of defining transhistorical "natural" regions. It showed no interest in the question of climate *change*. And yet the genre's elaborate discussions of the choice of a scale of analysis revealed the drawing of climatic borders to be an open-ended process, subject to perpetual revision. In this way, climatography created a space for perceiving climatic change not only across space but also over time. As we will see in part 3, the experience of drawing the borders of climatic regions propelled several Habsburg researchers to investigate evidence of climate change.

The fashion for writing climatographies had passed by the early twentieth century, and today the term, rarely encountered, has come to mean little more than a data set.[60] Yet the quest to communicate climatology "in word and image" has recently been renewed. In 2012, the Intergovernmental Panel on Climate Change (IPCC) published its first report to incorporate social science. It argued that the impacts of climate change "will inevitably need to be understood and responded to principally at the scale of the individual, the individual household, and the community." To illustrate this point, the authors offered a single "ethnographic vignette." They introduced "Joseph," an eighty-year-old Tanzanian, who, we are told, has "witnessed many changes." "What do 'changes' (*mabadiliko*) mean to someone whose father saw the Germans and British fight during the First World War and whose grandfather defended against Maasai cattle raids when Victoria was still Queen? . . . *What is 'climate change' (mabadiliko ya tabia nchi) to Joseph?*" This passage highlights Joseph's experience of shifting borders, both political and ecological, and suggests that he does not distinguish between natural and human drivers of change. In the context of the IPCC report, it poses the problem of mediating between expert and local knowledge.[61] Although "climate change" can be translated into Joseph's language, the concept carries no *meaning* for him.

It was also circa 2012 that writers in Europe and North America began to express an urgent need for literary innovation in order to make climate change meaningful to educated Western readers. Consider, for instance, the preface to the 2010 multiauthored volume *Climate Refugees*: "Our job is to tell stories we have heard and to bear witness to what we have seen. The science was already there when we started in 2004, but we wanted to emphasize the human dimension, especially for those most vulnerable." One review of *Climate Refugees* referred rightly to "the uncertain genre of the book," while a critic on the Amazon website expressed confusion: "I was expecting facts and figures, charts and graphs. In fact, I was a bit suspicious when I thumbed through this and saw all the gorgeous photos in this book: it seemed too luxurious for such a serious subject."[62] Here is an example of a self-conscious experiment in making human meaning out of scientific results, one that disrupted its readers' expectations by bending the rules of existing scientific and literary genres.

As this example suggests, the challenge of meaning-making identified by the IPCC is not simply a problem of translation. Nor is it new. The question of how to connect global models to "local stories" needs to be recognized as part of a long history of efforts to communicate discursively the human meaning of environmental information. This history includes some forms better known to literary scholars, like lyric poetry, travel narratives, nature writing, and futuristic fiction, and some more familiar to historians of science, such as cosmography, chorography, geography, natural history, medical geography, weather diaries, ship logs, and parish registers. Climatography is among the most recent of these genres, and it merits attention as a solution to the representational challenge first articulated by Karl Kreil: that of depicting climate, on the large scale and the small, maximally objectively and simultaneously subjectively, in its human significance.

CHAPTER 7

The Power of Local Differences

In 1884, Alexander Supan, professor of geography at the University of Czernowitz/Chernivtsi/Cernăuți summed up a lesson of the new dynamic climatology in his textbook *Principles of Physical Geography*: "It is, therefore, no exaggeration to say that the wind is the effective bearer of climate, and thus—since climatic conditions regulate organic life and with it human development—a cultural force of the greatest importance."[1] This chapter explains how the wind acquired this physical and cultural significance.

In designating the wind as the bearer of climate, Supan acknowledged and yet broke with a tradition of natural-philosophical explanation that dates back to ancient Greece. In the Aristotelian schema, climate is determined by the angle of incidence of sunlight at different latitudes, *klima* being the Greek word for slope or incline. Deviations from this "solar" climate were attributed to the winds that visited the location in question, each originating in a different location and carrying different qualities of air. Local winds and the variability they occasioned figured as disturbances overlaid on the simple geometry of climatic zones. Thus winds were incidental to *klima*, and yet they were significant within the ancient tradition of Hippocratic medicine. Knowledge of the typical winds at any given location was essential to maintaining good health. This tradition lived on in the nineteenth century. It was manifest, for instance, in the construction of wind roses, which visually summarized local statistics on the frequency of wind from each direction (figure 24).[2]

Supan alluded to the new significance that the nineteenth century had bestowed on winds. In the framework of dynamic climatology, winds were interpreted as products of encounters between contrasting air masses. Dynamic

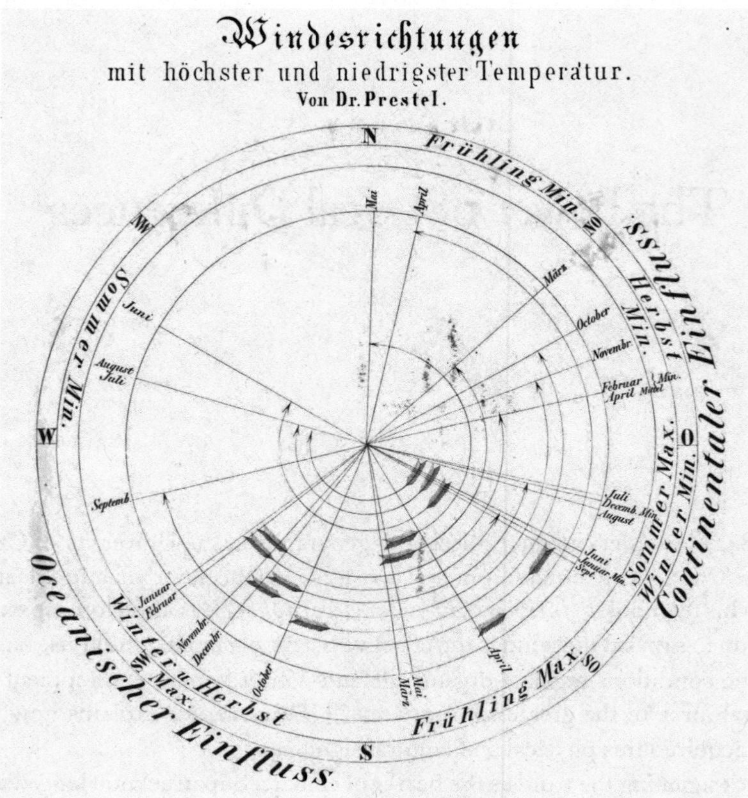

FIGURE 24. Thermal wind rose for northwestern Germany, 1861.

climatology posed questions such as, given a dry atmosphere initially at rest, what spatial contrasts of temperature and pressure would produce winds like those observed in nature? Once set in motion, how would such air currents be affected by the earth's rotation? Down the road, the theorist might try to incorporate the effects of moisture and friction into the explanatory framework. In this way, nineteenth-century dynamic climatology no longer treated winds as displaced air from a foreign clime, as in an Aristotelian framework. Nor did it approach winds in the manner of climate dynamics today. Indeed, this nineteenth-century way of posing questions might strike some readers as odd. Today, what seems to require explanation is not the onset of atmospheric motion, but departures from geostrophic flow—that is, deviations from an equilibrium state of motion in which the force arising from a pressure gradient is balanced against the Coriolis force of the earth's rotation, such that air flows along lines of equal pressure. This "quasi-geostrophic" way of thinking about atmospheric motions was not developed until the 1930s. It was driven, however, by questions that had been nagging nineteenth-century dynamicists for some time. Namely, what maintains the contrasts of temperature and pressure that give rise to energetic atmospheric motions? How is it that the atmosphere can support unstable conditions long enough for them to sustain strong winds? These were questions that Austrian researchers like Julius Hann and Alexander Supan helped to place on an international research agenda.

By the 1870s, governments and learned societies across Europe and North America had invested deeply in meteorological observations. In Britain, France, the Netherlands, Scandinavia, and the United States, the aim was, above all, to construct an advance-warning system for storms. These efforts, which rested on the synchronization of measurements via telegraphy and on the production of synoptic charts, resulted in a growing base of empirical knowledge about the strength and direction of winds in relation to the distribution of surface air pressure. Empirical rules for storm forecasting began to accumulate. One held that wind strength in a cyclone is proportional to the pressure gradient, or the difference between barometric readings at neighboring stations.[3] Another described the direction of winds in a cyclone, that is, a storm characterized by rotational motion around a center of low pressure.[4] For many naturalists at the time, these were simply handy rules of thumb for predicting strong winds.[5] For others, they held clues to the basic physics of storms. Only a few saw another potential: to apply this new empirical knowledge of pressure and winds to elucidate the global geography of climate. Those who rose to this last challenge tended to be scientists in the employ of Europe's great land empires—above all, Wladimir Köppen and Alexander Voeikov in Russia, and Julius Hann and Alexander Supan in Austria.

Hann developed his dynamical approach to climate as he painstakingly worked his way through the data of the growing station network of the ZAMG, which he joined as an assistant in 1867, becoming director from 1877 to 1897. He insisted that a detailed map of atmospheric pressure was "one of the most important foundations of the scientific understanding of the climatic conditions" of the region under study.[6] Here was a physical-mathematical solution to the Austrian Problem of precisely representing local variation while simultaneously revealing a higher unity. The form it took reflected the broader emphasis on phenomena of mixing and exchange in post-1848 accounts of the historical development of the Habsburg Monarchy and its economic and political future.

Thus the study of pressure gradients and the winds to which they gave rise seemed to hold the key to understanding the distribution of the earth's climates. This chapter also takes up the second part of Supan's claim: dynamic climatology was linked to new ways of thinking about human health and cultural development. The newly dynamic science of climate quickly became a topic of popular interest, in an age fascinated by stories of the progress of modern science. By the 1880s, schoolchildren as well as readers of German-language popular science journals and even provincial newspapers had ample opportunity to acquire a basic understanding of the new theories of atmospheric motion. Climatology provided tools of scaling with which people throughout the Habsburg lands could envision their place within imperial networks of circulation and exchange. Not until the turn of the century did scientists begin to question the physics behind this popular view.

THE WAYS OF THE WINDS

The Austrian research program in dynamic climatology can be said to have been launched in 1866, when Julius Hann, then an assistant at the ZAMG, overturned the reigning theory of the warm, dry mountain wind known as foehn. Hann had used thermodynamics, the new science of the relationship between heat and motion, to explain what happens as air is forced upward along a mountainside. As a parcel of air rises into regions of lower pressure on its way up the mountain, it does work by expanding. This process lowers the parcel's temperature as well as its specific pressure, causing condensation—and often precipitation. By the reverse process, as the air then makes its way down the other side of the mountain, it contracts and its temperature and specific pressure rise. This temperature rise will be greater than the temperature decrease on the upwind side, since the decrease was offset by the latent heat of condensation. This was a principle that was quickly understood to be applicable most generally to rising motions in the atmosphere. Hann had demonstrated the power of a new way of thinking about climatic phenomena, in terms of the interconversion of heat and motive power.

It was Hann's judgment that a global science of climate could only progress on the basis of more precise and detailed regional studies of the distribution of air pressure, a basic thermodynamic variable. The German-Russian climatologist Wladimir Köppen had made this point in 1874, when he pointed out that the Aristotelian use of the wind rose was flawed. Knowing the direction from which the wind blew was not enough to tell you the character of the wind. One also needed to know the surrounding pressure distribution, as he demonstrated with reference to data from Siberia.[7]

This was the motivation behind Hann's painstaking analysis of the distribution of pressure in central and southeastern Europe, averaged over the first thirty years of the operation of the ZAMG's network (1851–80). This was a daunting undertaking. Obtaining precise, standardized measurements of pressure over a wide region was not straightforward. Pressure differences between neighboring locations at the same altitude are far smaller in magnitude than differences of temperature; in fact, they were on the order of the systematic error of the barometric measurements of Hann's day.[8] Fortunately, the Austrian network had established the necessary conditions to produce suitable data. At each station the barometer was calibrated against a standard instrument, thanks to the inspection tours carried out every six years. And the elevation of each station was known with precision due to the geodetic measurements of the Imperial-Royal Military-Geographic Institute. What remained was the long and arduous task of averaging thirty years of data by hand.

Hann admitted that he had often spent a week or more deciding whether to adjust the average pressure at a single location by a tenth of a millimeter in either direction. "Many may well wonder if it is even worthy of a serious man to devote so much time and effort to such a minor result." Certainly, he wrote, there had been times when he doubted it was worth it. But anyone prone to such doubts for long, he insisted, was simply not fit to be a natural scientist, and he quoted Francis Bacon to that effect.[9] Just as Bacon had dedicated his empiricism to his queen, Hann's work was dedicated to governing a kingdom. In fact, Hann's *Distribution of Air Pressure* (1887) responded to the same representational challenge as did his chapter for the Austro-Hungarian Monarchy in Word and Image, which happens to have appeared that same year. In both cases, the goal was to keep fine-grained deviations in focus while constructing a total view of the empire. The isobaric maps and accompanying descriptions explained local peculiarities in relation to regional trends. Thus, for instance, the wintertime isobars showed an area of high pressure with its center on the south side of the eastern Alps. This corresponded to a "cold island" in the valleys, with a temperature increase or capping inversion above, which explained why warmer air from the south didn't penetrate into central Europe. The same phenomenon could be found in eastern Hungary and Transylvania. Meanwhile, the center of low pressure over the eastern Mediterranean and the Adriatic set up a pressure gradient that explained the strength of the downslope bora winds on the Dalmatian coast.[10] Hann looked forward to the day when the significance of such a map could be appreciated in relation to a complete description of the climatic conditions of Austria-Hungary. Only then would it be possible to use pressure differences "to explain the differences of wind conditions and their consequences."[11]

FIGURE 25. Postcard showing the weather house in the city park in Graz, 1898. Weather houses like these cropped up in city parks and public squares in spa towns throughout central Europe in the late nineteenth century.

DYNAMIC CLIMATOLOGY FOR ALL

Even as it was being worked out, the new dynamic climatology began to circulate to a broad audience. Under Hann's directorship of the ZAMG (1877–97), the number of observing stations in the network rose from 238 to 444.[12] Teachers, physicians, innkeepers, and telegraph operators were well represented among those who volunteered their time to take note of the state of the atmosphere at prescribed hours daily. Even those without access to their own meteorological instruments could participate in this ritual, thanks to the "weather houses" installed in parks and town squares (see figure 25). All the major Austrian spa and resort towns boasted such an edifice, "which tend to be of a very luxurious and tasteful design and are extremely popular with locals and visitors."[13] These volunteer observers and spa-goers were among the educated readers eager to hear of the latest progress in the sciences of weather and climate.

Hann's popular *The Earth as a Whole* (1872) introduced the basic principles of the application of thermodynamics to the atmosphere, explaining, for instance, the physics of sea breezes and the origins of the trade winds. A more application-oriented presentation could be found in the 1874 *Textbook of Climatology, With Particular Attention to Agriculture and Forestry*. It included the first climatic map of Austria-Hungary, which displayed the empire as a transition zone between "oceanic" and "Pontic" climates, between the West and the Orient, in which abrupt contrasts would immediately be smoothed into continuous transitions. On the large scale as on the small, the book explained, circulation was driven by "oppositions between warmer and colder neighbors"—a claim we will examine below.[14] Josef Roman Lorenz had almost completed the book when he was called to the agricultural ministry in Vienna; the last touches were left to Carl Rothe, a high school instructor in Vienna, who received the assistance of experts like Jelinek and Hann. Thus while many of the book's explanations incorporated the latest thermodynamics, other passages fell back on Dove's older account of a "struggle" between polar and tropical air, including his view of cyclones as *Ausnahme* (exceptions or deviations) within the general circulation.[15] In fact, the seventy-one-year-old Dove contributed the book's preface. One reviewer, attesting to the book's accessibility, deemed it just as useful for physicians as for farmers.

The press also helped to keep the public up to date on progress toward a dynamic understanding of climate. In 1880, for instance, a series of articles in the *Teplitz-Schönauer Anzeiger* (northern Bohemia) announced that "knowledge of the factors that determine air currents and their trajectory and speed has been quite a recent achievement. It was not long ago that we had no more correct idea of these relations than the ancient Greeks, who simply assumed that the allmighty Zeus had appointed one of his ancestors, the skilled sailor Aeolus, to be the guardian of the winds. . . . Now we know that the wind system of the Earth is governed on the whole by two dominant currents, which have their origin in the uneven heating of the earth's surface by the sun." From there the article went on to sketch the Hadley model of the general circulation (see chapter 8). Finally, the author explained that "the strength as well as the direction of the winds appear to be dependent on differences of air pressure and its distribution." A subsequent article in this series worked through an example, tracing the life cycle of a cyclone based on reports from the ZAMG.[16] In the *Wiener Landwirtschaftliche Zeitung* (a popular, illustrated, agricultural magazine), readers learned in 1885 how they could subscribe to the ZAMG's daily forecasting service. They also received a lesson in the "basic principles of modern meteorology," including the relationship between the distribution of pressure and the direction of winds. As the author explained, this would allow readers to decipher synoptic charts themselves, so that they "can form a judgment for themselves of the influence of the general weather situation on the local weather of their place of residence."[17]

By the 1880s, dynamic climatology had already found its way into at least one high school textbook. Students were taught to view their local weather as a link in a planetary chain of events: "For the most part, our weather is not determined by local conditions and circumstances, but rather by the course of the air-pressure minima and air-pressure maxima. The air-pressure minima originate in the Atlantic Ocean and mainly travel over Scotland and northern Europe. If such a depression center approaches us in central Europe, then we have south and southwesterly winds, and as these arise clouds cover the sky; west and northwesterly winds follow, from which the moisture falls as rain."[18] This text also conveyed the difficulty of predicting the outcome of such a synoptic situation: if the pressure minimum continued on its path, central Europe could experience a clear sky with winds from the northeast; but if a second minimum separated off from the first, the result could be storms in southern Europe, whether sirocco, foehn, or bora. By 1899, the subject of "isotherms, isobars, winds" had been incorporated into the physics curriculum of Austrian Realschulen, even if appropriate textbooks weren't always available to teach it.[19] In short, in the course of the last decades of the nineteenth century, an educated German-speaking public was gaining access to elements of a dynamic theory of climate.

COLD SPELLS, FROST SAINTS, AND NATIVE HUNGARIANS

Dynamic climatology was presented to students and newspaper readers not only as a signature achievement of modern science, but also as a bridge between science and folk wisdom. As we have seen, folk knowledge was built into the new atmospheric dynamics through scientists' engagement with local accounts of phenomena like foehn and mountain inversions. Presentations of dynamic climatology for a general audience highlighted this convergence between expert and lay perspectives.

For instance, one could expect to find climate-themed articles in local newspapers throughout much of central Europe during the second or third week in May. After the first warm weeks of spring, it often happened that the weather suddenly turned cold. This phenomenon was so familiar that an elaborate mythology had arisen around it. In popular speech, a temperature drop between the twelfth and fourteenth days of May was known in German as the *Eisheilige* (frost saints) or *Eismänner* (frost men), or the *strenge Herren*, the strict lords. In Czech, it went by the name *Pan Serboni* (Mr. Serboni), formed from the first syllables of the names of the saints associated with these dates: Pankrác, Servác, Bonifác. Hence the saying, *Pan Serboni pálí stromy*, Mr. Serboni withers the trees. In Polish, the phrase was *Pankracy, Serwacy, Bonifacy to źli na ogrody chłopacy*, suggesting that these saints were bad boys when it came to gardens. These frost saints were a source of great fear in central Europe, because they were capable of destroying entire crops at the very start of the growing season. Weather lore like this, passed down from one generation to the next, reminded farmers to take appropriate precautions. Many communities had devised tactics to protect their crops when a freeze threatened, most often by blanketing their fields in smoke.[20]

Whether or not this weather pattern occurs more frequently in mid-May than at other times of year has never been clear.[21] Already in the 1870s and 1880s, some scientists attributed the reports of a regular mid-May cold spell to faulty statistics and stubborn superstition.[22] As an 1887 article in the *Innsbrucker Nachrichten* put it, "Cases in which the effect is absent are forgotten, since, as Kepler already knew, one retains the occurrence, forgets the absence, since it's nothing special after all."[23] Nonetheless, much of the population of central Europe at the time was united in expecting a cold snap in mid-May, and newspaper editors aimed to address their concerns.

Hence the appeal of the frost saints as a research topic for scientists was, in part, the wide audience for results of any kind. The *Eisheilige* was also a tempting nut to crack for physical reasons. Here, after all, was a clear confrontation between air masses of different temperatures, as Dove had been the first to point out. Dove understood the frost saints as the last gasp of the polar current in its springtime struggle against tropical air. It carried such a chill, he said, because it blew from the region of melting ice in Labrador and Greenland.[24] An alternative explanation emerged in the 1870s, as scientists working in a dynamic framework became intent on explaining typical wind patterns according to average pressure distributions. The new theory (due to Wilhelm Bezold and W. J. von Bebber in Germany) started from the observation that, as winter passes into spring, land warms faster than water. Over a large landmass like the plains of Hungary and southeastern Europe, the warmer air will rise and create a low-pressure center at the surface. This low will allow cold air from the north to flow in, across central Europe, bringing a cold snap as it goes. Noting that these episodes were preceded by unusually warm temperatures in Hungary, Bezold nicknamed the cold spells "native Hungarians" (*geborene Ungarn*), a name that German-speaking scientists in Austria couldn't resist repeating.[25]

The theory of the "native Hungarians" became a hit with the popular press. Local German-language papers in Tyrol, Upper and Lower Austria, Bohemia, and Moravia carried articles that affirmed the reality of a phenomenon previously known only from popular lore. The *Linzer Tagespost* declared that explaining the *Eisheilige* was "one of the most difficult tasks of modern meteorology."[26] And the *Innsbrucker Nachrichten* reported that here for once was a case in which scientists had decided to take popular lore seriously.[27] In Transylvania, Ludwig/Lajos Reissenberger (1819–95) brought the dynamic perspective to his neighbors at the local Natural Scientific Society. Reissenberger was a Berlin-educated gymnasium teacher and meteorologist in Hermannstadt/Nagyszeben/Sibiu and had been a corresponding member of the ZAMG since its founding. He took an active role in organizing local scientific societies and in stimulating popular interest in meteorology. In his research, he took a particular interest in correlations between temperature variability and mortality. In taking up the question of the *Eismänner*, Reissenberger explained that what had been missing until recently was an understanding of how the pressure distribution governs flows of air—precisely the question on which Hann was working.[28]

These articles invited readers to consider their local climate from a synoptic perspective: to track the course of a cold spell as it swept across Europe from Sweden to Russia. As we will see in further detail below, the dynamic theory of climate offered nonscientists not only a compelling interpretation of a familiar phenomenon like late spring cold spells; it also provided a way to imagine central Europe as a physical unit, a space of atmospheric flows.

"He Died of Fresh Mountain Air, Bird Song, and the Scent of Roses"

Climatology captured the attention of many middle-class Habsburg subjects in the late nineteenth century as a means of taking control of personal health. Medical climatology placed heavy emphasis on the collection of empirical data, including the climatic characteristics and physiological effects of mountains, seacoasts and the open sea, steppe and desert. Textbooks in the field gave detailed accounts of the workings of meteorological instruments and insisted that the medical man must carry out his own climatic measurements. Climatology, in this view, was about firsthand observation, not theoretical study. Wilhelm Prausnitz, whose research and teaching at the Hygiene Institute in Graz included the health effects of indoor climates, insisted that "it is not possible to 'study' hygiene from a book. Hygienic research methods in particular must not only be seen but also tested out."[29] In 1901 the Austrian Society of Apothecaries took a field trip to the ZAMG, where members were fascinated by the profusion of instruments on display:

> It will certainly be of interest to everyone to know the climate of his place of residence and to pay closer attention to it. However, the climate can only be determined if one investigates precisely and at regular intervals the current state of the atmosphere—that is, the size and variation of air pressure, temperature,

> humidity, electrical and optical phenomena, as well as the air currents produced by the air pressure, the winds, the various and distinct forms of water vapor (clouds, fog, frost, dew) and the aqueous forms of precipitation (rain, snow, hail, sleet).[30]

While the pressure of the atmosphere may seem to be a factor that escapes direct human perception, changes in air pressure were widely believed to affect physical and mental health. This belief was supported by evidence collected by researchers of the ZAMG, who studied the effects of changing air pressure on the health of students, workers, and hospital patients.[31]

The results of medical climatological research were widely disseminated to both physicians and their patients. The *Österreichische Badezeitung* (later *Österreichisch-Ungarische Badezeitung*) launched in 1871 and published continuously for a quarter century. It was followed by the shorter-lived *Vierteljahrschrift für Klimatologie, mit besonderer Rücksicht auf klimatischer Kurorte*, the *Bade- und Reisejournal*, the *Illustrierte Fachzeitschrit für Kurorte, Hotels, Sanatorien, Reise und Sport*, and other periodicals with a similar orientation. These aimed to communicate the latest research on medical climatology to experts and nonexperts alike. As the *Vierteljahrschrift für Klimatologie* announced in its first issue: "The support and dissemination of our knowledge of climate, above all of its effects on human life and health, forms the charge of this quarterly journal. The scope and importance of this knowledge in its present stage of development more than justifies its compilation in its own periodical, and one intended not only for physicians but for educated readers in general."[32] Atmospheric dynamics was also introduced in reference works like Enoch Kisch's *Klimatotherapie* and Wilhelm Prausnitz's *Grundzüge der Hygiene*.[33]

This was an era when medical thinking was torn between environmentalist and contagionist explanations of disease. It's worth noting that the Habsburg state had reason to resist contagionism, since it implied the necessity of quarantine during outbreaks of cholera in southeastern Europe. Austria's commercial class lobbied against quarantines, as barriers to commercial exchange. Thus Austrian medical experts pursued public health alternatives to quarantines in the Balkans and the Levant—for instance, overseeing a trial program of sanitary reform and medical education in Constantinople.[34]

At the same time, ideas of what might constitute a "healthy climate" were shifting. By the close of the nineteenth century, Habsburg physicians agreed that the healthiness of a climate was relative rather than absolute. There was no single cure-all location. A given climate could be salubrious for some individuals but not for others, beneficial in some seasons but pernicious in others. As the medical director of the spa at Marienbad, Enoch Kisch (1841–1918) wrote in 1898, the last decades of the nineteenth century had seen a striking expansion in both the variety of diseases for which doctors advised climatic cures and the range of climates seen as potentially therapeutic. Earlier in the nineteenth century, taking a climatic cure had meant traveling to a "southern" land. Now, cold climates were almost as likely to be prescribed, even in winter.[35]

What's more, physicians often specifically recommended movement between one climate and another. For respiratory disorders, for instance, the best thing was "a change of climate," whether that might mean a "lengthy stay in the valley and in the mountains, in the south and on the coasts, in a mountain forest and the open sea." This advice built on the Hippocratic principle, *in morbis longis solum mutare* (in tedious diseases to change the place of residence). Variety rather than constancy of climate was recommended for a host of other diseases as well—scrofula (a skin disease associated with tuberculosis), diabetes, arthritis, heart and nerve ailments, "as well as various illnesses of the nervous system and the sexual organs." Exposing the body to multiple climates was said to serve the fundamental purpose of "enhancing organ function and improving total nutrition." In short, "change of climate is to be regarded as a common foundation of all climate cures."[36] Often, what a sick body needed was a change of air, any change. To be sure, this could cause the body strain, but after a few days the process of acclimatization was usually complete.

The most important factor to consider was the climatic character of a health resort *relative* to the patient's most recent abode. "Therefore what must be considered is not so much the absolute temperature of the climatic resort to which an invalid is sent, but the difference between the temperature from which he is departing and the one to which he will arrive." In this sense, there was a kernel of medical truth in a verse about the hypochondriac who "died of fresh mountain air, bird song, and the scent of roses." This school of thought was not without influence even among military men. Take, for instance, the Habsburg naval officer Karl Weyprecht, who led the Austro-Hungarian Polar Expedition of 1872–74. Weyprecht argued counterintuitively that his crew of sailors from the Adriatic was uniquely well prepared for their Arctic journey, having been seasoned for abrupt climate change by the inherent variability of their native climate.[37]

In this way, climate became a dynamic and relational concept, and the relocated body of the patient became a register of geographic difference. The relational character of climate was accentuated by medical textbooks of the day that explained climate therapy in terms of basic atmospheric dynamics. The new dynamic climatology taught that local conditions were not sui generis; they depended on prevailing winds and thus on the large-scale pressure distribution.[38] In this way, climate therapy furnished the patient with a kinesthetic experience of Austria-Hungary's natural diversity. Between the ocean and the steppe lay the many therapeutic climates of the Habsburg lands, endlessly diverse and yet in perpetual interaction with each other. "Thus it is the winds," wrote Kisch in Marienbad, citing Julius Hann, "that erase climatic borders and maintain neighboring climatic regions in constant communication."[39]

"STRIVING TOWARDS THE BALANCING OUT OF EXTREMES"

Dynamic climatology was quickly integrated into the geographical surveys of Austria-Hungary that publishing houses churned out with increasing frequency starting in the 1870s. In his widely read accounts of the climate of the Habsburg lands, Josef Roman Lorenz showed how local contrasts of temperature and pressure formed and were then "balanced out" by means of moving air currents: "The movements of the atmosphere derive from the same causes as those of the fluid envelope of the earth. In the atmosphere, temperature differences between horizontally and vertically adjacent layers of air are the stimulus to the drive towards the balancing out of extremes." Consider, for instance, Lorenz's account of the bora, a cold, dry wind that blows along the Dalmatian coast, on the Monarchy's southern periphery, where Lorenz had spent six years teaching high school and studying the coastal climate and its flora and fauna (see figure 26). The bora arose from the confrontation between two masses of air representing "the starkest opposition": a stationary mass of cold, dense air in the interior and the warm air on the Adriatic side of the Dinaric Alps. The strength of the bora depended on the magnitude of this contrast and the size of the air masses.

> If the opposition lasts for a while and is significant along a fair distance, then an inland current will flow for a period of time and draw as its replacement air masses from ever more distant regions to the north; in this way, in a Bora lasting several days, the temperature drops ever lower. . . . If the opposition is either merely local, or insignificant, then a narrow or weak inland wind will suffice for the equilibration, and a short local Bora or a moderate Borino will arise.

From a physical perspective, this analysis was rudimentary, neglecting the rising and falling motion of the air as it crossed the mountains. As a framework

FIGURE 26. The Dalmatian coast during a bora wind.

for geography, however, it was revelatory. Suddenly, the Dinaric Alps—long regarded as the dividing line between an oceanic climate and a continental one, between coastal civilization and mountain backwardness—no longer seemed quite so stark a barrier. The bora represented a sign of the true "interdependence" of regions across apparent borders.[40]

One of the most influential popularizers of geography in Austria-Hungary was Friedrich Umlauft (1844–1923). A protégé of Friedrich Simony, Umlauft was a gymnasium instructor who shared Simony's commitment to "whole-state" geography and the communication of scientific research to the public. The stated aim of his 1876 "Geographical-Statistical Handbook" of the Monarchy was to illustrate the interdependence and mutual determination of *Land und Leute*, or nature and culture. Austria-Hungary was characterized by "the harshest [*grellste*] oppositions with respect to physical conditions, population, and intellectual culture—which is why the Monarchy is rightly called a state of contrasts."[41] To do justice to such multiplicity without losing sight of the whole, Umlauft divided the territory according to Lorenz's system of climatic zones, based on rainfall and temperature. In this way, each region "maintains its position within the great climatic provinces of Europe," while "peculiarities are highlighted that are revealed under closer observation." In general, he explained, climate was the result of prevailing winds, and winds were the effect of a "balancing out" of differences of air pressure: "The distribution of air pressure, when disturbed by the uneven distribution of heat, are the stimulus to the drive towards the balancing out of extremes, to air currents." Umlauft applied these same images of natural flows when he turned to ethnographic description. He saw no hard-and-fast divisions among Austria's peoples, which included representatives of "all the main cultural groups of Europe." "Thus Austria's history converges with that of Germany, Hungary, and Poland, similarly to the confluence of different streams at different stages in a large riverbed, which then carries these waters along collectively." The Monarchy was thus a space of circulation and mixing, whether one attended to air, water, or human populations. "The nations mentioned do not occupy sharply defined and enclosed areas, but rather are interspersed in many regions. Thus in such border districts one often finds a uniquely mixed population. Indeed, nowhere in Europe can the intermixing of various nationalities be observed as strikingly as in our fatherland."[42] Umlauft's physical analogies served to naturalize his ethnographic observations.

While Dove and the later Norwegian school of meteorology chose images of "struggle" and "battle" to describe confrontations between divergent air masses, Austrian climatologists preferred the language of mixing, equilibration, exchange, and mutual dependence. Thus, as Dr. Kisch explained in his medical guide, winds were a force for "interdependence" and the "erasure of borders." To describe this interaction between contrasting air currents—and to circumvent the question of how their contrasts were maintained—Austrian climatologists even revived the Romantic concept of an *ausfüllende Bewegung* or self-fulfilling movement. Felix Exner, for instance, employed this archaic term in his otherwise highly technical 1925 textbook *Dynamic Meteorology*.[43] The concept of self-fulfilling movement was associated with the highly influential early nineteenth-century geographer Carl Ritter. It expressed a Leibnizian view of the cosmos simultaneously as a totality and an evolving configuration of moving parts. Ritter argued that the geographic relations among physical elements, as among human cultures, are always in flux, thanks to new technologies of observation, communication, and transportation. "What formerly was distant and unreachable, now approaches into closer contact, even into the realm of daily interaction." Thus for Ritter, the category of self-fulfilling movement included all manner of atmospheric and oceanic circulations and the organic responses they called forth, as well as the migrations of peoples and the intentional transformations of spatial relations wrought by human agency.[44] Analogously, an Austrian treatise in dynamic climatology posited that an air current with a component in the direction of the pressure gradient "strives to attenuate the contrasts; it is a 'self-fulfilling movement.'"[45] Readers of the day would have recognized this allusion to a Romantic cosmology of continuously regenerating variety. In this way, climatology lent physical plausibility to the Habsburg ideal of unity in diversity.

THE DYNAMICS OF LOCAL DIFFERENCES

In 1881, Alexander Supan published one of the first major monographs to apply atmospheric dynamics to the explanation of the climatic characteristics of the regions of the globe.[46] Alongside Coffin and Voeikov's 1875 *Winds of the Globe*, it was, in the words of Alfred Hettner, "the first approach to a physiological or genetic treatment of the climates of the earth."[47] Leaning heavily on Hann's observations and interpretations, Supan began by laying out the most recent conclusions concerning the relationship between winds and the distribution of air pressure. From there he moved on to an overview of the major wind systems of the Northern and Southern Hemisphere. Finally, the bulk of the book discussed each region of the world in turn, including tables of average wind frequencies, most of which he calculated directly from the station data, provided by Hann. In each case, he showed how the typical locations of primary and secondary pressure minima and maxima could be used to explain prevailing winds and, on that basis, known characteristics of the regional climate at different times of year—for instance, the typically warm winters along the Norwegian coast, or the extremely cool summers of Novaya Zemlya.

It was shortly after the publication of his treatise on winds that Supan began to reflect on the significance of Hann's insight for the discipline of geography more broadly. These were the years of a contentious struggle among geographers to define their field over and against the disciplines that increasingly encroached on geography's domain—sciences like geology, meteorology, economics, and anthropology. Geographers saw their discipline fracturing into narrow specializations. In the ensuing debates, Supan took a leading role, forcefully defending the unity of his discipline. His methodological pronouncements resonated well beyond Austria-Hungary, influencing future thinkers from Lenin to the Weimar school of geopolitics.[48]

At the German Geographers' Congress of 1889, Supan laid out his vision for the future of geographical research. The key to holding together the physical and human aspects of geography lay in raising the "special" or "chorographic" part of geography to the level of "chorology"—to go, in other words, beyond systematic description, toward causal analysis. The Austrian Problem was not far from his mind, as indicated by his critique of the Austro-Hungarian Empire in Word and Image for its failure to synthesize its multiauthored descriptions into a higher unity, to transcend chorography and achieve chorological insight. Supan proceeded to illustrate what he meant by chorology. Chorology was the study of the reciprocal relations between nature and man. In this respect, it rejected the environmental determinism of Friedrich Ratzel's anthropogeography. The first stage of chorological research was to mark out "geographic localities" that displayed homogenous conditions of orography, climate, vegetation, perhaps also fauna and minerals. Crucially, the influence of any given "geographic locality" on its human inhabitants was contingent on conditions in neighboring "localities." That is, how a human group adapts to its surroundings and exploits its local resources will depend on how those surroundings and resources differ from those of neighboring regions. As relationships of interregional interdependence grow, interactions between man and nature within each region will shift accordingly. Thus Supan's key insight was that natural conditions "guide the social development of their inhabitants in a particular direction"—not in any simple deterministic sense, but by fostering a relationship of interdependence, or potentially conflict, with another locality differing from the first in its natural conditions.[49]

This vision of neighborly difference underlay the research program that Supan prescribed for geography. It consisted in the study of the relationships *between* natural-human regions, their evolving interdependence. As he put it, "The power of neighboring geographic contrasts, which strive to balance each other out, is one of the most significant formative forces in the life of a nation." The mission of geography, Supan insisted, was to characterize these neighboring contrasts and to investigate the relations of dependence and conflict between neighboring societies to which they gave rise.

His choice to refer to these environmental contrasts repeatedly as *Kräfte*, forces, is telling. Supan had been among the first scholars to pursue climatology as the study of the atmospheric motions arising from gradients of pressure or temperature. Only six years later, in 1887, he was turning this program into an agenda for the study of politics and culture. As this suggests, imperial-royal scientists like Supan and Hann found the dynamic interpretation of climate compelling partly because of the analogies it suggested. Indeed, Supan relied on the pressure-wind relationship as the organizing principle of his volume on Austria-Hungary for the Länderkunde von Europa series (1889). Faced with the task of producing an overview of the multinational state, Supan, like Hann, seized on an interpretive method that recast difference as continuity. It was in the context of this overview that Supan penned the programmatic lines: "All living things blossom forth from the balancing out of neighboring contrasts. To ascertain these contrasts and describe their influence on men: this we see as our scientific responsibility."[50]

THE POLITICS OF EQUILIBRATION

To be sure, "the balancing out of neighboring contrasts" was only the roughest first approximation to a description of atmospheric dynamics. Yet it carried the authority of physical science. In yet another play on the terminology of force, Supan argued that his proposed method would give chorology more "wissenschaftliche *Kräfte*," scientific force.[51] He was prescribing for international geography most generally a research program that Austrian scholars had already adopted in the name of imperial unity. As we have seen, it had become a commonplace among imperial-royal scholars that the confrontation between dissimilar social elements set a process of development in motion. In the words of Alois Riegl, "when the unfamiliar meets the unfamiliar in a close and sustained relationship the process of development is set in motion."[52] What should now be clear is that this interpretation of imperial unity rested on an analogy developed among coordinated yet distinct disciplines concerned with the spatial distribution of natural and cultural resources, including climatology, geography, political economy, ethnography, and art history. It was a vision of the empire as a circulatory system, in which energy was released from a tension between local gradients.

Such was the powerful metaphor that coupled dynamic climatology to imperial ideology. When Habsburg scientists discussed the relationship between pressure gradient and wind, a cultural-economic analogy was implicit: difference creates circulation and thus cultural continuity and interdependence—literally, unity in diversity, or "the balancing out of neighboring contrasts."

The analogy was particularly resonant in the wake of the 1867 *Ausgleich* between Austria and Hungary, which granted Hungary domestic autonomy and recreated the empire as the Dual Monarchy. Typically translated as "compromise" or "settlement," the word *Ausgleich* was often used interchangeably with *Ausgleichung*, literally an "equilibration" or "balancing out." It was in this sense that Supan used the term *Ausgleichung* in his 1889 treatise on Austria-Hungary, where he wrote, both literally and metaphorically, of the balancing out of neighboring contrasts.

The slippage between *Ausgleich* and *Ausgleichung* was strategically employed by Habsburg-loyal writers to naturalize the 1867 status quo. The liberal statesman Gyula (Julius) Andrássy, instrumental in negotiating the compromise, lost his seat in the Hungarian parliament because he was perceived as overly sympathetic to Vienna. While out of office, Andrássy published a defense of Hungary's ties to Austria that drew on both historical and geographic arguments. Andrássy portrayed modern Hungary as a "small country" that could not survive independently. Austria was its natural partner because its borders were impossible to defend without Hungary's aid. The *Ausgleich* was the outcome of an increasing intensification of "differences" (*Gegensätze*) between the two countries: "Every human being, every organism composed of a group of human beings, can only survive through the balancing out of differences [*nur durch die Ausgleichung der Gegensätze fortbestehen*]." Andrássy thus presented the *Ausgleich* as the solution to a problem of disequilibrium. The goal remained to reestablish a state of balance: "Who can predict whether the old harmony will ever be reestablished, whether the oppositional drives will result in a new compromise, whether the means can indeed be found with which to bring the impending avalanche to a halt."[53] The pivotal term here was the German word *Ausgleich*. To nineteenth-century ears, it did not simply connote a diplomatic compromise. Far more vividly, it implied a physical process by which opposing forces were maintained in dynamic equilibrium.

THEORIES OF CIRCULATION

Dynamic climatology illustrated how diversity could be the motor of circulation—and how circulation could in turn "even out" the starkest of oppositions. This proved to be a fertile point of view for the field of political economy. It offered a vivid, physical analogy for the hope that Austria's natural oppositions would generate economic interdependence and therefore political unity.

The rise of dynamic climatology coincided with a spatial turn in central European political economy. This new departure leaned on the ideas of Johann Heinrich von Thünen (1783–1850), a north German landowner and agricultural improver. In 1811 von Thünen had attempted to derive the optimal geographic distribution of economic production. He assumed the existence of a single city with a single road leading out of it and a uniform natural environment. Based on the transport time from the site of production to the urban market, von Thünen posited that zones of agricultural production would develop as rings of increasing radius from the city center: first a zone of vegetable gardens, then zones of forestry, grain farming, and distilleries. Beyond a certain radius, agriculture would no longer be profitable, and the land would be useful only for hunting. Crude as this model may have been, central European scholars developed it as a tool for thinking through the expanding scale of trade in the nineteenth century. Their interest lay in the dynamic relationship between environmental, technological, or demographic change and the expansion or contraction of the economy.[54]

This line of inquiry led, for instance, to some of the first attempts to visualize Austria-Hungary's position within what the Vienna-based economic geographer Franz Neumann-Spallart (1837–88) called the "world economic organism."[55] From his start as an expert on Austro-Hungarian trade statistics, Neumann-Spallart moved on to developing methods for achieving an international economic "overview." To that end, Neumann-Spallart suggested that economics model itself on climatology. In both cases, the question was how to "represent statistically" "the economic situation of a state in its entirety over a certain time interval":

> This is a task that may be compared to the one that meteorology has to solve when it is supposed to determine the climatic character of a region. Just as the climate is in a certain sense the complex result of the interactions of a great number of interdependent elements, likewise what we call the economic situation [*wirtschaftliche Lage*] is the totality of a series of individual facts that express the degree of strength and health of the material life of a given population. In both cases . . . it is a matter of the analytical groundwork for the disaggregation of a holistic impression into its essential constitutive factors. However, meteorology finds factors such as air pressure, temperature, humidity, direction and strength of the wind, etc., which are genuine elements or factors of the situation, for which it possesses precise measuring instruments; and, by generalizing on the basis of a causal law, it can draw a conclusion from a single series of observations for all similar cases. Economic statistics, on the other hand, must rest content with surrogates for these natural-scientific methods.[56]

Otherwise put, climatology furnished political economy with an exemplary model of multicausal reasoning. Although economists did not have precise, causal laws to work with, nonetheless they too could analyze a complex state of affairs into its causally significant component factors.

Climatology also offered economics a model of spatial analysis, as Emanuel Herrmann argued in 1872. At that time, Herrmann was a docent at the Vienna Commercial Academy and an adviser to the imperial education ministry; from 1882 to 1902, he taught as a full professor of national economy at the Technical University in Vienna. Historians of economic thought have associated Herrmann with the subjectivist turn, because he explicitly sought to model economics on natural science. Yet Herrmann's model was not Newtonian physics but rather the empirical, geographical, historical, and statistical fields of natural history and climatology. Like a naturalist in pursuit of the diversity of life, Herrmann was fascinated by the variability of economic life across time and space. Like the naturalist, he sought to interpret the geography of economic activity in terms of interactions among general laws, local conditions, and historical trajectories. Indeed, he viewed the discipline of economics as continuous with evolutionary biology and human anthropology. Thus Herrmann noted that there was a telling affinity between von Thünen's circles of uniform economic production and Alexander von Humboldt's lines of equal temperature or "isotherms." "The lines of equal conditions of production in relation to an urban market form isotherms of a sort."[57] Herrmann noted that these two representational devices had been introduced to the world within a decade of each other.[58] He went on to construct an elaborate analogy between economic and climatic geography, likening "demand" to heat in its capacity to create "tropical" zones of economic growth.[59] Physical geography thus supplied Austrian economists not only with empirical data and statistical methods, but also with a new model for the spatial analysis of economic relations. Once von Thünen's rational approach had been adapted to allow for geographic variability, it suggested how economics could be transformed into a science: not as an abstract mechanics, but as an observational discipline modeled on physical geography.

Another exponent of this new spatial economics was Emil Sax, an independent-minded member of the circle around Carl Menger in Vienna. Like Neumann-Spallart, Sax set out to adapt von Thünen's analysis to the "new scale of global commerce" and analyze the impact of new modes of transportation. Before the coming of the railway and steamship, Hungary had been in the fifth or sixth von Thünen zone from Vienna. Subsequently, Hungarian cattle farmers now had to compete with those in Galicia; at the same time, it became easier to transport grain from Hungary to Vienna. So Hungarian farmers turned increasingly to grain, and the price of grain in the Alpine lands fell accordingly. As Sax interpreted these shifts, improvements in transportation had increased "the value of these natural regions" by raising the "marketability" [*Absatzfähigkeit*] of their products.[60] Thus the modern transportation network allowed Austria-Hungary to profit fully from the complementarity of its natural regions. In this vein, from 1900 to 1904, the liberal prime minister Ernst von Koerber promoted a vast program of economic development and integration, including a broad network of canals and railways—all intended, in his words, to "reduce the nationality strife" by paving a "road . . . free for the spiritual and economic development of the State."[61]

Liberals like Sax and Koerber were not alone in developing this spatial perspective on the economic life of the Monarchy. Karl Renner, one of the leading minds of the Social Democratic Party, disagreed with Sax about the railway's impact on Austria-Hungary, but he too emphasized the crucial significance of physical geography for the socioeconomic life of the Monarchy. The very possibility of imperial unity depended on the physical form of the territory.[62] More importantly, Renner echoed Sax on the value of natural diversity for the economic health of the state:

> To a superficial gaze it seems natural and adaptive that a commercial center lies at the center of a homogeneous region. Nothing is falser than this. Trade is the exchange of what a homogeneous region has in excess, in return for what it lacks; [trade therefore] always thrives on the region's periphery—that is, the space where contact is made between soils of one type and another, between one nation and another. . . . There, where mountains give way to plains, where the mouth of a river connects land and sea, where industrial land borders agricultural land, there the city arises.

Renner interpreted this geographical interdependence dialectically: "Dissimilarity of the parts and autarchy of the whole is the characteristic of all state formation, particularly of the large state. . . . Thus, opposites are incorporated into its being [*Dasein*] in order to be transcended [*aufzuheben*]." Applied to Austria-Hungary, this became an Austro-Marxist argument for the advantages of the supranational state in terms of the value of natural diversity: "Here are united not only agricultural and industrial land but also agricultural lands of the most different structures: forest, pasture, fields of rye, wheat, barley, beets, and animal feed, vineyards and orchards, land for horses and cattle."[63]

Although neither Sax nor Renner referred explicitly to dynamic climatology, all these analyses, economic and climatological alike, derived from the same "whole-state" discourse that assumed local contrasts to be a motor of circulation and, therefore, a force for unity. Climatic diversity, in particular, figured as the basis for a spatial division of labor that set trade in motion. As an article in the *Militärzeitung* from 1866 awkwardly expressed this atmospheric analogy:

> World trade in its broadest significance is the law of the flows that have as their medium the unevenness of nature and of cultural productions that are determined by climates and soils, and that have as their stimulus the compensatory [*ausgleichend*] efforts of man within the confines of his nature and needs—efforts whose geographical and historical origins, due to the infinitely complex composition of these needs, are almost as difficult to fathom as meteorological phenomena, which even today often have the appearance of hieroglyphs.[64]

It was not just that diversity propelled commercial exchange. Forms of circulation—atmospheric, economic, migratory—in turn served to smooth out stark contrasts. This was the image of the Monarchy evoked by the Ljubljana-based commercial geographer Franz Heiderich at the opening of the International Economics Workshop for economists and entrepreneurs, held in Vienna in 1910. Tasked with instructing the foreign guests on the "natural conditions" of Austria-Hungary's "economic life," Heiderich began with the obligatory description of the Monarchy's "pronounced geographic and economic contrasts." Yet nature and man had conspired to moderate these contrasts and stitch together the parts:

> By means of deposits from the rivers and the Ice-Age glaciers, the mountains have acquired gentler slopes along the plains, and the wind-blown deposits as well as the sediments from slowly shrinking bodies of water have further diminished the vertical difference in height and have connected differently formed regions. In this way Nature itself has erased the sharp tectonic borders and replaced them with gradual transitions and wide border zones. Across these flow, from one natural region [*Landschaftsgebiet*] to the next, cultural, economic, and political forms of life, at first distributed in colonies and then gradually coalescing. . . . The Monarchy can thus be regarded as a unity in a physical sense, its various parts solidly cemented together just like a giant breccia.[65]

Here Heiderich introduced two geological metaphors for Habsburg unity: one, the image of the Monarchy as a giant breccia—a rock formed of angular fragments, cemented together by the flow of debris. Its components retained their individuality, yet were bound together by a natural, irresistible process. The second image was a geological variant on a climatological metaphor: the equilibration of neighboring contrasts. In this case, gradients of elevation rather than pressure are evened out by a natural process of weathering. In this way, Heiderich cast empire-building as a natural process, akin to erosion, as unstoppable as the flow of wind and water.

Climate's economic significance was thus both literal and figurative. On the one hand, climatic oppositions set trade in motion. On the other, atmospheric circulation was an apt metaphor for the conciliatory effects of trade: "The exchange of goods and money is colorless like waves of air."[66]

Finally, the convergence of climatology and political economy in late imperial Austria cast new light on the future of industrial Europe. As Herrmann put it, modern man tended to think only in terms of immediate causes. But the earth sciences suggested a more appropriate scale for economic thought. The universe itself was a system of sustainable production, an "enduring economy [*Wechselwirthschaft*, literally crop rotation] of light, heat, gas, earth, waters." "The coal that warms our ovens was a verdant tree hundreds of millions of years ago, which, with so many others, was suddenly destroyed by a storm and washed over by the sea. The petroleum in our lamps derives from the fat of a fish. . . . But the earth must have compressed these stocks for millions or at least thousands of years like a protective container for us to be able to consume them unthinkingly today."[67] Even the milk and butter of a European breakfast must be seen as the product of millions of years of mammalian evolution. Alarmed at the rapid pace of resource exhaustion in the 1880s, Herrmann called for a worldwide organization to survey the earth's stocks and agree on their allocation.

*

In short, climatology can be said to have contributed a kinesthetic basis to the imperial ideology of unity in diversity. That is, familiarity with the observation of atmospheric phenomena and with the mapping of meteorological elements offered an intuitive, embodied way for scientists and nonscientists alike to imagine the space of the empire. The atmospheric dynamics communicated in atlases, newspapers, and medical guides taught Habsburg subjects to experience the wind as a force for the balancing out of oppositions. The idea that nature decreed the *Ausgleichung* of local differences lent support to the ongoing politics of *Ausgleich* in the late nineteenth century, as Bohemians, Galicians, and South Slavs each demanded their own "settlement"/"equilibration" with

Vienna. In the fields of medicine and political economy, appeals to the "balancing out of neighboring contrasts" grounded arguments for the benefits of the multinational state to the health and prosperity of its inhabitants. By the 1890s this simple model of atmospheric dynamics had captured the imagination of Habsburg citizens, expert and lay. It provided a vivid physical image of the emergence of unity out of diversity.

Much of the work of Habsburg climatology after 1900 went into refining this idealized picture of atmospheric circulation in order to understand how it resulted in the sensible characteristics of local climates. However, "the balancing out of neighboring contrasts" was but a first approximation to a description of atmospheric dynamics. As scientists today would put it, the pressure gradient only determines the winds under the simplifying conditions of geostrophic flow. Only in this case can the pressure difference be maintained and the wind remain constant, since the wind blows perpendicular to the gradient.[68] By the 1890s, however, the ideological force of this model made it seem self-evident. As we will see in the next section, it would take an outsider to question it.

A "UNIQUELY ODD FELLOW"

Of all the accomplished individuals working at the ZAMG circa 1900, Max Margules is, ironically, the only one whose name is commonly remembered today. Among atmospheric physicists, Margules is best known as the author of the "tendency equation," a cornerstone of early computerized weather forecasting, and the inventor of the concept of available potential energy, long central to the work of climate modelers.[69] Some textbooks of atmospheric physics will tell you that Margules worked in Vienna circa 1900, that he was known as something of an oddball, and that he met a tragic end. In fact, he left behind few clues to illuminate the mysteries of his life. The irony of Margules's fame today is that in his own lifetime he was a liminal figure in the circles of imperial-royal science. A Jew in a casually anti-Semitic academic world, he was never a contender for a high-ranking post. Socially awkward by some reports, Margules has gone down in history as a loner, as someone who worked, to quote one textbook, in "intellectual isolation."[70] There is indeed evidence that he was an intensely private man. The archives hold no more than a few terse manuscripts on physical chemistry, the topic he took up after abandoning atmospheric physics in 1906. Even his published papers are dense with equations, short on words. And yet the claim that Margules was isolated from his milieu remains to be evaluated. His story must be pieced together from government records and the posthumous recollections of his colleagues.

FIGURE 27. Max Margules (1856–1920).

Margules was born to a Jewish family in the primarily Jewish town of Brody, in eastern Galicia, in 1856. He moved to Vienna for his last two years of gymnasium, where he lived in the Jewish quarter of Leopoldstadt. He studied physics at the University of Vienna and then, in 1879–80, at the University of Berlin. The chair of physics at Berlin was then Hermann von Helmholtz, who had recently returned to the study of atmospheric discontinuities. Margules was not well prepared to contribute to this inquiry, having trained as a mathematical physicist with a focus on electromagnetism. Nor had he experienced the leisurely childhood in the Alps that had set many of his colleagues on the path to a geoscientific career. Nonetheless, he turned his focus to the science of the atmosphere.

For twenty years, interrupted only by his stay in Berlin, Margules served as an assistant, adjunct, and then secretary at the ZAMG. He was hired in 1877 in part to work on the institute's yearbook, which had fallen far behind its publication schedule due to the arduous labor of reducing data from the fast-growing station network. Margules succeeded so well that by 1885 the yearbook for 1883 was in press. Likely due to his knowledge of a Slavic language, he was put in charge of maintaining communication with the weather stations in the eastern and southern portions of the Monarchy.[71] As we saw in chapter 4, the ZAMG's network was unevenly spread across the crown lands, and its early directors made it a priority to increase the density of observations from regions like Galicia, Bukovina, and Dalmatia. Margules made this goal his own. He expressed satisfaction at seeing stations established in poorly represented areas.[72] By 1888, he was responsible for reviewing the observations submitted by all the stations of the network and preparing them for publication. He was also in charge of inspecting stations in Galicia, Bukovina, Dalmatia, and Bosnia and Herzegovina—to which was soon added Austrian Silesia, Upper Hungary, and Transylvania.[73] These inspection tours took Margules to the distant reaches of the empire, from where he reported on the quality of observers and observations. Back in Vienna, he took pains to stay in contact with observers in these parts.[74] Margules also took charge of reducing vast amounts of raw data.[75] Not one for collaboration, he would nonetheless occasionally surprise colleagues with a sheaf of measurement values, on which observations relevant to their research had been highlighted with his signature red pen.[76]

Although Margules is thought of as a "fundamental" researcher who worked "in isolation," his research questions were firmly embedded in the program of imperial-royal science. His most lasting contributions began as an interrogation of the central metaphor of Habsburg climatology: the potential of local differences to power an integrative circulation.

INTERROGATING THE CENTRAL METAPHOR

What Margules realized in the 1890s was that existing scales of observation failed to capture the phenomena of relevance to a quantitative evaluation of this model of atmospheric motion. The stations of the ZAMG's network were irregularly spaced, and even at its densest in Carinthia there was no more than one station for every three square miles.[77] This made it impossible to track phenomena like squall lines, with dimensions of roughly one hundred kilometers. And so Margules defined a new scale of observation. That is, he constructed climatology's first purpose-built *meso*scale observing network, comprising four stations arrayed at a sixty-kilometer radius from Vienna.[78] Observations from this network would be used to determine the relationship between observed pressure gradients and the force of a squall—that is, a localized storm or violent gust of wind.

Here was an infrastructure suited to studying the question at the top of the institute's research agenda: the strength of winds to be expected from what Supan had called the "balancing out of neighboring contrasts." The data from the stations' barometers and anemometers indicated that bigger pressure differentials did not, in fact, correlate with stronger winds—"not even approximately." Margules began to suspect that pressure contrasts were not the driving force behind atmospheric motions—that they were a "mere cogwheel in the machine," as he would later put it. Here was empirical evidence that the model of a circulation powered by pressure differentials didn't quite work. Refining it became the new agenda of Habsburg climatology.

In his theoretical work, Margules sought to understand this situation from first principles. Take the example of his "tendency equation," still taught in introductory courses in atmospheric physics today and a governing equation of many computerized climate models. The tendency equation relates a change in pressure to the movement of air. Margules derived it in 1904 from basic considerations about the incompressibility of air and the relationship between air pressure and altitude.[79] The pressure of the air at any point is determined by the weight of air above it. Margules's tendency equation (equation 1) gives the relationship between the change in pressure at a given point and the wind blowing toward or away from that point (neglecting friction and the rotation of the earth). The first term is the rate of change of air pressure; the second and third terms are the divergence of air in the horizontal plane; and the fourth term is the air's vertical motion. The equation says that as air flows horizontally away from a point, the pressure at that point will fall, unless balanced by a vertical inflow of air.

$$(1)\quad \frac{\partial p}{\partial t} + \frac{\partial (pu)}{\partial x} + \frac{\partial (pv)}{\partial y} + g\mu_h w_h = 0.$$

Historians of meteorology have described the equation as an early attempt at weather forecasting. The goal would be to *predict* a rise or fall of the barometer from observation of the wind field. The equation showed Margules that a small error in the measurement of the wind field will skew the forecast significantly. In the 1940s, this problem came to be dealt with by means of "quasi-geostrophic" theory. This approach, developed by Jules Charney, used vorticity, a measure of rotation, to arrive at approximate calculations of the divergent flows that worried Margules. These, in any case, tend to be small outside of the tropics. Back in 1904, however, Margules's analysis reinforced the skepticism with which he and his colleagues at the ZAMG already viewed the job of weather forecasting. As Felix Exner would argue, mathematical models of atmospheric processes were of use for explaining, not forecasting.[80] Margules was unequivocal: forecasting, he said, was "immoral and dangerous to the character of the meteorologist."[81]

How then should we interpret Margules's work toward the tendency equation? Was this an attempt to prove the practical impossibility of forecasting? Or was he perhaps after knowledge of a different kind? Consider next a very similar expression that Margules had published three years earlier. In this case (equation 2), he began by balancing the work performed by an expanding parcel of air against the work exerted by the surrounding air pressure. Equation 2 allows for a similar calculation as equation 1, assuming the change in pressure to be small and the movement of air to be only in the horizontal plane.

$$(2) \quad \frac{1}{2}(V^2 - V_0^2) = RT\frac{p_0 - p}{p_0} + \frac{RT}{p_0}\int(\partial p/\partial t)\,dt\,.$$

It is telling that equation 2 states the equivalence in the opposite order as equation 1: now a change in wind speed is to be calculated from a pressure gradient. In other words, the interest is not in forecasting a clear or stormy day. Instead, the question is how much motion is generated when air flows from an initial location to a final location at which the pressure, though lower, is not constant. Rising pressure will produce a greater final wind speed; falling pressure will produce a lower final wind speed. Here, the motivation, as the title of the 1901 paper indicates, is to understand the "Energy of a Pressure Distribution." In other words, equation 2 is an expression of the motive force produced by what Margules's contemporaries referred to as "neighboring contrasts."

THE ATMOSPHERE'S STORE OF ENERGY

The second key concept for which Margules is remembered today is that of "available potential energy" (APE), which played an important part in the first general circulation models of the 1960s and 1970s and remains central to the analysis of the instability associated with regions of the atmosphere where temperature varies abruptly (baroclinic zones). We can define "potential energy" as the energy stored in a system; "available potential energy" then designates the fraction of that energy available to do work—that is, to generate motion. In the atmosphere, this is only a small fraction of total potential energy. Margules showed that APE (or what he called "available kinetic energy") could be calculated as the difference between the potential energy of an initial configuration of a gas and that of a final state in which the potential energy of the gas is

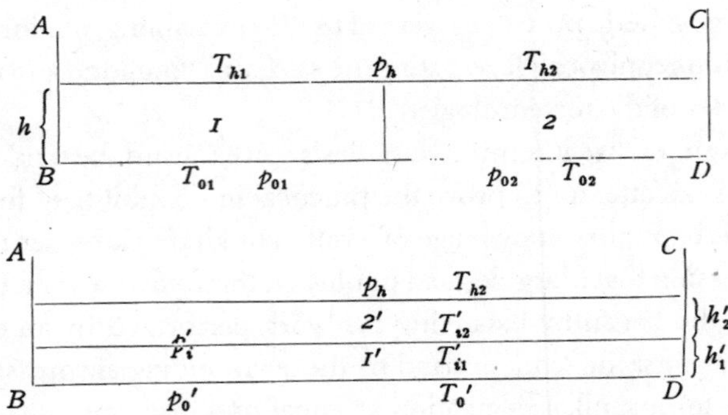

FIGURE 28. Calculating available potential energy: initial and final states of a chamber of gas with dividing wall removed.

reduced to a minimum, without adding or subtracting heat. A very simple case is illustrated in figure 28. The initial state consists of air at different temperatures separated by a vertical wall. The final state is layered horizontally: gas with higher pressure and higher temperature above, lower pressure and lower temperature below. It is instructive at this point to note the difference between APE and total potential energy. Imagine a horizontally stratified atmosphere like this stable final state. It would still have gravitational potential energy, since the molecules are above ground height. However, the system can do no work: it cannot move to a state of lower potential energy (all molecules at ground level) without removing heat.[82]

The significance of APE is that it allows one to calculate accurately something new and important: it measures the energy stored in different states of the atmosphere and *available* to generate motion. This gave Margules a way to test the presumed power of neighboring contrasts. Suppose a room five meters high is divided in two by a sliding wall. Assume that the air on one side is at a higher pressure than the air on the other side, with a difference of ten millimeters of mercury at ground level. Gradients of this size are often observed during strong winds. What happens when you remove the sliding wall? According to Margules's calculations, the redistribution of air would generate a wind of only 1.5 meters per second: no more than a gentle breeze. No matter how big the room is, the answer is the same. Now consider the case in which the room is again divided in two, but the two sides are maintained at different *temperatures*: 0°C (32°F) in one half, and 10°C (50°F) on the other. Remove the sliding wall, and you'll have a wind of .67 meters per second, somewhat lighter than a light breeze. Not very impressive, to be sure. However, if the height of the room were six thousand meters, the height of mid-level clouds, the speed achieved would be twenty-three meters per second—close to a storm-strength wind.

Margules was challenging the reigning theories of cyclone formation. Hann's thermodynamic theory of the origin of foehn had inspired an idea that, ironically, Hann would fight hard to disprove: the thermal theory of cyclones. This held that the churning energy of a cyclone derives, like the foehn's warmth, from the latent heat released by a rising current of moist air. This view came easily in an era obsessed with the new industrial engines, since it cast steam in the role of motive power. As Gisela Kutzbach has recounted in a masterful study, the thermal theory of cyclones became orthodoxy in the 1860s. But Hann had never bought it. He had focused instead on the motive power of the pressure distribution, but had been unable to explain what maintained that distribution. Margules went further. Both his observations and his theoretical calculations had suggested that pressure gradients were not responsible for storm winds. So what was?

Inspired by studies of inversions and other stark temperature contrasts in the atmosphere, Margules hypothesized that strong winds could be produced by horizontal gradients of *temperature*. He reasoned that such temperature differences create a situation in which small air currents might displace parcels of lighter (warmer) air into colder regions, and parcels of heavier (colder) air into warmer regions. These parcels would then experience a strong gravitational force tending to restore them to their original positions. According to this interpretation, storm winds are the result of conditions of instability in the atmosphere that accrue gradually until they are suddenly released by a small movement of air. Margules introduced the fiction of the sliding wall because, as he admitted, he was unable to explain how such unstable conditions were maintained in the free atmosphere. He concluded that the forces "set free" when an unstable equilibrium is disturbed "are greater than those corresponding to the largest horizontal pressure gradients that have been observed in the atmosphere."[83]

Upon publication in 1903, Margules's theory was overlooked by many and met resistance from some who made the effort to read it. Thus his colleague Trabert had to defend him for having ignored vertical air currents.[84] Particularly controversial was Margules's claim that the latent heat of condensation contributes little to the energy of most storms.[85] Sir Napier Shaw contended that Margules had set an arbitrary limit on the vertical convection of warm, moist air, the driving force of the cyclone according to the thermal theory.[86] In fact, Margules's fundamental contribution to atmospheric dynamics would not be recognized outside of the German-speaking world for several decades.[87]

In 1954 the pathbreaking American meteorologist Edward Lorenz seized on Margules's concept of APE as the quantity that would make it possible to track the flow of energy in the atmosphere.[88] Lorenz tweaked the concept in two ways. First, he applied it not to an individual storm, as Margules had, but to the atmosphere as a whole (which could more properly be regarded as a closed system, in which a fixed mass of air would be redistributed within a fixed volume). Second, he renamed it. "Available potential energy" made clear, in a way that Margules's "available kinetic energy" did not, that the energy in question was stored in the form of a tension in the atmosphere. Lorenz then showed how APE could be used to track the exchange of atmospheric energy between large scale and small, "between zonal winds and the eddies." In this way, it became possible to show that larger eddies—that is, cyclones—play a vital role by transferring enough angular momentum to the zonal flow to compensate for the energy dissipated as friction. The concept of APE thus helped to confirm a picture of the general circulation that Julius Hann had already begun to imagine in qualitative form circa 1900, as we will see in the next chapter.

A TRAGIC END

For his hard work and "excellent knowledge," Director Hann repeatedly petitioned the ministry to promote Margules and give him a raise.[89] In 1890, Margules was made an adjunct, and in 1901, he was promoted to the institute's secretary, the first to hold that post. Many years later, he expressed gratitude to Hann for all that he had learned from him.[90] Then, in 1897, Josef Maria Pernter succeeded Hann as the ZAMG's director. Pernter was a staunch and politically active Catholic conservative, a Tyrolean patriot, and very likely an anti-Semite. Ficker recalled: "Anyone who got to know Pernter without knowing who and what he was, would never have guessed that this lively, combative South Tyrolian had been captured by science. . . . This product of a Jesuit education looked more likely to be a politician or a pugnacious cardinal than the director of a scientific institute. And if one had to write the life story of this extraordinary man, one would have to reach the conclusion that political and religious controversies occupied him at least as much as the problems of science." In Ficker's judgment, "Anyone who knew Pernter and Margules would find it lamentable but not incomprehensible that these two men could not bear each other."[91]

Pernter must have made Margules's life at the ZAMG unbearable. It became all too clear that he would never rise above the position of assistant. Two years after publishing his concept of APE, Margules quit his post as secretary of the ZAMG and abandoned meteorology for good. He explained to the education ministry that he had fallen into conflict with his colleagues, pointing out as well that he had been passed over repeatedly for promotions.

Director Pernter agreed to his request for early retirement, adding that Margules was an *eigenartigen Sonderling*, a "uniquely odd fellow." He was "overly sensitive," perceived himself to be always under attack, and did not respond well to efforts to improve his situation.[92]

Other colleagues described Margules far more fondly. Ficker wrote:

> I had the great good fortune to arrive at the ZAMG when Margules was still active there—and had the even greater good fortune to develop a closer relationship with him. I remember very well . . . how he looked me over with his gray eyes, how he then said, 'I've read your Föhn study. I have here a couple years of data charts from the Sonnblick and a valley station that I've prepared for you. You can always come and tell me what you've found. But I won't give you any advice. And he stuck to that while I put together the study of the transport of cold air masses over the central Alps. That by the way was the first meteorological investigation that was carried out with a view to his new ideas. Only after it went to press would he give criticism. He sure wasn't a teacher. Once he said to me, "You should really deal with the theoretical side now. You'll get the math soon enough!"[93]

From Felix Exner, we have the following account of Margules's tragic fate:

> In the last years, when I had the privilege to visit Margules from time to time, I found in him an enlightened, amicably disposed wise man, with no trace of bitterness, who had given up on all the joys and vanities of the world and who was leading the life of an urban recluse. I never left him without being deeply impressed by the greatness of his soul. In the last years of his life there literally was no one with whom he would communicate on a regular basis; he was a bachelor, living alone in a small, undecorated flat without house help. Margules, who preferred independence and freedom to everything else, wrote to me once that he had almost nothing to eat and asked if I was still alive. Afterwards, patrons were found both in and outside Austria, who sent him some food; and still it was not an easy task to persuade him to accept it. And so Margules died a hungry death in complete consciousness, unwilling to become a burden to others and to take anything that was not his due.[94]

As much as Margules's colleagues valued his contributions, they could not or would not welcome him into the social world of imperial-royal science. Margules died with the empire, another faithful Jewish servant of Franz Josef.

And yet, in the years around the turn of the century, Margules had made the Austrian Problem his own. He had worked to expand the imperial observing network both extensively and intensively and had nurtured relationships with its observers. He was not, as is often supposed, a "lone wolf."[95] But he had retained enough intellectual autonomy from the ideology of imperial-royal science to put its central metaphor to an empirical test. Other Austrian researchers would follow his lead.

CHAPTER 8

Planetary Disturbances

Habsburg contributions to dynamic climatology elucidated phenomena at scales ranging from the planetary down to those of agriculture and human health. Far from being disconnected investigations, these studies built on each other, working toward an understanding of interactions between phenomena of such disparate dimensions. To this end, between 1903 and 1921, researchers affiliated with the ZAMG, including Margules, Schmidt, and Defant, developed two essential tools of scaling, described in this chapter: small-scale fluid models of atmospheric motion, and a quantitative measure of turbulent motion, applicable to flows of any dimensions.[1] Putting these together with Margules's concept of APE (chapter 7), they were able to estimate the contribution from turbulent eddies to the flow of heat and angular momentum between equator and poles. This added up to a revolutionary idea. Cyclones and smaller eddies no longer appeared as "local disturbances" superimposed on steady planetary currents. Rather, these disorderly motions came to be seen as essential components of the atmospheric system.

MODELS OF THE ATMOSPHERE AND ITS "DISTURBANCES"

The historian of meteorology Hans-Günther Körber has suggested that the opening to a "dynamic" science of the atmosphere came with the adoption of the Copernican system in the seventeenth century. For the first time, winds could be explained in terms of the movement of the earth. Galileo, in fact, cited the easterly winds in the tropics as evidence of the earth's rotation.[2] And yet seventeenth- and eighteenth-century natural philosophers remained wedded to the Aristotelian schema of solar climatic zones. In 1686, Edmund Halley interpreted a familiar wind pattern as an effect of convection: the warmth of the sun causes air to rise at the equator, while colder air rushes in beneath to fill the vacuum. This sets up a circulation that corresponds roughly to the observed "trade winds": surface-level, equatorward movement of air in the tropics. In 1735, George Hadley modified this model to take into account the rotation of the earth. For the atmosphere to rotate along with the earth, its speed will have to be faster at the equator than at higher latitudes. For this reason, air moving from the tropics to midlatitudes high above the surface in the Northern Hemisphere will have a relative velocity to the east; on its return to the tropics, the surface-level wind will be directed to the west. Accordingly, the trade winds are observed to blow toward the southwest in the Northern Hemisphere. In the nineteenth century, this model was further elaborated by the American physicist William Ferrel (1817–91). Ferrel realized that as the upper-level Hadley flow approaches midlatitudes (moving toward the northeast in the Northern Hemisphere), it will begin to cool and sink, and friction between the air and the earth's surface will slow it down. In order to maintain the earth's constant rotational velocity, Ferrel assumed a counterbalancing midlatitude surface-level flow toward the poles, directed toward the east in the Northern Hemisphere, with an equatorward flow toward the west at higher altitudes (see figure 29). This model remained standard until the early twentieth century.

Ferrel's model of the general circulation coincided with his theory of the origin of cyclones. According to Ferrel, a cyclone was a vertical circulation driven—like a steam engine—by the heat released when water vapor condenses. Warm air cools as it rises, causing condensation; this releases latent heat, which causes the air to expand; the expansion creates a surface depression, such that air rushes in below, setting up a circulation. Thus Ferrel viewed cyclones as closed systems. According to his model of the general circulation, cyclones and smaller eddies can be viewed as "local disturbances"—that is, as disorderly motions superimposed on the zonal circulation and responsible for its "irregularities."[3] An alternative theory of cyclones and of their relationship to the general circulation was proposed by Heinrich Dove in Prussia (1803–79). Dove understood cyclones as a confrontation between polar and tropical air currents, a notion that bears only a superficial resemblance to the theory of the polar front associated with Scandinavian meteorologists of the early twentieth century. What Ferrel, Dove, and nearly all their contemporaries had in common was their failure to specify the interactions between regional

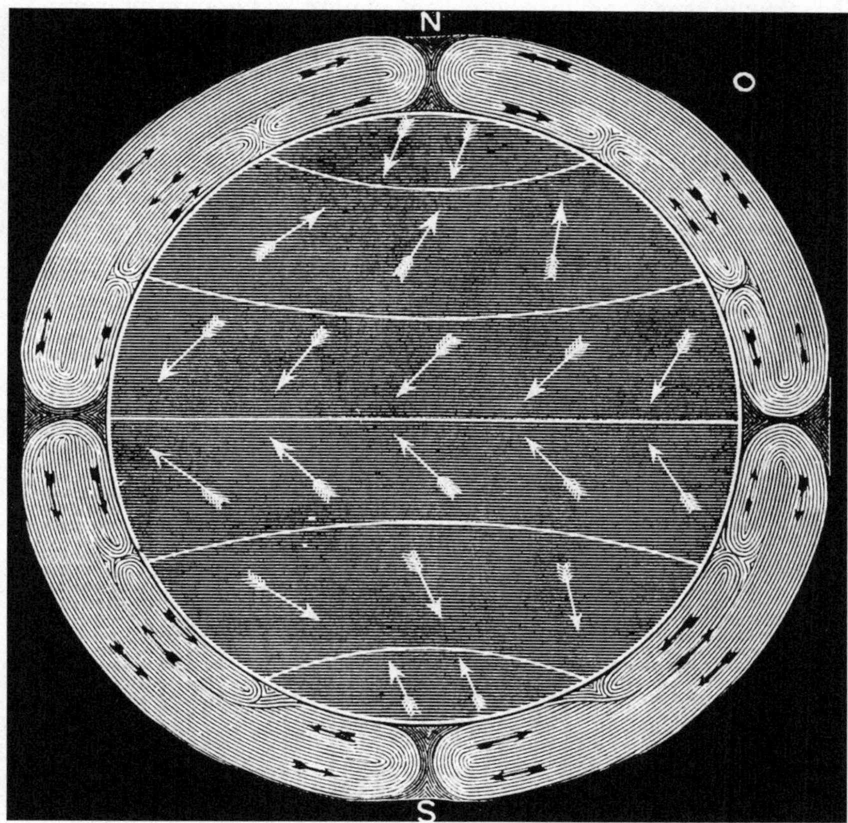

FIGURE 29. Diagram of the general circulation of the atmosphere, by William Ferrel, 1859.

and planetary phenomena. Since Ferrel believed that the energy of cyclones was generated by the latent heat of condensation, planetary-scale currents were irrelevant to their formation. Dove assumed, on the contrary, that the energy of cyclones derived directly from global, equatorial currents; and for that reason he took no interest in the mechanism of energy transfer between the zonal flow and eddies. In short, nineteenth-century models of the general circulation treated phenomena of the dimensions of a cyclone (that is, on the order of one hundred to one thousand kilometers in diameter, lasting a few days) as incidental deviations from the hemispheric circulation. They ignored the exchange of energy between atmospheric phenomena at different scales.

By the 1870s Julius Hann had rejected both Ferrel's theory that cyclones are caused by the latent heat of condensation from rising air and Dove's view that cyclones arise from the confrontation of polar and tropical air currents. Instead, he argued that cyclones grow from the release of tension between regional differences of air pressure, which in turn are maintained by the planetary-scale circulation. Here was a potentially revolutionary view of the significance of local disorderly motions in the context of the circulation of the atmosphere from equator to poles and back again.

THE ATMOSPHERE'S "SOMEWHAT TANGLED MOVEMENT"

Most scientists in Hann's day had little motivation to reconcile their explanation of cyclones with their model of the general circulation of the atmosphere. In an era of naval expansion and empire-building, storms garnered far more attention than planetary-scale processes. And since storms continued to be viewed as mere local disturbances, there was no motivation to incorporate them into a theory of the atmosphere as a whole.[4]

This changed with the work of the Scottish meteorologist Alexander Buchan (1829–1907). In 1869 Buchan published "Mean Pressure and Prevailing Winds of the Globe," which suggested that cyclones exemplify a fundamental relationship between airflow and pressure distribution. Buchan also demonstrated the value of barometric maps for understanding the climates of the world. Inviting controversy from a German-speaking audience, Hann judged that Buchan's contribution was more significant than Dove's in showing the way forward toward a genuinely "scientific" climatology.[5]

Hann's investigations into the distribution of air pressure and temperature over central Europe led him to a pathbreaking interpretation of the formation of cyclonic storms. On the basis of synoptic maps, Hann showed that cyclones tend to form when a pressure low is surrounded by high-pressure centers on both sides. The regions of high pressure, he supposed, supply the air needed to maintain cyclonic motion. Here he assumed that a small difference of pressure, acting over a long distance, could be the cause of a powerful rotation. He showed this to be plausible by means of admittedly crude calculations, which treated the air as a rotating disk and estimated its kinetic energy accordingly. Hann continued to refine this argument, and in the 1890s he was able to draw on new data from the first generation of high-altitude observatories, an innovation that he had tirelessly promoted. Hann's conclusion, in a nutshell, was that the convective theory of storms confused cause and effect.[6] It treated the initial temperature difference between rising and sinking air as the *cause* of the motion, as in a steam engine. Hann insisted that this contrast was instead an *effect* of the motion. He pointed out that such a closed circulation would quickly peter out without the addition of heat. Against the thermal theory, Hann argued that storms were not "exceptions" to the general circulation; they were small-scale effects of the same forces that drove the atmosphere on the largest scale. Though aperiodic when analyzed individually, regularities appeared when they were considered in aggregate: when plotted as a monthly average, the relationship between pressure gradient and wind could be seen to follow the empirical relationship known as Buys Ballot's law. As open systems, cyclones could be understood to feed off low-pressure air masses over the ocean. From this perspective, it was easy to see why the frequency of cyclones at midlatitudes increased in winter, when the temperature contrast between equator and poles was at its starkest. As Hann put it in a visionary passage in 1890, "Cyclones and anticyclones are only partial phenomena in the general circulation of the atmosphere."[7]

In this way, Hann insisted on a transformative shift in the scale of analysis. In the poetic formulation of the Harvard geomorphologist William Morris Davis, Hann had shown cyclones to be "driven eddies in the somewhat tangled movement of the general circumpolar winds."[8] From the dimensions of the eddy itself, Hann had expanded outward to take in the surrounding pressure field and the circulation of the atmosphere at large. Cyclones and anticyclones were not closed systems unto themselves, but components of the general atmospheric circulation. Indeed, according to Supan, the victory of Hann's point of view in the storm controversy was the result of this shift in scale. Previously, Supan observed, storms had been perceived as "unperiodic and local phenomena"—that is, they had appeared irrelevant to what Dove described as the planetary-scale contest between polar and tropical currents. Hann's achievement was to reconcile an account of the formation of storms with the emerging understanding of the relationship between pressure distribution and winds. "In a word, what appeared to be an exception became the rule: the phenomenon of general and periodic atmospheric disturbances to equilibrium in our latitudes are composed of a series of small-scale phenomena, to which had previously been attributed a local and unperiodic character."[9] Hann had revealed the significance of what had previously appeared a mere "detail," a "disturbance," by viewing it as a part within a greater whole.

Hann's colleagues termed this new theory of cyclogenesis "dynamic," by which they meant that it explained cyclones with reference to forces generated by the rotation of the earth. Two decades later, the Austrian meteorologist Wilhelm Trabert would lament that the term "dynamic" had become so widely applied to theories of the atmosphere that it hardly carried any specificity.[10]

This is a charge that we will return to at the end of this chapter, when we consider how twentieth-century scientists recounted the origins of "dynamic climatology."

LOCAL DISTURBANCES

If climate is defined as average weather, then the variability of weather in time should lie beyond the scope of climatology. But this is *not* how Hann and his colleagues defined climate, as we have seen. Variability was essential to their concept of climate—as is the case today, in the era of rapid anthropocentric warming. Indeed, as early as 1872, Hann identified the irregularity of the climate of the midlatitudes as a feature worthy of study. "The climate of the temperate and polar zones are thus governed by the alternation between opposing wind directions, and thus far no one has been able to discover a rule for this alternation. The climate [*Witterung*] of the tropical zone bears the character of permanence, the climate of the extratropical zone that of utter irregularity and variability [*Regellosigkeit und Veränderlichkeit*]."[11] Hann was disputing Dove's axiom that the atmosphere circulates in two continuous currents. As he and his colleagues pointed out, this assumption was contradicted by every sudden gust of wind. That is, we have direct experience of the unsteady character of winds ("Böigkeit," "Unstetigkeit"). It is visible, for instance, in the swaying of a tall field of grass or the quivering of a tree's branches and leaves.[12]

By 1891, on the basis of data from the new high-altitude station on the Sonnblick, Hann had reached the conclusion that irregularity was a fundamental feature of the general circulation:

> The stationary state of the atmospheric circulation, in the manner dictated by the theory [i.e., the theory associated with Dove, assuming a direct circulation between equator and pole] never obtains in reality. . . . The large temperature differentials and their temporal displacement near the earth's surface in higher latitudes of the wintry hemisphere, the resulting spatial retardations and accelerations of the upper currents, the variable resistance, etc. constitute a continuously changing source of disturbances. In the rapidly rotating upper-level currents, these must give rise to eddies, which then also set the lower layers of air in motion. . . . How can one suppose that the atmospheric circulation proceeds entirely without local disturbances, given the many instigating factors that arise from the temporally and spatially changing temperature gradients and the different coefficients of resistance.[13]

PLATE 1. Kremsmünster and its surroundings, with the astronomical tower at the center, painted by Adalbert Stifter, ca. 1823–25.

PLATE 2. *Venedigergruppe* (Venediger group), Eastern Alps, painting by Friedrich Simony, 1862. The subject of this painting is a section of the Hohe Tauern reaching over ten thousand feet and which Simony described as displaying "unusual symmetry." The accompanying text draws the reader's eyes to the geological structure of the mountains in the background. But one should also note the foreground details, particularly the pilgrim gazing up at the devotional image on the pine tree. Simony's vantage point for this scene, the village of Neukirchen, had long been a pilgrimage destination.

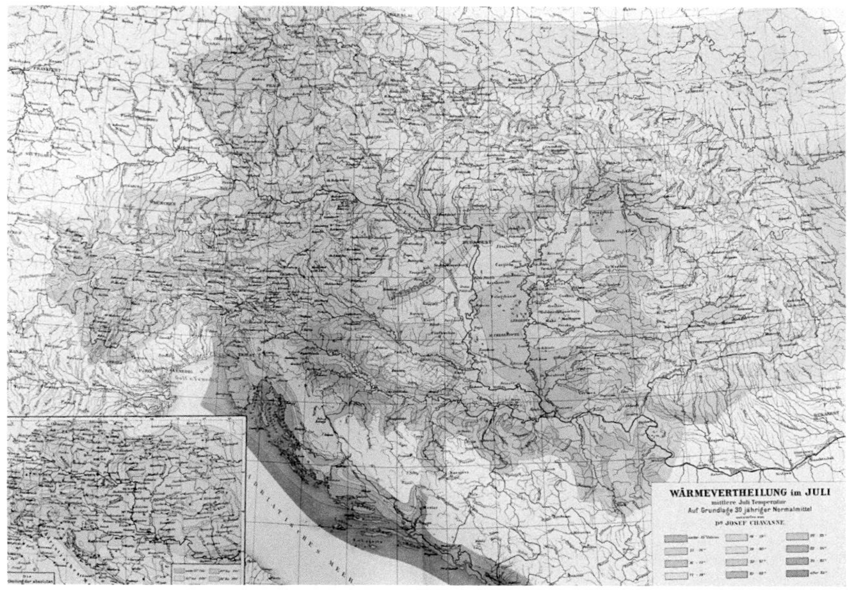

PLATE 3. "Distribution of Heat in July," by Josef Chavanne. In *Physikalisch-statistischer Handatlas von Österreich-Ungarn.*

PLATE 4. Vegetation of the puszta, the Hungarian steppe, painted by Anton Kerner, ca. 1855–60.

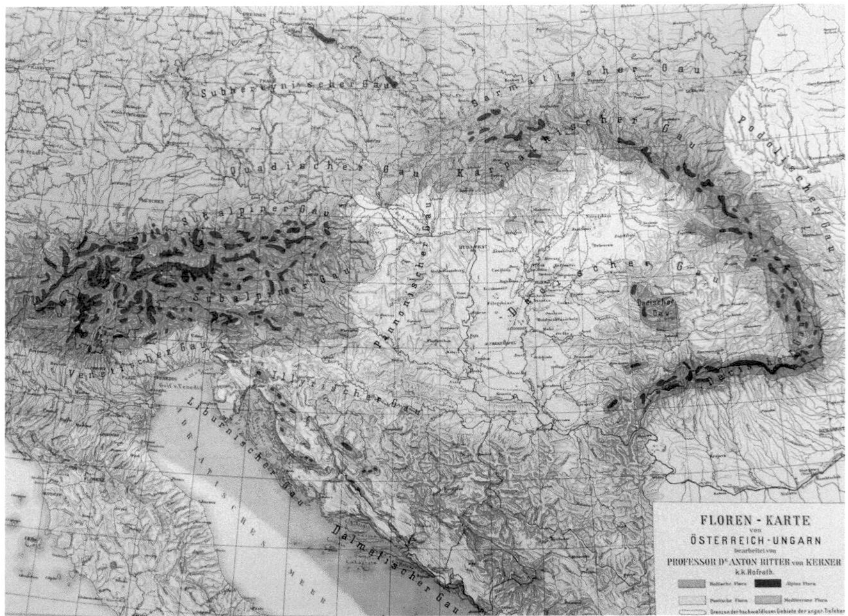

PLATE 5. "Floral Map of Austria-Hungary," by Anton Kerner von Marilaun, 1888. The map portrays the Habsburg territory as the intersection of four distinct flora: Alpine (*red*), Baltic (*green*), Pontic (*yellow*), and Mediterranean (*pink*).

It would be another thirty years before scientists would be able to substantiate Hann's hypothesis on the basis of fundamental physics. Why then did Hann find this picture of energy flow in the atmosphere so compelling?

Hann's discussion of "atmospheric disturbances" in his 1901 *Lehrbuch der Meteorologie* suggests an answer. Climate, he observed, is an averaged distillation of experience, while weather is a momentary "total impression" (*Totaleindruck*) of the "interactions" (*Zusammenspiel*) of all atmospheric elements at once. We cannot rightly speak of the weather of a week, a month, a year—indeed, of any period longer than a day. Even the term *Witterung* (atmospheric conditions) "itself designates an abstraction; weather is a real condition [*reeler Zustand*], an individual event [*Akt*] singled out from the ever-changing series of atmospheric phenomena." As soon as we begin the process of averaging the readings of our instruments, we have moved away from the representation of weather, toward climate—that is, toward a picture of the average atmospheric conditions at a given location. In this sense, "weather" can give a misleading impression of "climate." What we encounter as weather is a disturbance of the stationary state of the atmosphere. It is for this reason, Hann suggests, that it took so long to elucidate that stationary state. "The atmospheric circulation actually takes place only in the form of disturbances, which unfold in such a way that their average effect is responsible for the transport of air that corresponds to the schematic image of the general atmospheric movements of the higher latitudes. *Here*, for that reason, success has come so late and only partially in the discovery of the hidden law in the fast-changing phenomena."[14] Hann's point was that this picture of the atmospheric circulation is but one local perspective, that of the inhabitant of the midlatitudes. For there is another region of the earth where climate looks much the same as "weather." In the tropics, one has a fair chance of experiencing "average" conditions directly at any given moment, because the range of variation is so narrow.[15] In this passage, Hann effected a double reversal of scale. First, he revealed the provincialism of the purportedly global perspective of the European scientist. Second, he suggested that what appear to be insignificant details within that global perspective might in fact be vital to the whole.[16]

This process of scaling had led Hann to the insight that disorderly motions were essential to the transport of energy between equator and poles. And yet he had no way to quantify and test this hypothesis. Without a measure of the energies involved, he could not accurately judge the significance of atmospheric "irregularities."

MEASURING TURBULENCE

How well the best of our flying birds know to adapt themselves to all the small variations of air currents, by the positioning of their wings and tail![17]

Wilhelm Schmidt

Interest in the irregular motions of liquids and gases exploded during the First World War, due to the push to design more efficient aircraft and submarines. But a basic understanding of turbulent motions in the atmosphere had already emerged in the last two decades of the nineteenth century, as Olivier Darrigol explains in his elegant history of fluid dynamics. Helmholtz took up the study of atmospheric turbulence in the 1880s. While hiking in the Swiss Alps, he had seen a thin layer of clouds in the sky in the form of strips of whirling eddies. He surmised that these disorderly motions were the result of the confrontation of two layers of air of different densities. He could see how larger eddies spun off into ever-smaller whirls, forming what would today be called a fractal pattern. From this observation, he realized that atmospheric turbulence could be responsible for damping the force of winds in the upper atmosphere. Indeed, without this resistance, the earth's rotation would accelerate winds to unimaginable speeds.[18] The English physicist Osborne Reynolds would soon name this pseudo-frictional effect of turbulence "eddy viscosity." In 1922, the meteorologist Lewis Fry Richardson captured this phenomenon in a twist on a verse by Jonathan Swift: "Big whirls have little whirls that feed on their velocity, and little whirls have lesser whirls and so on to viscosity—in the molecular sense." The pressing question was how to describe this phenomenon mathematically.

An empirical answer to this question would depend on identifying turbulent processes like those occurring in the atmosphere, yet readily measurable. In other words, what are the conditions of physical similarity between planetary-scale phenomena and devices that can be manipulated in a laboratory? A clue lay in the work of Osborne Reynolds in England. Reynolds used long, narrow troughs of water to study the dynamics of flowing liquids. He made turbulent motion visible by injecting a thin stream of colored dye into water flowing at an adjustable speed. In the 1880s, Reynolds learned from his experiments how to predict when a stream of fluid would shift from orderly to disorderly motion; that is, he developed a criterion for the transition from laminar to turbulent flow.

In doing so, Reynolds introduced a powerful new approach to scaling. He argued that it was a logical error to suppose that the character of fluid flow through a pipe depends "in any way" on the "absolute size or absolute velocity" of the flow in question.[19] It should therefore be possible to study large-scale turbulence via small-scale models, provided that one holds a certain ratio constant. That ratio has come to be called the Reynolds number, and it is proportional to the product of velocity, density, and the radius of the pipe, divided by viscosity. In principle, flows through channels of the same shapes and proportions but different sizes will display the same degree of turbulence if they share the same Reynolds number. The Reynolds number thus indicates what remains invariant across scales, and it is therefore a powerful tool of scaling. Reynolds himself was keen to answer a version of the question that already preoccupied Hann: namely, to measure the quantity of energy transferred between larger- and smaller-scale motions—in Reynolds's terms, "the vis viva consumed by internal friction" and "converted into heat."[20]

There remained a problem with applying this approach to meteorology. The flow of a fluid through a relatively narrow trough or pipe depends on the height and width of the walls. How would this compare to the flow of air in an unbounded space like the atmosphere? Measuring phenomena as complex and fleeting as the disorderly motions of the free atmosphere demanded new tools and practices of scaling.

EXPERIMENTAL CLIMATES

In introductory classes today on atmospheric physics, students might have the good fortune to encounter the "dishpan" demonstration. Fill a cylindrical vat with water and place a block of colored ice at the center to simulate one of the poles. Now set the vat on a rapidly rotating turntable and watch as the ice melts and a wave pattern appears. Look closely and you will see the hemispheric circulation of air between equator and pole. You might even see spiraling whirls start to form: the genesis of cyclones.

What Peter Galison calls "mimetic" models held an appeal to scientists in the early twentieth century that is hard to recapture in our own digital age.[21] By translating atmospheric or geological processes into the workings of human-built devices, models made it possible to study impossibly large, distant, or exceedingly gradual processes. Models allowed scientists to "observe" processes at work high in the atmosphere or deep inside the earth. For scientists who studied processes too complicated to be described by linear differential equations, models offered a necessary simplification. They also gave earth scientists a rare opportunity to conduct "experiments," manipulating the parameters of their model and the forces acting on it as they could never do for a planet. They could test hypotheses about the forces responsible for phenomena like cyclones, tornadoes, or the convolutions of geological strata. Among the atmospheric conditions that could be mimicked in the lab were small-scale turbulence, baroclinic instability (i.e., the movements induced by horizontal temperature gradients), the Hadley circulation, and Rossby waves. For many scientists, working with models also had an aesthetic appeal. As they found themselves spending more and more time crunching data from station networks, models supplied the real-time, multisensory, full-bodied knowledge they missed.

Yet model-building was not to every geoscientist's taste. As Naomi Oreskes has shown, there is a tradition of skepticism with respect to models in geology dating back to the eighteenth century. Some scientists refused to accept evidence from models that contradicted their theories. Others granted models merely heuristic value; that is, models could demonstrate the plausibility of proposed causes, but could not prove that those causes operated in nature.[22] Despite the enduring and sometimes acrid character of these debates, scientists gave little attention to the criteria that might define a model's validity as an analogue for planetary processes.

So it was that when Wilhelm Schmidt (1883–1936) took up atmospheric model-building as part of his dissertation research in 1910, he had the sense that "pure experiments" were "very rarely made in meteorology." By "pure," he meant those that have a goal other than "procuring observations."[23] In other words, a true experiment is one that tests a causal hypothesis. From a historical perspective, he was right. James Thompson had described how one might construct a model of the general circulation in the 1860s, but had gone no further. In the 1850s, a Berlin physician and amateur meteorologist named Friedrich Vettin had built a series of model tornadoes. However, Vettin's work was little known among meteorologists, allegedly because Heinrich Dove had intimidated him into silence.[24] Schmidt believed that meteorology had much to gain from honing an experimental method, precisely because atmospheric phenomena were so complicated and depended on so many different factors. As a student of Franz Serafin Exner and Ludwig Boltzmann, Schmidt had been trained to consider nature's fluctuations as phenomena in their own right and to work to describe them mathematically.[25]

The first hypothesis that Schmidt hoped to test was that a squall begins when a mass of colder air collides with a mass of warmer air. Despite Margules's efforts, Schmidt judged that here "mathematics fails, or at best is able to give only distant approximations." The success of the experiment would depend, first, on a precise formulation of the hypothesis, and, second, on a set-up that simplified the conditions to be found in nature, yet without departing too far from them. "Thus one must ascribe demonstrative force to every

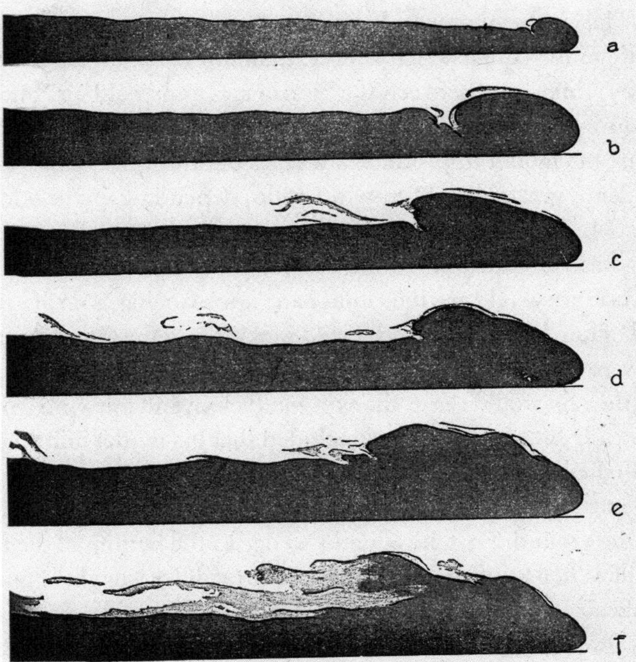

FIGURE 30. Propagation (from left to right) of the "head" of Wilhelm Schmidt's experimental storm (with the difference between warmer and colder fluid ranging from 0.5°, a, to 35°, f).

investigable and testable analogy, if the conclusions derived from it hold up against *the scale of nature*."[26]

The apparatus for Schmidt's experiment was a trough measuring 181 by 31 by 4 centimeters, capped with glass plates on each end lengthwise. Forty centimeters from one end was a divider, set at an angle. The longer end of the trough was filled with fluid, while a denser fluid was added to the shorter end, not quite up to the top of the divider. These fluids of different densities were meant to simulate air at different temperatures. (For a mesoscale system like a squall, it was not unreasonable to omit the force of the earth's rotation.) When the divider was removed and the fluids mixed, a characteristic eddy form appeared, which Schmidt described as a "raised head" (see figure 30).[27] Here was a way to test Margules's theory that cold air intrudes on warmer air in a wedge-shaped mass. In order to investigate this miniature squall more closely, Schmidt suspended sawdust in both fluids, illuminated it with sunlight, and photographed the mixing process. The image made the streamlines visible, clearly showing the rotatory motion induced by the bolt of cold air. The velocity of flow at each point was proportional to the length of the mark made by the moving sawdust in the photographic image, illustrating the relative wind speed on each side of the squall. What's more, Schmidt could adjust the temperature (density) difference between the "warm air" and "cold air" and thereby modify features of the squall, such as its velocity of propagation. That is, he could adjust his model to produce a weaker or stronger cyclone (a "head" with a smaller or greater height-to-width ratio) depending on the temperature difference. Schmidt took this process of scaling even further by asking how this rearing head would appear from the perspective of a terrestrial observer. First, the observer would see the cloud bank approaching from afar, followed by the wall of rain. Then the wind would pick up and the temperature would drop; if it were summer, this might be followed by thunder and lightning. But quite soon the rain would cease, the sky would clear, and the wind would calm to a cool, steady breeze. Schmidt concluded that his model mimicked actual squalls both qualitatively and quantitatively.

Schmidt next designed two modifications of this experiment to imitate conditions on the ground. First, he added a wedge to the bottom of the trough to serve as a hill. When a puff of smoke was sent through the trough, mimicking the influx of cold air, it again formed a "head." In this case, however, the "head" rose higher as it propagated than it had with a flat surface, and its size diminished as it moved uphill. When the smoke was instead forced downhill, the height of the head fell as it propagated, but the speed of propagation increased. Schmidt interpreted these results as confirmation of Trabert's empirical observation that "every landscape favors the formation of storms in a given direction and inhibits it in the opposite direction," depending on the angle of the ground's slope.

In a second modification, Schmidt observed the effect of an influx of cold air when the initial conditions involved a surface inversion: colder air below, warmer air above.[28] One half of the trough was now filled a few centimeters high with a dense, lightly colored glycerin solution, then filled nearly to the top with water. Into the other half of the trough, separated by a divider, Schmidt added a darkly colored solution of a density in between those of the other two liquids. In this case, the head was smaller and flatter than in the previous experiments, and one could see the formation of a middle layer. These results bore comparison to Margules's empirical description of the *Auflecken* or "lapping up" of a surface inversion. Schmidt reasoned that the arrival of a cold air mass at a higher altitude would create a stronger horizontal pressure gradient on high that would "squeeze out" the cold surface air. The observer on the ground would experience a gradual warming, but with almost no wind. In conclusion, to those who might say that his models oversimplified reality,

Schmidt countered: "The allegation that nature is in every case much more complicated is no more true of experiment than of every thought, every abstraction. Cognition is always a path from the simple to the complex."[29]

THE COMING OF WAR

The Great War brought a new urgency to questions of aerodynamics and weather, as a result of the introduction of two new weapons: fighter planes and chemical gas. In Austria, civilian meteorology was uniquely well positioned to respond to the military's needs. Meteorology had a dedicated professorship at each university and was a required subject for candidates for a teaching certificate in physics, mathematics, and geography.[30] In 1915 Felix Exner was promoted to director of the ZAMG and put in charge of the weather service of the Austro-Hungarian forces, which took up residence in the institute's villa on the Hohe Warte. Several of the younger staff members were called to the front, but Wilhelm Schmidt and Albert Defant remained behind, the former in his capacity as secretary of the ZAMG, the latter as an adjunct in the weather service.[31] Research unrelated to warfare ground virtually to a halt.

On the home front, censorship was strict, and food was in short supply. With the death of Franz Josef in November 1916, after seventy years on the throne, the dynasty's legitimacy was in jeopardy. Nationalists began to clamor for the federalization of the empire, and in October 1918, the new emperor, Charles I, acquiesced. When peace was finally declared, approximately 1.2 million Austro-Hungarian soldiers had died, while nearly 2 million had been interned as prisoners of war, many of whom were struggling to make their way home in the midst of Russia's civil war.[32]

The researchers of the ZAMG believed their science had proved its practical value in wartime. They expected it to be suitably supported by the postwar state. But war's end brought an economic crisis to the new Republic of Austria. Exner and his coworkers were struggling to find enough food to eat and coal to warm their homes. Hardest hit was Max Margules. On the rare occasion when an acquaintance laid eyes on him, his physical deterioration was worrying. Food had been hard to come by during the war, and the postwar inflation had left the retired meteorologist with little to spend. Still, Margules refused to accept assistance. In 1919, the Austrian Meteorological Society awarded him its Hann Medal. Margules declined the check that came with it, insisting that the money could be put to better use elsewhere.

In these economic conditions, government funding for basic research was not to be had. What's more, with the breakup of the empire in 1918, the scientists of the ZAMG could no longer count on access to observations from beyond Austria's new borders. At that point, the virtue of working with scale models became a necessity.

It was only then that Schmidt realized that a rival group of researchers had been developing similar tools of scaling. In 1907, the University of Göttingen established an institute for aerodynamic modeling and testing under the direction of Ludwig Prandtl, and its first wind tunnel began operation the following year. Prandtl was an engineer by training, educated at Munich's Polytechnic University. Based on his experimental studies, he had already begun to work out a mathematical description of the flow of air close to the edges of objects and the conditions under which vortices would form.[33] However, Prandtl did not consider the application of his results to meteorology until the early 1920s.

In 1926, when Wilhelm Schmidt first contacted Ludwig Prandtl about their shared interest in atmospheric modeling, he complained of the "meager means available to us in Austria." Prandtl was by then director of the new Kaiser Wilhelm Institute for Fluid Dynamics (Strömungsforschung) in Göttingen. Schmidt had recently applied to the Notgemeinschaft der Deutschen Wissenschaft (Emergency Association for German Science) to support a "larger program for more precise studies of turbulence in the free atmosphere."[34] Here was a mode of fundamental atmospheric research that seemed custom made for the reduced conditions of the postwar ZAMG.[35] And yet, how much more could be accomplished with the sophisticated apparatus of Prandtl's laboratory! Prandtl's studies of atmospheric turbulence were likely fueled in part by the hang-gliding craze that took postwar Germany by storm. But his bigger bait was the new work of Vilhelm Bjerknes on the dynamics of cyclones. Like Schmidt and Exner, Prandtl was skeptical of Bjerknes's description of cyclones as waves in a surface of discontinuity between polar and tropical air. If such a surface were stable enough to support waves, what would cause these waves to develop into cyclones? And if the surface were unstable, how could it be considered a permanent feature of the atmosphere? Bjerknes dreamed of reducing atmospheric motions to analytic equations, but this approach could not be applied to a nonlinear process like eddy formation, as Exner pointed out. He and Schmidt turned to experiment "out of the need to make progress by other means."[36]

TEMPEST IN A TEAPOT

Indeed, it was Exner who designed the classic "rotating dishpan" model of the general circulation. He first discussed it in print in 1923, at the height of

FIGURE 31. Felix Exner's rotating-dishpan model of the general circulation, photographed from above, 1923. Tongues of cold water radiate out from the ice at the center and form vortices akin to cyclones.

Austria's inflationary crisis. Exner filled a cylindrical vat with water, positioned a block of ice at the center, set it rotating, and applied heat to the rim to produce a starker temperature gradient between equator and pole. Photographs captured the eddying forms that Exner likened to cyclones (see figure 31). Another project accomplished on the shoestring budget of the postwar ZAMG was Exner's "artificial tornado." This was a rotating vat containing nothing but air, warmed along the rim. As the air above the rim warms and rises, it flows toward the center, and a tall column of eddying air forms, which Exner used smoke to visualize. Crucially, Exner could intervene in this process at will. By blowing through a pipe aimed at the center and bottom of the vat, he could trigger the formation of a new eddy.

By manipulating these whirls up close, Exner came to feel he had penetrated to the heart of cyclogenesis. "If one observes this phenomenon frequently, a process becomes clear—as simple as it may be, it is the essential one in the formation not only of these vortices, but in cyclones in general."[37] This essential feature was the barrier that the tongue of cold air posed as it flowed outward into the more rapidly rotating air closer to the rim. The result, Exner inferred, was the formation of a center of low pressure behind the cold air, which gave the air a rotational motion. What's more, Exner found that if he eliminated the temperature gradient between the center and rim of the rotating vat, the eddies vanished.

On the basis of these experiments, Exner objected to Bjerknes's characterization of cyclones as waves in a stable circumpolar surface of discontinuity between polar and equatorial air. This seemed to Exner irreconcilable with the observation, originally due to Hann, that cyclones were essential components of the general circulation of the atmosphere. How could a wave take part in a closed circulation? Instead, cyclones represented the "unfurling [*Aufrollen*]" of an unstable surface.[38] At the same time, Exner used these experiments to refute Wegener's recent assertion that cyclone formation was driven by a contrast between air masses in two different states of motion.[39] The rotating dishpan demonstrated that a temperature difference was necessary to instigate a rotational motion. As the model demonstrated, rotation began when the warm westerly wind was presented with a barrier of cold air. The questions Exner raised about Bjerknes's model would not be answered until the late 1930s, when Carl-Gustaf Rossby elucidated the conditions under which stable planetary-scale waves could form at the boundary between polar and tropical air.[40]

LABORATORY LANDSCAPES

Schmidt and Exner found inspiration for their laboratory models in living landscapes. In lakes, riverbeds, and sand dunes, they saw natural analogues to the processes unfolding in the atmosphere. These landscape features recorded, less ephemerally than clouds, the characteristic forms of motion that arise when contrasting layers of air, water, or earth flow past one another. For instance, Exner's 1921 study of dunes and riverbeds opened by conceptualizing the riverbed as a trace, a record of the ways in which water and sediment had shaped each other as one flowed over the other. Likewise, dunes could be thought of as the trace of interacting flows of air and earth.[41] Landscapes were semipermanent registers of fluid dynamics.

As a professor of climatology at Vienna's Agricultural University, Schmidt's research concerned the interactions of soil, water, and air. Back in the laboratory at the ZAMG, Schmidt began to study the effects of the earth's rotation on the course of rivers and their beds. It had long been assumed that riverbeds, like air currents, had a natural tendency to curve due to the rotation of the earth (Baer's law). This would explain the rightward bends of many rivers in the

Northern Hemisphere. It was tempting to compare this effect to Buys Ballot's law for atmospheric rotation, but Schmidt suspected that this analogy was misleading. There were essential differences, he argued, between the conditions of flow for water in a river and for air in the free atmosphere.

In pursuing this question, Schmidt was evaluating the analogy between river formation and atmospheric dynamics. What were the conditions of similarity? He turned the question into an exercise in dimensional analysis. The gradient force is much stronger in the case of water falling under the force of gravity, he noted, than when air is driven by a difference in pressure. However, the frictional force is also far greater in the river's case, where the flow is bounded on three sides, than in the free atmosphere. Taking friction into account, Schmidt estimated that the deflecting force of the earth's rotation was 390 times greater when acting on the atmosphere than when acting on a river. On these grounds, he doubted that a river's curves were an effect of the Coriolis force. To test this reasoning, Schmidt designed a motor-powered rotating turntable. Atop this turntable he placed a flat-bottomed vat filled with damp sand, the surface of which sloped in one direction. He then poured water slowly over the sand in one location and observed its course. On this basis, Schmidt claimed that the combined effect of rotation and erosion was a sharp leftward bend of the flowing water—"in precisely the opposite direction as one has typically assumed and even in some cases believed to observe."[42]

Exner, however, was not convinced. He devised his own variation on this experiment, using a mixture of sand and silt. He gave the material a gentler slope than Schmidt had, dug a straight canal before adding the water, and added a drain at the opposite end. In this way, Exner obtained results that agreed better with observations of natural rivers. What he found was that the rotational force pressed the flowing water toward the right side of the canal, so that the surface of the water sloped to the right, and the fastest-moving part of the stream was not the middle but the right-hand side. This rise in velocity increased erosion on the right side of the bed, which created a slight bend in the river toward the right. At a certain point in the water's course, however, the gravitational potential of the slope outweighed the centrifugal force, and the water's course then bent to the left, downhill. Thus the cumulative effect was a riverbed that displayed curves to the right followed by sharp leftward bends when the flow was heavy. This explained both Schmidt's observations and the common wisdom contained in Baer's law.[43] In fact, Exner's observations also squared with the terse theoretical analysis of river meandering that Albert Einstein had published the previous year, apparently unbeknownst to Exner.[44]

Today, geophysicists continue to debate the causes of meandering in rivers, but tend to emphasize local effects of turbulence and gravity over the influence of the earth's rotation.[45]

Like his observations of dunes and meanders in nature, these experiments brought home to Exner the sheer complexity of the phenomena involved. He remarked on the sensitivity of the results to conditions such as the amount of water flowing and the composition of the soil. This was not a reason to give up—on the contrary, complexity implied the limits of mathematical analysis and the indispensability of models.

What Exner sought was a more explicit condition of similarity between small scale and large. In order to compare lab results to observations of the Danube, for instance, one had to consider the rotational velocity of the turntable in comparison to that of the earth at the latitude of Vienna. A more realistic model would also require a material other than sand or silt, one that was only as porous as the natural riverbed. Finally, Exner turned to dimensional analysis to compare the effect of erosion in the laboratory to that observed in nature. Making simple assumptions about the mathematical form of the forces at work, he derived an equation for the average velocity of flow that applied equally to laboratory models and full-scale rivers. Here was a plausible condition of similarity.[46]

Schmidt had plans to apply this rotating turntable to his prewar experiments on the inflow of cold air, but he was strapped for resources. By 1926, Prandtl, by contrast, had built an entire rotating room, a "carousel," as his daughters fondly called it.[47] It was cylindrical in shape and three meters wide by two meters high.[48] "It seems to me," Prandtl wrote Schmidt, "that your experiments and mine are not in competition with each other. But I would welcome establishing an ongoing contact between us, such that each of us would inform the other of any experiments in this field, in order to avoid any unnecessary duplication of labor." In one case, Prandtl was kind enough to offer the services of his laboratory for the precise determination of the coefficient of resistance of a pockmarked copper sphere that Schmidt intended to use to study the role of turbulence in generating air resistance. Prandtl, for his part, was unaware until then that related research was under way in Vienna. He had known nothing of Exner's earlier experiments with the rotating dishpan until Schmidt referred to them in a subsequent letter.[49] Nonetheless, the Göttingen engineer shared some of the same motivations as Vienna's geophysicists. "One keeps hearing," Schmidt complained to Prandtl, "that the storm fronts originate in the collapse of such unstable layers. Purely topologically, that is not possible." Prandtl had expressed this same doubt at the 1922 meeting of the German

Physical Society.[50] More fundamentally, Prandtl shared the hunch of Schmidt, Exner, and Defant that atmospheric physics had thus far neglected the fundamental role of turbulence. Schmidt suspected that the instability of the polar front was a turbulent effect of the earth's rotation, and this was among the hypotheses that his rotating table was meant to test. "You can see where this is leading: to the formation of systems of eddies of given dimensions, similar to those we see on earth."[51]

The question remained: how could researchers model large-scale turbulence in the free atmosphere at the far smaller dimensions of the laboratory? In their private exchanges, Schmidt and Prandtl confessed their uncertainty about the epistemic status of their models. Schmidt wrote of the need for further experiments to model large-scale circulation, but expressed his concern that it was not possible to reach similar levels of turbulence in the laboratory. "The one great difficulty lies indeed in that one must actually keep the turbulence proportionately low. Not enough attention is paid to this when one is dealing with more tangled [*maschere*] flows."[52] Prandtl agreed: "With these experiments on the imitation of the large-scale circulation, the question of scale naturally concerns me as well. But I will be content for the time being if only a qualitative analogy comes out of it." Schmidt, by contrast, did not abandon the hope of hitting on criteria that would ensure a quantitative as well as qualitative similarity between laboratory model and free atmosphere. In a letter to Prandtl in April 1932, Schmidt wondered "if one might be able to find a simple criterion for turbulent versus laminar flow for the free atmosphere as you have managed to find in your experiments. It's a matter namely of deciding the question I've already mentioned: whether the free atmosphere exhibits processes that the meteorologist would still treat as laminar, while the physicist would consider them turbulent, or even perhaps the reverse." Schmidt was questioning the very definition of turbulence in the free atmosphere. The laboratory physicist's usual criteria did not necessarily apply, and it was possible that the Reynolds number was "something entirely different" for atmospheric flows. Back in 1917, Schmidt had raised the possibility that "a new form of movement" would be discovered in the atmosphere, one that was related to but not identical with "turbulence" as it was known from laboratory studies.[53] Schmidt and Prandtl agreed that unusually strong forces and a high degree of precision would be necessary in order to draw quantitative conclusions about the free atmosphere from laboratory experiments. Indeed, Prandtl calculated that any laboratory model would need to achieve ten to fifty times the force of gravity to generate a density profile comparable to that of the earth's atmosphere.[54]

This quest to construct laboratory models of atmospheric dynamics found little appreciation outside of central Europe in the 1920s. At that time, the growth of military and commercial aviation was driving demand for reliable weather predictions. Most meteorologists were intent on applying the new concepts of fronts and air masses to more or less practical schemes of forecasting. Lewis Fry Richardson famously imagined a forecasting "factory" with sixty-four thousand workers. Tellingly, he included in this proposal a disparaging caricature of "an enthusiast" in the basement "observing eddies in the liquid lining of a huge spinning bowl."[55] And yet the Austrians' persistent recourse to mimetic experimentation was not a mere eccentricity. It was, in part, a critique of the rush to reduce the complexity of atmospheric processes to solvable equations.

Ironically, mimetic models of the atmosphere only began to draw the attention of leading international researchers circa 1950—just as the first digital computers made possible quantitative simulations of atmospheric dynamics, apparently rendering analog methods obsolete. In 1947, Carl-Gustaf Rossby, the Swedish-born atmospheric physicist and chair of the new department of meteorology at the University of Chicago, launched a series of laboratory experiments on large-scale turbulent mixing. After his departure for Stockholm in 1950, several of his colleagues and students at Chicago continued for several years to hone these models.[56] They were inspired by their rediscovery of Exner's papers of the 1920s: "Exner's work with a rotating disk was pointed out to us by Dr. O. R. Wulf and a decision was made to repeat and extend Exner's experiment particularly from the standpoint of making numerical measurements."[57] By the early 1950s, similar experiments were under way at MIT and Cambridge University, and their results received theoretical attention from Edward Lorenz, among others.[58] Fultz's group also made a detailed study of the history of this mode of experimentation. They came to suspect that it had played a bigger role in the history of meteorological thought than had ever been acknowledged. This appeared to them an "extremely interesting, and valuable" topic for "a historian of science."[59] Of course, these were the very years in which numerical weather prediction was getting off the ground at Princeton. Mimetic models were quickly overshadowed by the promise of unprecedented calculating power.

Today, these laboratory models of whirling atmospheres remind us of what computation alone cannot achieve. Computer climate models have grown so staggeringly complex that even experts cannot understand their operation at an intuitive level. One of the world leaders in climate science today, Isaac Held, has therefore called for the construction of a hierarchy of models, from complex to simple, much as biologists study simpler organisms in order to understand more complex ones. In that light, mimetic models may have a new role to play. As Held suggests, "Laboratory simulations of rotating and/or convecting fluids remain valuable and underutilized."[60]

QUANTIFYING "EXCHANGE"

By 1913, Schmidt had begun to envision a generalized mathematical approach to disorderly motions in the atmosphere.[61] He aimed to do for fluid dynamics what an earlier generation of Austrian physicists had done for thermodynamics. With their contributions to the kinetic theory of gases, Ludwig Boltzmann, Josef Stefan, and Franz Serafin Exner (Felix's uncle) had provided a statistical description of the emergence of order out of disorder. Echoing the elder Exner, with whom he had studied, Schmidt explained in 1916 that "turbulent motion bears the stamp of the random, the incalculable—this is true, if one tries to follow the processes down to the smallest detail; not true, if one simply proceeds on the basis of the laws of its effects."[62] The analogy between the statistical treatment of thermodynamics and Schmidt's approach to turbulence ran deep. The differential equation describing turbulent mass exchange took the same form as the equation for heat conduction in the kinetic theory of gases. Turbulence, however, is a much more efficient means of mixing warmer and colder air than ordinary conduction.

In 1917 Schmidt introduced a mathematical definition of *Austausch* or "exchange."[63] This concept built on the recognition, due to Helmholtz and Reynolds, that turbulence plays out in the atmosphere at multiple scales at once: in the motions of molecules, in the synoptic-scale motions we call storms, and in the planetary-scale circulation of the atmosphere as a whole. There must, therefore, be a way to measure what is transferred between smaller and larger eddies by means of this disorderly motion, whether mass or another property such as momentum, heat, or even impurities suspended in the air. And it should be possible to do so even without any detailed knowledge of the paths of flow. Schmidt reasoned further that this measure will depend only on the eddying motion, not on the composition of the fluids or gases, nor on the properties being transferred.[64] He derived the *Austausch* coefficient directly from considerations of the conservation of momentum of a fluid or gas flowing horizontally with no mean vertical component of motion. The coefficient of "exchange" is then the momentum transferred from an upper layer to a lower layer in a given cross-sectional area. In terms more familiar today, the *Austausch* coefficient is proportional to the product of mixing length, density, and local average vertical velocity. Mathematically, Schmidt's *Austausch* coefficient is analogous to the coefficients of diffusion, conductivity, and viscosity in kinetic gas theory, but it depends on the dynamic rather than molecular properties of the gas. Thus it varies with atmospheric conditions, namely, wind shear and thermal stability.[65] According to Schmidt's measurements, the *Austausch* coefficient is 1 kg m^{-1} sec^{-1} under calm conditions and in layers of the atmosphere close to the ground. It averages approximately 90 kg m^{-1} sec^{-1} at the top of the Eiffel Tower, and is as high as 140 kg m^{-1} sec^{-1} in the region of the north Atlantic upper-level trade wind.[66]

Schmidt's notion of *Austausch* filled a conceptual hole in the atmospheric science of his day. It soon became widely used, and it remained common well into the 1930s, particularly in agricultural climatology and "microclimatology" more broadly, as well as in empirical measurements of atmospheric turbulence.[67] Schmidt was recognized as one of the pioneers of the study of atmospheric turbulence, along with his Austrian colleague Albert Defant and G. I. Taylor and Harold Jeffreys in Britain. Schmidt applied the *Austausch* concept to a startlingly wide range of phenomena; with one stroke, he synthesized much of the research on atmospheric (and oceanic) turbulence of the previous half century. Commentators praised the broad, synthetic scope and fundamental significance of his contribution. "The new ideas concerning the 'Austausch,' introduced to a large extent into meteorology by Wilhelm Schmidt himself, have shown themselves to have such a wide applicability within almost all branches of this science, that the knowledge here conferred on the reader must be regarded as fundamental for the conception of meteorological phenomena in general."[68] Ludwig Prandtl expressed the "great pleasure" he took in seeing the "nice and clear results that have already come out of your investigations."[69] *Austausch* was only gradually displaced as a measure of turbulence in atmospheric physics as Prandtl's concept of "mixing length" caught on.

The *Austausch* concept opened up two directions for research at once. First, when applied to movements of air perpendicular to the ground, it was of great practical value. Note that vertical motions are the means by which layers of warmer air higher in the atmosphere mix with cooler layers lower down. A region of the atmosphere with a lower *Austausch* coefficient implies a local climate that is more independent of larger-scale atmospheric conditions. In this respect, the *Austausch* coefficient provided "important foundations for the characterization of the 'local climate.'"[70] For instance, valleys were marked by a smaller degree of *Austausch* than mountain peaks and ridges. This explains why valleys experience greater disparities between temperature highs and lows both diurnally and annually. Likewise, the climate of a forest clearing (low *Austausch*) could be considered "independent," while that of an open field (high *Austausch*) was "dependent." Always fond of phenomenological interpretations, Schmidt noted that a hunting dog sniffing at the ground knows intuitively where the *Austausch* coefficient is lowest. By the same token, plants growing close to the ground live in a "different climate" than taller plants. Because evaporation depends on the degree of atmospheric mixing, rates of evaporation are higher for taller vegetation. *Austausch* thus tracked climatic characteristics that were of vital significance to living things.

In 1925 Schmidt published *The Exchange of Mass in the Free Atmosphere and Related Phenomena*. He described it as an "attempt" to present "a scientific field of study still under construction to a wider circle of readers and in a more easily accessible form." He addressed the book to a range of applied scientists, primarily meteorologists, oceanographers, geographers, physicists, botanists, and agriculturalists. Indeed, one reviewer judged that the book should be "of interest to all concerned with meteorology, oceanography, climatology, atmospheric electricity, and even botany."[71] True to the "word and image" tradition, Schmidt de-emphasized equations, instead furnishing numerical examples and tables of quantitative results. In his conclusion he reported satisfaction at having been able to present "tangible images" (*greifbaren Bilder*) of turbulent motions in the atmosphere.[72]

The *Austausch* framework had many practical applications. It could give farmers insight into the process of evaporation (as Schmidt had already indicated in his 1916 study), while it offered fisheries an understanding of the distribution of plankton in the oceans. *Austausch* coefficients could also be used to predict the local climatic impact of human activities such as deforestation and urban growth. Such questions had been on Schmidt's mind for decades. Back in 1905, he had measured the "impurity" of city air, defined by the number of dust particles revealed by an Aitken condensation chamber.[73] These concerns remained central to his scientific work up until his sudden death in 1936. At that time, he had just completed a book on climate in "human surroundings," coauthored with an expert on industrial hygiene (see the conclusion to this book).

Still, Schmidt's interest in *Austausch* was not exhausted by its potential to serve local, practical needs. The concept was equally applicable to global investigations. Shifting his attention from vertical to horizontal transfers of energy, Schmidt was able to study planetary-scale processes with the same mathematical tool he had developed for the local scale. The *Austausch* framework makes it possible to quantify turbulent exchange between equator and poles. In this way, it enabled calculations of the planetary-scale ramifications of climatic fluctuations.[74] Already in 1917, Schmidt had reasoned that the energy transferred horizontally via turbulent eddies was a more significant factor in the maintenance of the general circulation than dissipation via molecular friction. This was the conviction Helmholtz had reached in the 1880s, but now Schmidt had shown how to give it precise mathematical form. It was, Schmidt hoped, a step toward understanding "the energy and water budget of the earth as a whole."[75]

THE GENERAL CIRCULATION

In 1921, these threads of research into turbulent cross-scale energy exchanges were woven together into a vivid new image of the atmosphere. In that year, Albert Defant published the first theory of climate variability based on atmospheric dynamics.[76] Defant brought the full arsenal of Habsburg scaling techniques—mathematical, experimental, and rhetorical—to bear on what he termed a "theory of climate fluctuations." The result was the first account of the general circulation that attributed a causal role to phenomena on the spatial and temporal scale of daily life—phenomena like ordinary storms, which previous theorists had overlooked as mere "local disturbances."

It happens that the British physicist Harold Jeffreys arrived at the same conclusion five years later, but by a very different path. It is helpful to begin by looking at Jeffreys's approach, because the contrast with Defant's is revealing. Jeffreys reasoned from first principles. He began by deducing the winds to be expected from a given distribution of temperature in the earth's atmosphere. He arrived at equations that predicted winds of the right order of magnitude, but all blowing from east to west. This result was contradicted by well-known belts of westerly winds in each hemisphere. What was missing from his calculations, he reasoned, was friction. But friction alone would not generate westerlies. The lesson of the exercise was that no solution to this mathematical model was possible in which the pressure distribution was symmetrical and alone determined the winds. That is, the general circulation must include a small ageostrophic component. This highly formal argument led to the conclusion that the only possible pressure distribution that could maintain the general circulation was one built up of large-scale eddies or cyclones at midlatitudes. From this new perspective, cyclones could not be said to "represent either an instability of the general circulation, or oscillations about a steady general circulation." Rather, cyclones were the only possible form that the atmospheric circulation could take at midlatitudes. Throughout this analysis, Jeffreys's style was formal and terse, and the force of his conclusion rested on abstract mathematical arguments.[77]

Defant's approach was a world apart. A native of South Tyrol, he had been educated at Innsbruck and Vienna and had been appointed professor of cosmic physics at Innsbruck in 1919. In tackling the problem of the general circulation, he drew on a range of scaling practices developed by Habsburg researchers over the previous half century. In his introduction, he invited the reader to contemplate "the earth as a whole." How could a mortal observer possibly arrive at a visual image of the general circulation of the atmosphere? What would such a planetary-scale, time-averaged picture of atmospheric motions look like? Echoing Hann's *Textbook of Meteorology*, Defant noted that the question would be easier to answer for an observer residing in the tropics. Elsewhere, an image of the planetary-scale circulation can only be constructed on an appropriately long time scale. Defant then engaged in an act of scaling as bold and imaginative as Adalbert Stifter's poetic juxtapositions of large and small. He proposed to consider "the atmospheric currents of the temperate latitudes as a marked phenomenon of turbulence on a grand scale." To make this intuitive, Defant appealed to the reader's everyday experience with turbulence. He conjured up "a large, wide river" in which the movement of water "is not flowing, but rolling, whirling, and pulsing." A reader at the time might have been reminded of Exner's and Schmidt's studies of meanders. Averaging the flow of water over time would produce an orderly distribution of velocities for a given cross-section. Yet this distribution "does not exist in actuality, but is only an abstraction from the disorderly flow of the water." Just so, the orderly, time-averaged picture of hemispheric air currents was a fiction. What was the reality?[78]

Defant next asked the reader to picture an "ordinary air current." It was common to feel small eddies or "restlessness" (*Unruhe*) in such everyday winds. Who would describe those small turbulent effects as anything but "insignificant disturbances"? And yet recent research by Wegener had suggested that wind patterns would be entirely different without this small-scale turbulence.[79] By the same token, studies of the general circulation had previously ignored cyclones and anticyclones, dismissing them as mere "disturbances." Of course, viewed from a human perspective, cyclones were too big—too extensive and too long-lived—to seem like random fluctuations. However, as Defant's calculations showed, the lifetime of cyclones stood in the same spatial and temporal proportion to the general circulation as did the turbulent motions characteristic of a wind's "restlessness" to the wind itself. Researchers had failed to recognize the true character of cyclones for much the same reason that they had discounted this restlessness—"because it impinges too forcefully on the course of weather in our latitudes, and because at the dimensions that man has introduced as practical, the size of the disturbed region and the duration of the perturbation are very large."[80] Scientists had attended to the human scale at the expense of the study of phenomena that either fell short of or exceeded it.

The other key ingredient in Defant's analysis was Schmidt's concept of *Austausch*. He proposed to characterize the degree of turbulence of the atmosphere at midlatitudes in the Northern Hemisphere by means of the *Austausch* coefficient, A, calculated as an annual average at a given location. Defant gave his own derivation of this quantity for the case of horizontal exchange—that is, for turbulent flow from south to north. This required estimating the relevant parameters. He took the characteristic length to be the average diameter of cyclones and anticyclones, "which are indeed the turbulent elements of our disorderly motion," and he used anemometer data to estimate the wind speed and thus the momentum flux. This led to a value of 10^7 kg m^{-1} sec^{-1} for A in the region of "central Europe." He then proceeded to a second calculation of A by means of a proportionality argument due to Felix Exner. Recall that Exner, in his study of meanders, had derived conditions under which one could expect the turbulence of two fluid systems of equal density to be geometrically similar. Specifically, a faster flow of larger dimensions would need correspondingly greater viscosity to produce similar motions. This provided a way to solve for the "virtual" viscosity, or *Austausch*, of the midlatitude circulation. Given measured values for the size of turbulent eddies in ordinary air currents of a given velocity, and known values for the diameter of cyclones, this simple scaling argument likewise led to a horizontal *Austausch* coefficient at midlatitudes on the order of 10^7 kg m^{-1} sec^{-1}. Defant noted that this was a large value. As we've seen, coefficients of vertical *Austausch*, even at the top of the Eiffel Tower, were at least five orders of magnitude smaller. It was so large that one could not deny the key role that turbulence must play in the transport of heat between equator and poles. Here was confirmation of a principle that Helmholtz had intuited long before: namely, what keeps the earth's winds from accelerating beyond the fastest gales on record is not friction with the earth's surface but the eddy viscosity generated by large-scale turbulence. Defant drew the conclusion that the general circulation outside of the tropics was "the largest of the random motions that we have known of until now."[81]

As this suggests, Defant rested his claim to knowledge of the general circulation on a very different basis than Jeffreys did. While Jeffreys began with mathematical proof, Defant opened with a set of vivid verbal images. While Jeffreys remained at the scale of the earth as a whole, Defant asked readers to consider dimensions ranging from the microscopic to the synoptic. Defant demonstrated how turbulence affected the natural world at every scale, from the motions of gas molecules to eddying winds, from the flow of water in a narrow pipe to a major river. Indeed, the relativity of scale was a methodological precept for Defant. Echoing the ethos and language of Habsburg naturalists from J. E. Purkyně, to Eduard Suess, to his own mentor, Julius Hann, Defant showed how naive humans would be to view the world exclusively at dimensions corresponding to the perceptual capacities of our species.

THE STABILITY OF THE EARTH'S CLIMATE

From there Defant took up a question that Jeffreys hadn't even considered. Namely, what does the essential role of turbulence in the general circulation imply about climate change? That is, Defant intended to use the *Austausch* framework to assess the impact of a local climatic disturbance on the state of the atmospheric circulation as a whole. Just how stable was the earth's climate?

This was a question inspired by long-standing debates over the significance of historical evidence of climatic variations (chapter 9). Despite the common impression that, in Defant's words, "the temperature and pressure conditions of the earth represent a stationary state," recent research had revealed the existence of both periodic and aperiodic climate fluctuations. These had been studied in depth by the climatologist Eduard Brückner, Defant's colleague in Vienna, whose 1890 *Climate Fluctuations since 1700* Defant cited in his conclusion. Brückner had identified a thirty-five-year cycle of cold and warm periods on the basis of historical documents, but he had not been able to explain its cause. James Croll had attributed such shifts to astronomical causes—variations in the shape of the earth's orbit around the sun and in the tilt of its axis, plus the wobble of its axis known as precession.[82] Following John Tyndall and Svante Arrhenius, on the other hand, Defant posited that at least some of this variability must be of "terrestrial origin." He supposed that that such variations could arise from a local disturbance that precipitated changes in other regions of the atmosphere. What then did the *Austausch* framework imply about the manner in which such a perturbation would propagate through the atmosphere?

Defant took a concrete example. Suppose there exists an oscillation of temperature in the tropics with an eleven-year period (corresponding to the observed cycle of solar activity). What effect would this have on the climate of the rest of the earth? Assuming a constant coefficient of exchange, Defant's calculations predicted that the disturbance would spread to higher latitudes without a significant change in period, but with rapidly decreasing amplitude and a slight phase delay. The problem with this result was that, according to observations, the strongest climate fluctuations occurred at middle and higher latitudes, not in the tropics. Defant resolved this contradiction by pointing out that the assumption of a constant coefficient of *Austausch*, A, was unwarranted. On the contrary, he suggested, large fluctuations of temperature in the middle and higher latitudes should correlate with large fluctuations in the coefficient of exchange. In this way, Defant suggested, by conceptualizing the general circulation in terms of *Austausch* and allowing the coefficient A to vary with time—by treating it as a parameter rather than a constant—it was possible to arrive at a comprehensive "theory of climate fluctuations."[83]

Defant showed that the variability of atmospheric temperature with time at a given latitude could be written as a function of two terms, one depending on the variation of A with latitude, and the other depending on the intensity of solar radiation and the radiative properties of the atmosphere (which Defant defined in terms of a "coefficient of emission," the value of which he admitted was highly uncertain). This led him to argue that climate fluctuations could be attributed most generally to three factors: 1) changes in solar activity, 2) changes in the composition of the atmosphere, and 3) changes in the degree of dynamic exchange between equatorial and polar regions, as measured by A. For instance, a rise in solar activity would increase the temperature difference between poles and equator; this would increase A, which in turn would balance out the temperature gradient. Drawing an analogy to a steam engine's governor, Defant concluded that "over the larger time intervals in which exchange [*Austausch*] occurs, the circulation of the atmosphere behaves like a regulator." And yet he went on to admit that things might not be that simple. There was geological evidence that these fluctuations had been more significant earlier in the earth's history. Moreover, it had become clear that "the absorptivity of the atmosphere is subject to modifications as a result of changes in its composition (for instance, as a result of volcanic eruptions, etc.)." Presumably, the greenhouse effect of industrial carbon dioxide was contained in Defant's "etc." If a change in the composition of the atmosphere were to raise the "coefficient of emission," Defant assumed that the result would be a stronger north-south temperature gradient; this would ramp up the turbulent exchange, which in turn would even out the distribution of heat. The original temperature distribution would return, although the atmosphere's emissivity would remain elevated, as would the coefficient of exchange. Defant assumed that atmospheric absorption would eventually return to its lower initial level. At that point, the heightened atmospheric exchange would carry more heat to the poles, which in turn would decrease the exchange between equator and poles; in short, the earth would return to its original state. Again, turbulence would act as a "regulator" to maintain the stability of the general circulation. Defant acknowledged, however, the uncertainty of two key factors: first, the intensity of the climatic fluctuation as experienced at higher latitudes, and, second, the amount of time necessary for the system to equilibrate: the first might be small and the second short, "as in the small climate fluctuations of historical times, or both large, like the climate fluctuations stretching over many millennia that geology has discovered in the Ice Age [*Diluvialzeit*]."[84]

Viewing the general circulation as a turbulent phenomenon thus had significant implications for interpreting the historical variability of the earth's climate. It convinced Defant that these variations must be due at least in part to terrestrial rather than astronomical causes. In this sense, the new framework raised the distinct possibility that a significant shift in atmospheric composition could cause climate variability of great magnitude and long duration. However, Defant made no explicit mention of the evidence put forward by Tyndall and Arrhenius to show that human activity could alter the composition of the atmosphere. Readers might well have been left with the impression that earth's turbulent atmosphere was self-stabilizing in the face of nearly any disturbing factor.

Indeed, Defant burnished his picture of the atmosphere with providential brushstrokes. It was atmospheric turbulence at midlatitudes that

> sends the necessary amount of entropy to higher latitudes, regulates the balancing out of temperatures from one season to another, facilitates the development of the temperature conditions at higher latitudes that alone allow men, animals, and plants to live and flourish, ensures the even distribution of the sun's radiation over the whole earth, and thereby prevents that in the equatorial regions everything withers from heat and at higher latitudes, by contrast, everything freezes from cold. *On a planet that is hardly lacking in marvels, it is one of the most wondrous natural phenomena of all.*[85]

With this passage, Defant placed his study firmly in the Austrian Catholic tradition of physico-theology, which lavished scientific attention on the most unassuming parts of the divine plan. His goal was not simply a "fundamental equation," but rather a picture of "the earth as a whole," as he had promised readers at the outset. Yet his conclusion was ambivalent. Although Defant invited his audience to marvel at the stability of the earth system, he had also shown them the conditions under which it could potentially suffer a serious, long-term shift in climate. Here was a framework in which it was possible, in principle, to solve for the planetary-scale consequences of human-scale disturbances.

CONCLUSION

By the first decade of the twentieth century, atmospheric science had become "dynamic." What that meant, however, was not yet clear. In a 1908 article in the *Meteorologische Zeitschrift*, the newly appointed director of the ZAMG, Wilhelm Trabert, insisted that "dynamic" had become a "buzzword" among atmospheric scientists, its meaning obscured by overuse.[86] The phrase "dynamic *meteorology*" came into use in German, French, and English in the late 1870s. While appropriate methods were up for debate, the label itself seemed unproblematic: it designated the project of explaining atmospheric motions mathematically, in terms of thermodynamics and hydrodynamics. By contrast, the phrase "dynamic *climatology*" only emerged two decades later, and its meaning could never be pinned down.

The first use of the phrase "dynamic climatology" is attributed to Tor Bergeron, one of the leaders of the Bergen school of meteorology, in a 1929 address to the German Meteorological Society.[87] At the time, meteorology was reinventing itself as a mathematical science built on physical principles, and climatologists felt left behind. Bergeron spoke of a "renewal" (*Erneurung*) of climatology. The title under which this speech was published, "Richtlinien einer dynamischen Klimatologie," suggests the normative thrust of Bergeron's history of the subfield, since *Richtlinien* might best be translated as instructions or guidelines. Bergeron initially defined dynamic climatology as the application of hydrodynamics and thermodynamics to the atmosphere. But he also prescribed a method: identify self-contained atmospheric systems, describe their *Strömungsglieder*, or components of flow, and locate the origin of these components in the general circulation of the atmosphere. This qualitative approach gained enormously in popularity during the Second World War, when processing data statistically proved too slow for "operational needs." However, this method was more often described as "synoptic" than "dynamic" and was widely criticized for the subjective character of its judgments of "typical" conditions.[88] Bergeron further narrowed the meaning of dynamic climatology by drawing his examples exclusively from frontal analysis. He made it seem as if a dynamic perspective had been introduced to climatology solely for the purpose of storm forecasting. On the other hand, Bergeron's critics were no more faithful to history. For example, Sergey Chromow in Moscow attempted to show that Heinrich Dove had invented dynamic climatology half a century before the Bergen school, "in so far as Dove's polar and equatorial currents can be seen as a precursor of the components of large-scale atmospheric flow."[89] Yet Dove had no inkling of a thermodynamic and fluid mechanical explanation of these processes. Two decades later, geographers were still hashing out the true meaning of dynamic climatology. The Working Group for Dynamic Climatology of the World Meteorological Organization took two years to agree on a definition, but they were quickly accused of having conflated dynamic and synoptic approaches. Indeed, the chair of the working group admitted that "I did, in fact, originally see the light of true dynamic climatology and recognized its essential premises; but . . . later lost the way."[90]

The definitional problem stemmed in part from the gap in status between climatology and meteorology at midcentury, particularly in the United States. There, all eyes were on numerical weather forecasting. Where this work aimed at long-term forecasts, it might well have been described as "dynamic climatology." But its developers shunned the label of climatology, which had already acquired an old-fashioned ring and an association with mindless empiricism. In the words of Kenneth Hare, who held a PhD in geography, "Treason doth never prosper. What's the reason? If it doth prosper, none dare call it treason. Substitute 'climatology' for 'treason,' and the point is made."[91] In this terminological confusion we can begin to see why it has taken so long to grasp the historical significance of dynamic climatology.

And yet this controversy distracts from a far more important shift in meaning that took place in the first two decades of the century: a transformation in the very meaning of "climate." Climate was now a concept applicable at multiple scales, from the local up to planetary. As recently as 1895, in the definitions of Wladimir Köppen, the general circulation had marked the limit where climatology edged into meteorology.[92] In showing how to apply a climatological tool like *Austausch* on the scale of the atmosphere as a whole, Schmidt, Exner, and Defant had laid claim to the general circulation as an object of climatological research. Indeed, they had given substance to the notion of the "planetary climate," the *Weltklima*. One might object that studies of the general circulation had always been climatological, in so far as, outside of the tropics, they had treated time-averaged states of the atmosphere. But only with the work of Hann and his collaborators did these studies become reflective about what was being averaged away. Just as dynamic meteorology was defined in 1885 as the study of disturbances to atmospheric equilibrium, dynamic climatology emerged in imperial Austria circa 1900 as the study of disturbances to the earth's climate system. It thus became possible for the first time to investigate the global effects of local climatic change. Climate had become a multiscalar, dynamic concept.

Central to this transformation were the representational practices of imperial-royal science. The rise of dynamic climatology in imperial Austria depended on innovative techniques of scaling, from tabletop models of the planetary atmosphere, to analogies between the hemispheric turbulence and the palpable "restlessness" of ordinary winds. The mathematical vocabulary of *Austausch* and APE was the physicist's correlate to the visual and literary techniques examined in the preceding chapters, all of which aimed to achieve a synthetic overview without effacing difference at the smallest scales. In the work of Hann's group, the goal of climatology remained, as Hann had defined it, to produce "a maximally life-like picture" of atmospheric phenomena. Hence the wonderfully evocative analogies, poetic descriptions, and appeals to intuition that we find in the writings of Schmidt, Exner, and Defant. Words and images were more than illustrations of equations. They led into a realm of complexity to which analysis was inadequate, but which might yet be grasped via physical intuition.

3

The Work of Scaling

CHAPTER 9

The Forest-Climate Question

When Sigmund Freud sought a metaphor to describe the psychic effects of the civilizing process, he likened it to the draining of a swamp. The id, the seat of the primitive drives, was like a wild landscape, as yet untamed by human hands. The displacement of id by ego, the adult self, was akin to reclaiming wasteland for agriculture. Freud asked, "But have we a right to assume the survival of something that was originally there, alongside of what was later derived from it? Undoubtedly."[1] In the realm of the psyche, he argued, the id makes its presence known through experiences of the uncanny, encounters that can be interpreted as reminders of the seething impulses of the id. The ecological analogy is suggestive. In the realm of the psyche as in the nonhuman landscape, the civilizing process can have unintended consequences. In the case of the swamp, the forces that previously ruled this landscape have merely been pacified, not eliminated. The unpredictable floods that plague a land reclaimed from swamp are uncanny in their own way. Before the First World War, one of the more prominent discourses on civilization's discontents in Austria concerned the devastating environmental effects of agricultural cultivation. And one of the features of that discourse sure to strike a man of Freud's interests was the contested status of memory. How accurately might inhabitants remember the land that was "originally there"?

In a widely circulated 1878 pamphlet by Freud's Vienna colleague Friedrich Simony, Freud could have read the following testimony from a northern German forester:

> What you report about the water sources struck a chord with me [*ist mir aus der Seele gesprochen*]. In my capacity as the first forest official of the local province, formerly of the kingdom of Hannover, I have spent twenty five years hiking through what was a once densely forested valley, now sparsely wooded . . . and on this basis I can say how extraordinarily the water conditions have changed. . . . The springs and groundwater were once [*einst*] sufficient for alders and beeches to grow.[2]

Evidence of climate change rested on memories. These were memories that everyone living in the region could be expected to share, memories of a past that could be gestured at rather than measured, a past evoked by the simple term *einst* or "once."

And yet, even at the time, certain commentators noted the inconsistency of memory when it came to questions of climate. Eduard Brückner, later a professor at Vienna, famously observed that in dry periods, people claimed that deforestation was responsible for desiccation, but in times of floods it was often claimed that deforestation increased rainfall.[3] As the 1905 *Handbook of Forest Politics* observed, until the middle of the nineteenth century, evidence of anthropogenic (human-induced) climate change rested on "historical comparison," namely, on local observations linking human interventions to natural disasters or to a general degradation of agricultural conditions. "This train of thought, which continues to lead people astray today, has something inherently seductive, because it is tied to facts that are displayed to everyone's eyes and therefore exert the full force of proof on public opinion. In contrast, the public faces the tremendously difficult task of dividing causes and effects, and excluding the exaggerated conclusions and errors deriving from subjective impressions." As the author went on to elaborate, "public opinion" (*die öffentliche Meinung*) was prone to certain errors when it came to climate change: a tendency to downplay the intensity of past phenomena in comparison to present ones and a willingness to ascribe any and all "adverse natural phenomena" to the accelerating pace of forest destruction—even in an era when forests were successfully being preserved.[4] Another expert dismissed as "legendary" reports that deforestation had desiccated the karst lands of southeastern Europe: "All the reports that are widespread among the public today—that today for the first time masses of rubble from the Isonzo are being carried into the plain, that now in Dalmatia olive trees no longer flourish as in the time of ancient Rome, that the soils have all been swept away, that it used to rain more often, that there is no morning dew because of the scarcity of forest-moisture, and that springs have run dry because of deforestation—are all legends [*Sagen*] that people tell each other without investigating what there is to them."[5] Or, as the *Neue Freie Presse* put it, "The masses are indeed so easily swayed, they are content with belief in the place of knowledge; the former is much more convenient."[6] Late nineteenth-century Austrian debates over climate change were thus caught between an eighteenth-century ideal of rational debate and a fin de siècle anxiety about the suggestibility of the masses. The *Handbook of Forest Politics*, for instance, by no means dismissed the relevance of subjective experience. On the contrary, it noted that the evidence for climate change produced by scientific experts "furnished nothing surprising, in that it corresponds perfectly to human sensation and conjecture."[7] One wonders: what room was there for evidence that failed to corroborate public opinion?

This was the question at stake in the 1870s when a few bold scientists set out to counter a widespread belief about the climatic impact of deforestation. The late Habsburg Monarchy witnessed a series of clashes over the reality of anthropogenic climate change, debates that will sound familiar, at first glance, to twenty-first-century readers. Such debates, then as now, are necessarily negotiations between the spatial and temporal frameworks of scientific evidence and those of collective memory. They are, in the terms of this book, exercises in scaling.

Scaling, as we have seen, is a bodily experience, tied to movement through space and time. Yet, as Husserl observed, it is also a social process. The "near" is defined in part by what is familiar, by one's own community, while the "far" corresponds to what is foreign. What we learn of the "far" often comes more from other people than from firsthand experience.[8] When we consider scaling across dimensions of time, this social aspect is especially conspicuous. Our access to the past, to what is distant in time rather than space, typically comes from memories other than our own. Scaling thus becomes a process of commensuration between dimensions marked out by scientists' tools, on one hand, and by communal identities and collective memory, on the other.

A HURLY-BURLY OVER CLIMATE CHANGE

Historians looking for the roots of our present ideas about anthropogenic climate change have dwelled on the long life of what they have called the "desiccation theory," the idea that clearing an area of its vegetation could result in diminished rainfall. This was typically a theory of local rather than global change, although deforestation was the leading cause of increasing levels of atmospheric carbon dioxide until the twentieth century. Richard Grove has traced the origin of this theory back to sixteenth-century debates over the consequences of European imperialism in Latin America. As he and others have shown, efforts to demonstrate the beneficial climatic influence of forests were linked to critiques of the impacts of European colonization on indigenous environments and societies. From its roots in encounters between Europeans and indigenous peoples from the islands of the South Pacific to the plains of North America, by the eighteenth century the desiccation theory had convinced many people of the climatic benefits of forests.[9] This theory has drawn the interest of environmental historians as a sign of a lost ecological consciousness; as a form of ecological anticolonialism in Latin America; or, on the contrary, as a means of legitimating European imperialism in Asia and Africa.[10] However, with a few exceptions, much of this effort to get at the meaning of the desiccation theory neglects the evidence that this was less a theory than a question.

In fact, in much of central and eastern Europe in the nineteenth century and into the twentieth, the so-called desiccation theory went by the name of the *Wald-Klima-Frage*, the forest-climate question, and equivalents in other languages. As one forester remarked in 1901, "Seldom before and likely in the future has a question been debated and addressed from so many sides and so relentlessly as the forest-climate-question."[11] The putative climatic benefits of forested land were vigorously debated throughout this region, as David Moon has chronicled for Russia.[12] Why was the question so charged in this part of the world? Conditions on the ground are part of the explanation. Compared to western Europe, much of central and eastern Europe is prone to dryness and extremes of temperature. In addition, the Austrian and Russian empires were both more heavily forested in their centers than on their peripheries; hence the strong impression made on imperial officials by the absence of forest in the grasslands of southern Russia and central Asia, as in the karst lands of the Balkans and the Hungarian steppe.

In the Austrian lands, the debate over anthropogenic climate change played out in the imperial parliament in Vienna, in provincial assemblies, and in the press, particularly in the growing ranks of forestry and agricultural newspapers. These were forums not only for experts but also for "countless simple country folk, who do not tire of bringing their own perspectives and experiences to public awareness."[13] As Vienna's liberal daily, the *Neue Freie Presse*, observed in 1879, "No question has been debated as frequently, as widely, and at times as fiercely as that of the influence of an area's forest-cover on its climate and water sources."[14] It was a *Wirrwarr*, a hurly-burly, in the words of Eduard Brückner, one of the scientists working for the Austrian state whose work productively engaged it.[15]

The formulation "forest-climate question" invited comparison to other hot topics of the day, such as "the woman question" or "the Jewish question." As the historian Holly Case has recently observed, this locution was a strategy for elevating what might have been a local social issue to a matter of debate for the new international public sphere.[16] It is a sign that we as historians must attend to the politics of scale. At stake was the authority of local, imperial, or international governance. As ordinary citizens grappled with this question, they scaled between the vernacular terms in which they framed their natural and social environment and the framework presented by dynamic climatology. As a result, a remarkable shift becomes apparent in popular discourse. For the first time, nonscientists began to refer to "climate" not as a characteristic of a local environment but as a dynamic system of planetary dimensions.

THE POLITICS OF FORESTS

In the Habsburg Monarchy, as in other nineteenth-century European states, forest policy weighed the interventionist tradition of the eighteenth-century "police state" against the ascendant principle of private property. Rights to the forest had been a particularly tense issue in the Austrian lands since the emancipation of forced labor in 1848. This act deprived peasants of the use of their lords' forests and pastures. The result was an increase in what landowners perceived as "wood theft," even as peasants believed they were simply taking what was rightfully theirs. The first half of the nineteenth century also saw the founding of the empire's first forestry academies, beginning with Mariabrünn in Lower Austria in 1805. In the absolutist spirit of the eighteenth century, these institutions trained foresters to be loyal state servants, not researchers. After 1848, the rising class of professional foresters accused peasants of illegally allowing their animals to graze in the forests and wantonly felling trees. As evidence, they pointed to a growing timber shortage, though historians have noted the lack of evidence to support that claim.[17] Foresters also insisted that woodlands offered a range of public benefits: they increased rainfall, moderated temperatures, shielded the surroundings from winds, and protected against floods, avalanches, rockfalls, and erosion. Invoking the medieval tradition of the *Bannwald*, or protected forest, nineteenth-century foresters called on the state for legal protection.[18]

The first empire-wide law for the protection of forests was signed in 1852, giving the state the right to intervene in privately owned forests in the interest of the "public good." The intentions behind this law remain open to debate. In part, major floods in 1848 and 1851 may have inspired the move to protect forests as flood breaks. Historians have noted, though, that the law aimed to maintain productive forests for the sake of industry, most of which was still state owned; in this sense, it was oriented more toward the *use* than the protection of forests.[19] Although the law remained in effect throughout the life of the Monarchy, it was rarely enforced.

By the 1870s there was agitation from associations of foresters for a stronger measure of state intervention, inspired partly by the example of France. The Second Empire, under Napoleon III, enacted a law for the "reforestation of the mountains" in 1860, in reaction to major floods in southern France. Prominent scientists like Antoine Becquerel lent their authority to this policy by attesting to the climatic benefits of forests.[20] Austrian officials later visited France to study the implementation of these measures.[21] Thereafter, the forest-climate question was repeatedly discussed in the pages of the *Journal of the Austrian Society for Meteorology*. However, publications on the topic for a general audience were still rare.[22]

Then, in 1872, a devastating flood in western Bohemia prompted speculation that deforestation had left the region vulnerable. Bohemian foresters argued that depletion of the forests was to blame for abrupt swings between drought and flooding. Reporting on the official investigation of the flood damage, the botanist Emanuel Purkyně tried to counter speculation that forest destruction had brought a decrease in rainfall.[23] Subsequently, the Bohemian Diet considered but failed to pass a law requiring afforestation.

Nonetheless, the forest-climate question continued to simmer and became a frequent subject of debate in the popular press. The topic was raised at the first International Congress of Agriculturalists and Foresters, which met in Vienna in 1873 in conjunction with the World Exposition. The discussion barely skimmed the surface, but the congress did issue a call for an international observing system to study the climatic influence of forests. Meanwhile, in the exhibit halls, the public could scrutinize displays of forest-climatological measurements made in Prussia, Bavaria, and Bohemia.[24] In the face of rising concern, the imperial government began to consider a change of course. A reorganization of the empire-wide forestry administration was announced by imperial decree in 1873, giving oversight of state forests to the ministry of agriculture; five years later, a new imperial forest law was drafted but never enacted.[25]

A new wave of concern arose when severe floods struck the Alpine lands in 1881. Approximately fifty-one people lost their lives, fewer than in Bohemia in 1872. Nonetheless, as one historian of forestry has argued, this tragedy prompted a qualitatively new response from Vienna. For the first time, the imperial state acted as a humanitarian force.[26] In doing so, the state was able to take advantage of the railway and telegraph networks, as well as of its newly integrated network of professional scientists. In the wake of the catastrophe, the state commissioned scientific experts in Tyrol and Carinthia to investigate the flooded areas and determine the most significant causal factor, whether abnormal rainfall or excessive deforestation.

Further support for forest protection came, surprisingly, from large landowners. In the 1848 era, owners of large wooded estates had resisted state intervention into their practices of forest management. But socioeconomic conditions had changed considerably since then, and it was no longer clear what kind of policy best served the interests of this elite. Many landowners came to see a stronger forest law as working in their favor. They likely hoped to free themselves from peasant demands for traditional rights of forest use, to profit from higher prices for wood, and to exploit the forests for their own manufacturing enterprises.[27] At a dramatic meeting of the Association of Austrian Foresters in 1883, Prince Karl von Schwarzenberg appealed to members of the landed aristocracy to support a new forest law, stronger than that of 1852. He played to their fears of wood speculators and of a swelling proletariat. His argument was that stricter state protection would harm only small landowners. Once middling farmers were forced to sell their land, the proprietors of large estates would be free to buy them at low cost. Even better, the nobility would thereby prevent this land from falling into the hands of capitalists. They could simultaneously rescue the forests from clearance and save peasants from proletarianization. Using implicitly anti-Semitic language, Schwarzenberg argued that large estate owners were "in a position to take up the battle against big capitalists" and were thus "compelled to acquire small parcels of land in order to prevent the mass liquidation of farms [*Güterschächterei*, an allusion to Jewish ritual slaughter]."[28] Schwarzenberg helped to seal an alliance between the landed aristocracy and professional foresters against small landowners.

So it was that, in the last quarter of the nineteenth century, the Reichsrat confronted a series of proposals to strengthen forest protection. In the course of these debates, elected representatives repeatedly testified to the beneficial influence of forests on the surrounding climate. The Galician aristocrat Stanisław Mieroszowski attested that the forest served as a climate "regulator," since it "tempers extremes of cold in the winter and moderates extremes of heat in the summer; attracts atmospheric precipitation to itself in great quantities; collects humidity in its underground reservoirs; and returns it gradually, especially in summertime, to the streams and rivers."[29] Sometimes the "fact" of the forest's climatic influence was cited instead as a critique of the aristocracy and their land-use practices. Thus a representative who identified himself as a Galician farmer accused the large landowners of a "pillage economy" (*Devastationswirtschaft*), the effects of which were not only a timber shortage. "Entire woodlands, which once adorned our land and protected it from harsh winds, fell victim to this barbaric economy. (Agreement.) I don't need to mention, that this has the most unfortunate consequences for the climate."[30] It would seem that the question was closed: forests were to be protected for the sake of a healthy and fertile climate, in the name of the public good.

THE LEGAL QUANDARY

However, the confident tone of these statements belies the complexity of the matter at hand. As a legislative issue, what mattered was not simply whether forests modified the climate of their surroundings, but how wide an area was affected. The spatial extent of the effect was decisive, because imperial forest legislation contained a pregnant ambiguity. The Forest Law of 1852 allowed the state to intervene in privately owned forests in cases where it was necessary in order to protect "the public good." But how was one to know, in any given case, whether "the public good" was at stake? How were the public and its interests to be defined? Likewise, the law of March 1883 requiring flood protection in Tyrol divided responsibility among the imperial government, the crown land, and the *Interessenten*, or interested parties. But who were the parties affected by the alleged climatic impact of deforestation? The problem, as the foremost historian of Austrian forest policy observes, was that "the concept of the interested parties was not clearly delineated."[31]

Gradually, legal scholars and policy makers recognized the difficulty of defining the public affected by the clearing of any given forest stand. As the author of an 1898 history of Austrian agricultural policy put it, "we are indeed missing any more precise specification of what might count as a 'public interest,'" and he wondered how the climatic effect of deforestation was to be weighed.[32] In 1884 the ministry of agriculture, acknowledging this ambiguity, called for expert investigation "with the utmost rigor" into the effects of forest clearing on "agricultural conditions in general." Forestry experts were called on to determine the "repercussions" to be expected from forest clearing "from a climatic and atmospheric perspective for the agricultural conditions of the afflicted province or district."[33]

The gradual recognition of this ambiguity is evidence that a shift was occurring in environmental thought. Consider the language of the debate between Austrian liberals and foresters. In the view of many liberals, the forest was indeed a "public good," but the implication was not that it therefore required protection.[34] On the contrary, this meant that everyone should be free to use the forest as they saw fit, without restriction, regulated only by the market. The forest, liberals said, should be as freely accessible *as air*. Foresters objected to this comparison on the grounds that deforestation spoiled the local climate. They argued that forests must not be treated like air, subject to unlimited use by all, because forests were a limited and vulnerable resource, while "the air as a public good cannot be misused, cannot be polluted at will."[35] Indeed, David Ricardo had argued that no price can be set on air, water, or the pressure of the atmosphere, because these resources are boundless; they do not require a choice to be made between use and preservation.[36] This would seem to be a simple and self-evident distinction: forests were a limited resource; air was not. The irony is that the foresters of the 1870s were arguing precisely that deforestation threatened *the air*. Nonetheless, they persisted in contrasting the bounded forest with the boundless air. Arguably, they did so because they could not quite imagine how something as uncontained as the atmosphere could be vulnerable to local human actions.

What was needed was a framework for evaluating the *scale* on which the atmosphere was threatened by deforestation. Think about the range of dimensions in play: from the microscale of plant transpiration, to the mesoscale winds passing through or over a forest canopy, to the macroscale patterns of the global atmospheric circulation. In principle, the dimensions of the climatic effect would determine which level of governance within the Austrian half of the Monarchy was responsible for forest protection—whether municipalities, the diets of the crown lands, or the imperial administration in Vienna—or whether this was rightly a private matter. The question was, how far beyond the borders of the forest (and how far into the future) is a forest capable of modifying the surrounding atmosphere? This legal question propelled a research program for forest climatology.

THE RISE OF A RESEARCH PROGRAM

Until the 1870s, the small body of scientific literature on the climatic effects of forests had been based almost entirely on anecdote and speculation. The literature generally treated "the forest" in general, failing to distinguish between different types of forest and different geographic conditions. The most common references dated back to the 1820s and 1830s. Most prominent was the French military officer and hobby naturalist Alexandre Moreau de Jonnès, who contended, on the basis of comparative geography, that a healthy climate required neither too little nor too much forested land.[37] In Austria in 1831,

Gottlieb von Zötl, an assistant professor at the forestry academy in Mariabrünn, had made the case that forests lower the surrounding temperature by blocking winds and absorbing sunlight.[38] Zötl's claims were repeated by others, but no concerted effort was made to verify them empirically. By the same token, textbooks on "agricultural meteorology" for forestry students tended to be overviews of meteorology as a whole, diluted for easy consumption, rather than targeted discussions of phenomena relevant to agriculture.[39] An 1853 article in a Vienna agricultural newspaper lamented this oversight: "Why is it that we are so often mistaken in the evaluation of a local climate? Simply because for so long farmers and foresters did not write down enough of the observations that they made of it."[40] In 1869, when Julius Hann addressed the forest-climate question in the third volume of the *Journal of the Austrian Meteorological Society*, he admitted that a farmer could still "embarrass" a climatologist by asking whether forest clearing would threaten his harvests by altering rainfall patterns. At present, the climatologist was "unable to confirm these theoretically well-founded fears with any precision."[41]

The publication in 1873 of preliminary results from Ernst Ebermayer's experimental forest in Bavaria heralded a new era.[42] Three years later, at the 1876 International Statistical Congress in Budapest, scientists considered a proposal for the international coordination of research into the forest-climate question. The proposal was submitted by Josef Roman Lorenz von Liburnau, who had been appointed three years earlier as an advisor to the newly expanded agricultural ministry in Vienna. In the course of his career, during postings in Salzburg, Fiume, and Vienna, Lorenz had published extensively on a range of natural environments, from the moors of Upper Austria, to the coast of the Adriatic, to the Danube river valley. He had won the authority to speak as an "imperial-royal scientist," an expert on the territory of the Monarchy most broadly. For the Congress's section on agricultural meteorology, Lorenz drew up an ambitious research program to collect meteorological data relevant to agriculture. But who would take responsibility for producing this data? Lorenz hoped to convince his international audience that the burden should fall to meteorological institutes.[43] He hoped to learn from his international colleagues whether state institutes of meteorology were prepared to take up research on agricultural meteorology and climatology, or whether these would fall "of necessity, in the last instance" to agricultural ministries.[44]

Privately, Lorenz reported that the discussion of his proposal in Budapest had been no more than preliminary. "The related discussions were, as is almost always the case at this conference, quite rushed and perfunctory; one is always trying to be done with it at the designated time, because again some festivity is about to begin. The value of such conferences evidently lies not in what one accomplishes there, but in the preparations that such conferences inspire and which always have a lasting value or even lead to significant proposals."[45] Still, the congress issued a brief resolution calling on "the governments of all states" to establish observing stations for agricultural meteorology that would be responsible for forecasting, phenological observations, and studies of the climatic influence of deforestation and reforestation.[46]

Unlike the statisticians, attendees at the Association of German Foresters in Eisenach in 1876 could not be faulted for a lack of interest in the forest-climate question. Lorenz came to this congress prepared to redirect the typically "wild" discussion of the question around eight scientific propositions. On this international stage, Lorenz introduced a strategic shift in scale. "In order to understand correctly the circulation of water, it is necessary to realize that the earth and its atmosphere share a store of water of a fixed size—the circulation of which, however, brings a very unequal distribution to different places at different times." He then pointed out that existing research had demonstrated the forest's climatic influence *within* its borders, but had left open the question of "the influence of the forest on the climate of the *more proximate and more distant* surroundings."[47] The key question was the spatial and temporal scale of the phenomenon, depending on the type of forest and its geographic location. Lorenz underlined that any climatic influence of forests on their surroundings would have to be *dynamic* in nature. It wouldn't do to assume, as foresters previously had, that the forest's effect was as simple as "living quarters are colder over an ice cellar and warmer over an oven; *sapienti sat*."[48] Rather, the forest's effect could only be explained dynamically, in terms of the deflection and modification of air currents as they passed through or over a forest. Lorenz was arguing that the question had to be analyzed not only in its regional dimensions, but also from the perspective of the *planetary* water budget—any regional effect would have to be consistent with this global model. He was introducing foresters to a new way of thinking about climate: the perspective of dynamic climatology as it was then being worked out at the ZAMG.

To Lorenz's frustration, however, these were not the terms in which foresters wished to discuss the matter. As a German newspaper described, "The discussion did not, however, move in the direction of exact research, but rather mainly supplied countless and repetitive examples of the kind that are hardly suited to decide the question."[49] With time running short, the assembly reached the conclusion that too few of the assembled foresters could even make sense of Lorenz's points. All Lorenz achieved was a resolution calling for further discussion at a later point. Defeated, he wrote a colleague that perhaps the foresters would at least concede that their conviction "still has too little exact evidence."[50] It might be tempting to call Lorenz—in today's terms—a climate skeptic and a merchant of doubt, if it weren't for the fact that he was right: the existing evidence was entirely inconclusive, and no scientific consensus had been reached.

According to Josef Wessely, the director of the forestry academy at Mariabrünn until its transfer to Vienna in 1875, "the entire congress was hogwash." Wessely was already sixty years old at the time. Like Lorenz, his career had been highly mobile, taking him from South Tyrol to Carniola to Moravia and even the Banat—and affording the opportunity to gain expertise on the forests of each region. With the state's liberal turn in the 1860s, Wessely had begun to advocate expanding Austria's export of timber. To that end, he sought to inform the public about the nature and distribution of "Austria's treasure trove of wood" in each and every crown land. Based on his observations and calculations, Wessely claimed that Austria possessed an "excessive uninvested capital in the forests," which "could be harvested at any time without detriment to its long-term sustainability." He therefore viewed the outcry over the climatic impact of deforestation as a threat to Austria's forest industry, and he had no patience for the unscientific manner in which the matter tended to be discussed by foresters. As he reported privately on the 1877 forestry congress, the foresters weren't interested in "ascertaining the truth," but rather in "filching a cloak of rational judgment for their homegrown prejudices."[51] The ensuing discussions tended to be "total hogwash." Wessely privately hurled insults at his opponents: one was the "giant swindler," another "the knight with the sad face," and another "the mercenary"; of one he wrote that he couldn't rightly say if he was a "Jesuit, pedant, or ass-kisser." It was also obvious to Wessely that no one had bothered to read his proposal, since they hadn't taken the brochure with them and couldn't have read it during the debate. Wessely was left feeling alienated from what he perceived as an anti-intellectual culture: "how long have I experienced from so many sides—entirely undeservedly, in my opinion—hate, slander, anger, and persecution. So that I often want to slap myself for being a miserable Austrian, or at least for not having used my power to emigrate to a land where those who dare to deviate just a few degrees from the accepted line are less mistreated than in our Heimat, on the so-called blue Danube."[52]

GOING PUBLIC

Nonetheless, Lorenz persevered. His 1878 book *Forest, Climate, and Water* brought his dynamic perspective on the forest-climate question to the German-speaking public. An American author called this book "the best popular discussion of forest influences by the most prominent scientific investigator of the subject."[53] It appeared as a volume of the series the Forces of Nature: A Natural-Scientific Popular Library, which was priced low and advertised in venues including women's magazines. In this widely accessible book, Lorenz taught readers to differentiate environmental factors according to the spatial scale on which they might affect the climate.

He used the term "modifying factors" (*Modificatoren*) to designate conditions that contributed "a very significant alteration to the pure solar climate." These modifying factors "have a very varied range. Some are capable, if not of completely changing the climate of a large part of the world, then at least of very clearly influencing it. The effect of others reaches only a small parcel of land or even only very narrowly delimited points." He deemed it "sufficient for the correct interpretation of climatic phenomena" to differentiate three scales of modifying factors. First, the "global" factors, such as the shapes of the continents and oceans and the major ocean currents. Second-order factors included orography, exposition, waterways, and vegetation. Crucially, however, Lorenz observed that these same factors made up the category of third-order and smaller factors. It all depended "precisely on the standpoint of the person judging. For a small village, an individual parcel of forested land, even an individual garden or farmed field has a particular shape, exposition, water supply, and distribution of vegetation, as for an entire country."[54] In other words, the scale of observation was decisive.

In the final chapter of *Forest, Climate, and Water*, Lorenz addressed the implications of dynamic climatology for legislation. The goal, as he saw it, was to avoid two possible extremes: either completely prohibiting timbering or leaving it to the mercy of the free market. It was misguided to divide the world into "friends" or "enemies" of the forest, *Waldheuler* or *Waldschinder*. Instead, legislators should follow the principle of intervening only in cases where the public good needed to be defended: "It is thus a matter above all of recognizing in which respects and in which cases the forest influences the public good or the interests of specific neighbors." He admitted that this "influence" could never be determined with certainty. Yet one could draw a distinction between cases where timbering was economically necessary and others where it was avoidable; further distinctions could be made by estimating the scope of the climatic effect, since "the sphere of influence of the forest has very variable dimensions." In his view, preventing the overall expansion of farmland and pasturage in central Europe was unfeasible, given current rates of population growth. "For the preservation of mankind—for whose sake, of course, the whole issue has been raised and the legislation set in motion—an expansion of agriculture and pasturage is indispensable."[55] This was the crux of Lorenz's position: if the law was to serve the public good, the benefits of forest protection would have to be weighed against the public's need for economic development.

Still, adjudicating any given case would require judging the dimensions of the forest's climatic influence. Lorenz therefore led his reader through several examples of "effects of different dimensions," in order to consider "what might be concluded legally and in relation to enforcement." If wind patterns indicated that the impact of deforestation would cross national borders, then effective laws could only be created through international bodies. The movement for bird protection furnished an example of success in this respect. In contrast, in the case of the clearing of a small parcel of forested land, the matter was generally a private one. The neighbors had been fortunate to live in the proximity of a forest, but they did not have a right to continue to enjoy its benefits. The situation was different, however, if deforestation would render the neighbors' land infertile. Then it became necessary to consider the interest of future generations and their right to be able to feed themselves. Thus the legal decision to permit forest clearing or not would need to be made case by case, taking into account the scale of the expected climatic effect. If clearing were not permitted, the forest owner was to be paid a subvention by the administrative body corresponding to the "interested parties," whether municipality, district, or crown land.[56]

Reviewers clearly grasped the point. Writers for the liberal *Neue Freie Presse* and the *Landwirtschaftliche Zeitung* appreciated the correction Lorenz provided to "general" opinion. Both reviewers lauded Lorenz's book as a model of writing for a "popular" or "lay" audience, which taught readers how to use a scientific framework to reach their own practical conclusions. Both further praised Lorenz for having countered prevalent exaggerations with a realistic assessment of the scale of the problem. "It is gratifying to read . . . especially that the popular exaggerations of the consequences of deforestation have been *reduced to their proper size* [*auf ihr richtiges Maß*], without in the least denying the significance of the demonstrable climatic influence of forests."[57] Likewise, the *Landwirtschaftliche Zeitung* acknowledged Lorenz's authority in evaluating the relative significance of the many physical factors in play: "Anyone who sets out to write about the relationship between forest, climate, and water in an original way, which properly illuminates the matter, *must have surveilled so many forces of nature in their mutual influence, and acquired such a facility for judging natural phenomena*, as until now few minds have been capable of doing."[58] As these lines demonstrate, Lorenz had won a new kind of authority, one associated with the work of scaling between phenomena of disparate dimensions.

Lorenz's accomplishment in this public debate was to expand the scale of the concept of climate. Mid-nineteenth-century theories of anthropogenic climate change had hinged on a regional definition of climate. Climate was assumed to be a property of a region, a unit that was more than local yet less than hemispheric. It was this interpretation of climate that made plausible the claim that clear-cutting should be addressed at the imperial level. Once this claim was subjected to legal analysis in the 1870s, the precise extent of the suspected climatic effect began to matter. The competing levels of authority within the multinational state drove empirical inquiry into the dimensions of atmospheric phenomena. Likewise, the resources to investigate this question depended on the structure of the multinational state. The investigation was shaped, first, by the state's capacity to coordinate research across widely divergent ecosystems, and, second, by its corps of imperial-royal scientists trained to think synthetically and comparatively across those locales. Throughout his career, an imperial-royal scientist like Lorenz von Liburnau honed his ability to analyze a system at multiple scales: to identify, for any given dimensions of observation, which forces must be calculated exactly and which could be neglected or approximated. This is what atmospheric physicists today call scale analysis, and it remains an essential part of their work.

A CACTUS IN THE BOHEMIAN FOREST

So far we have followed the story of the forest-climate question primarily as it can be pieced together from the public record: parliamentary debates, newspapers, conference proceedings, scientific journals. Now we come to a remarkable stash of letters in an archive in Prague. They might be compared to the email exchanges among climate scientists that were exposed in the "Climategate" scandal of 2009. That is to say, this correspondence takes us behind the scenes of the public debate, revealing the strategies adopted by scientists under public attack. These documents shed light on another aspect of the Habsburg politics of scale: the making of the "imperial-royal scientist."

The letters belong to an individual who left barely a trace in the public record. Only the most careful reader of the transcript of the parliamentary debates would recognize his name. Emanuel Purkyně (see figure 32) was the eldest son of Jan Evangelista Purkyně, a physiologist known as one of the founding intellects of the Czech "cultural awakening," the early nineteenth-

FIGURE 32. Emanuel Purkyně (1831–82).

century movement promoting popular Enlightenment, the refinement of the Czech language, and scholarship on the history, culture, and welfare of the Czech nation. His mother was the daughter of a professor of physiology and anatomy at Berlin. He had grown up in Breslau, where his father taught physiology, and did not live in Bohemia until the age of nineteen, when the elder Purkyně was offered a post at the University of Prague. In Breslau, Emanuel Purkyně was exposed to a cosmopolitan, humanistic, Pan-Slavic patriotism, a program his father would continue to advance throughout his career.[59] Emanuel trained under his father as a plant physiologist, later joined by Julius Sachs and Ladislav Čelakovský, the orphaned sons of two of his father's scientific colleagues. Upon arriving in Prague, Emanuel was found to be so deficient in the humanistic disciplines that he was forced to complete two years of gymnasium before enrolling at the university.[60] In 1855, he became a professor of natural science at the forestry academy in Weißwasser/Bělá pod Bezdězem, Bohemia, an institution founded in that year on the initiative of large landowners.[61] This was one of only four forestry schools in the Habsburg lands at the time; the others could be found in Lower Austria, Moravia, and central Slovakia. Fortunate as he may have been to land an academic position, Purkyně stuck out at the forestry academy like a cactus in a pine forest.

Purkyně's interests were those of a plant physiologist and geographer, not a professional forester. His true passion was the microscopic investigation of plant anatomy, but he was repeatedly called on to investigate other topics of greater interest to the Bohemian public. In this respect, he was part of the efflorescence of plant geography in the Czech lands in the 1850s. His research was published primarily in Czech, often in his father's journal *Živa*. His early work included, for instance, a study of the dependence of vegetation in peat bogs on a complex set of environmental conditions, and the conditions favorable to the spread of a certain species of tree.[62] He also compiled a catalog of the flora of Bohemia in the commission of the Bohemian National Museum, and for over a decade he investigated the causes of regional differences in agricultural yield for the Czech National Museum in Prague.[63] It was in the course of this research that he began to doubt the reliability of existing measurements of climate in forested areas.

To explore this question, Purkyně carried out what was likely the world's first systematic microclimatological research, in the garden of the Bohemian National Museum and on the city outskirts, starting in 1857. He used thermometers installed on varying terrain, both shaded and exposed, at different heights above the ground, and at various depths in the soil. In this way, Purkyně pioneered the study of climate in the lowest reaches of the atmosphere, the climate near the ground.

As Purkyně put it, "I saw that on a limited area numerous climates could be found next to each other."[64] As a botanist and physiologist, Purkyně recognized the significance of these measurements: here were the climatic conditions that made life possible. He effectively founded the subfield of microclimatology, with its vital applications to agriculture and human health, decades before researchers in Germany and Russia: before the forest-meteorological observations begun by Ebermayer in Bavaria in 1866, before the agricultural climatology of Voeikov and Dokuchaev on the Russian steppes in the 1880s, and well before the better-known work of Rudolf Geiger at the turn of the century.[65]

It is tempting to say that Purkyně was following his father's injunction, in the first issue of *Živa*, to *relativize* the human scale. Human needs could not be the only measure of significance in "infinite nature," his father had written. For the younger Purkyně, this principle entailed a critique of observations carried out unthinkingly at spatial dimensions convenient to human activity. Measurements made at eye level, anywhere it was easy to place a thermometer, in areas where scientists happened to live—these were not evidence that could be used to gauge the long-term stability or variability of climate. Climate displayed different degrees of variability at different scales.

Purkyně noticed, in particular, that measurements of temperature and rainfall are affected dynamically by their *exposure* to different air currents. Exposure depends on the natural and built environment and height above the ground. Greater exposure to winds—for instance, on a high peak or tower—could significantly reduce the amount of precipitation measured. Purkyně pointed, for example, to the dramatic difference between a series of measurements of rainfall taken from an ombrometer in the courtyard of the University of Breslau and those made from the height of the university's tower. On this basis, he charged that it was invalid to base a climatic comparison between forest and city on rainfall measurements from the Prague observatory, where windy conditions artificially reduced the amount of precipitation recorded.[66] On this basis, Purkyně argued that the data being collected to settle the forest-climate question was virtually useless. His microclimatological perspective implicated all the observations that had thus far been cited as proof that forests increased rainfall.

More generally, Purkyně came to see that the spatial and temporal scales of climatic phenomena are interdependent. A "local climate" is typically identifiable only in the short periods when winds are calm and little mixing of air occurs; long-term average measurements partially conceal its existence. Thus measurements of climate across time can be thrown off by variations in climate across space. In order to establish directed climatic change, Purkyně realized that it would be necessary to build a very dense network of observing stations, evenly distributed in space, with instruments at a consistent height above the ground. All existing networks were severely skewed to the distribution of human populations. The abiding lesson was that researchers needed to correct for the variability of climate on the small scale, rather than comparing observations made at different heights and under different conditions of exposure to sun and wind.[67]

At the same time, Purkyně also demanded a larger scale of analysis for the forest-climate question, one that took into account the planetary and continental determinants of regional and local climates. Particularly effective in communicating this spatial argument were his large-format maps of global forest cover. These were meant to demonstrate the dependence of rainfall above all on wind patterns arising from the distribution of mountains and bodies of water, rather than on plant cover.

Indeed, Purkyně's visuals proved far more compelling to his contemporaries than his dense prose. As he put it privately, "I was often treated hostilely because I expressed an opinion that was diametrically opposed to that of the elites. Only striking, mostly visual cartographic and graphical representations finally helped to win something of a victory."[68] Although these graphics proved impossible to reproduce for publication, his colleagues took care to describe them. A report of a meeting of the Bohemian Forestry Society in 1879 noted that Purkyně displayed

> a large wall map of the entire earth, on which rainfall zones and amounts of rainfall were indicated by means of hatching; and forests, steppes, and deserts by means of colors. He used it to show how, for northern continents, the amount of rainfall diminishes from the ocean to the center [of the continent], while the forest cover increases from the coast to the interior. . . . [He] pointed further to the rain-free zones in the regions of the trade winds [in the tropics] . . . and demonstrated how little the rains fall in the regions where humidity is produced.

Next he showed a map of North America, which illustrated how the distribution of rainfall depended on mountains, ocean currents, and sea and land winds. Finally, he turned to maps of climatic "anomalies," deviations from a long-term average. In the case of a small area like Bohemia, he explained, it was "much more difficult to represent clearly on a map the course of clouds and dry winds," but even here he could show that "frequently the eastern or southern half of Bohemia behaves differently than the western or northern, which depends only the dominance and extent of certain winds."[69] Purkyně was teaching his audience to see climate as a dynamic problem that could only be analyzed with regard to the global circulation of the atmosphere.

Purkyně also stressed the need to extend the temporal scale of analysis. He used both historical documents and geological observations to demonstrate that rainfall patterns in Bohemia could not have changed substantially in the past millennium. According to the historical record, Bohemia had been plagued by floods and droughts even when thickly forested. In short, in order to assess the forest-climate question, it was necessary "to follow the distribution of forests and steppes in all parts of the world, then the climate of individual lands and the causes of local variations of climate, to illuminate the general meteorological laws and the modifications of them due to mountains and the extent of land and ocean in individual countries," as well as to collect "all known historical facts." On the basis of evidence of all these kinds, Purkyně arrived at a wholly original interpretation of the forest-climate question.[70]

The essence of his argument was that talk of climate change hid far wider and more deeply rooted problems. Regions said to be suffering from climate change were often, in fact, afflicted with a wholesale socioeconomic and ecological crisis. In many cases, the woes attributed to climate change were the result of "a misguided choice of agricultural specialization and of settlement location, and an over-use of existing vegetation through over-population and over-loading with cattle, etc."[71] Purkyně insisted that it was dangerously reductive to treat reforestation as a cure-all for problems that were fundamentally socioeconomic in origin. He pointed to the example of France under Napoleon III, where the state had seized on reforestation as the one-size-fits-all solution to social ills, with the backing of scientists like Becquerel. The public had been led to "hope for future salvation from a large-scale afforestation, which is unfeasible without a modification of the demographic and economic circumstances, and any real redress will only be delayed, and the public and government misled to sit with their hands in their laps, instead of working with their heads and hands." He went so far as to blame a "superstitious" faith in the power of forests for the absence of other forms of state support—measures that might actually be effective against disasters, like building levies or relocating homes to less vulnerable areas. Purkyně concluded that the question of reforestation "could not be decided in general; these are questions of an entirely local nature."[72] In this way, already in the early 1870s, Purkyně became the first scientist to make a case for rescaling the forest-climate question.

SWIMMING AGAINST THE CURRENT IN BOHEMIA

Purkyně's position was, to say the least, a surprising one for a faculty member at a forestry academy. The forestry profession at the time stood firm behind the principle that forests brought climatic benefits. As Purkyně would learn the hard way, skepticism brought few rewards in this branch of science. In letters to his colleague George Engelmann, a German-émigré botanist in Saint Louis, Missouri, Purkyně painted himself as a martyr to empiricism: "I do not deceive myself. My manner of swimming always against the current will win me no friends—at least not for ten or twenty years, when it will be too late."[73]

Privately, Purkyně attributed his iconoclastic ways to the influence of a man known to his followers as the democratizer of German science. The naturalist Otto Volger was a radical democrat and German nationalist who had fled to Switzerland to escape prosecution in the aftermath of 1848. A decade later, he became the founder of the Free German Foundation for Science, Art, and Public Education (Freie Deutsche Hochstift für Wissenschaften, Künste und allgemeine Bildung, or FDH), a Frankfurt-based institute aiming to bring natural knowledge to the public and to counteract the fragmentation of the sciences. Volger presented himself as a disciple of Johann Wolfgang von Goethe—less the Goethe of *Werther* and *Faust* than the Goethe of the *Farbenlehre*: the brazen naturalist laying siege to the fortress of Newtonian color theory. (Purkyně's father, Jan Evangelista Purkyně, had also been a great admirer of Goethe and an ally in the construction of a "subjective" theory of color.) It was in this spirit that Volger purchased Frankfurt's Goethehaus and moved his institute into it. To his opponents, Volger was known as a contrarian: an enemy of Plutonism when it was the dominant theory in geology; a proponent of the fixity of species in the face of Darwnism. But to his admirers, Julius Hann among them, Volger was a model of scientific integrity.[74]

It might appear ironic that Purkyně took Volger, a German nationalist, as his hero. Yet nationalism for both Purkyně and Volger (and for many veterans of 1848 in central Europe) had more to do with class than ethnicity: hostility to the aristocracy was at the core of their identity. In practice, their nationalism was closer to what we would call scientific internationalism, since it promoted cooperation across the narrow political borders of post-Napoleonic central Europe. On the occasion of the death of a mutual friend and fellow naturalist, Purkyně wrote to Engelmann:

> Good old Leonardi is dead. He possessed something all too rare in our time, a deep feeling for friendship and a genuine striving for clarity in everything concerning Nature. He saw before him only one Nature, not departments and guilds. I am indebted to him, since I was in close contact with him from 1850 to 1860, and even if he could not convert me to Krause's philosophy (for all philosophy is utterly foreign to me), still I learned a great deal from him. In particular, he introduced me to Volger, at which point I became a new man. I learned that one must see and think for oneself in everything and that mere scholarliness is of no help to a natural scientist. Much of Goethe's spirit must have remained among the Frankfurters, or perhaps Goethe himself was only a favored incarnation of the Frankfurter Spirit.[75]

Purkyně thus aligned himself with Volger and Goethe, and with their tradition of freethinking, public-spirited natural history. Purkyně would have the honor of becoming a member of the FDH in 1881, just a year before his death.[76] Perhaps it was a consolation to him at that lonely time.

Purkyně also faced a powerful rival in Bohemia. In 1873, the mathematician František Josef Studnička (1836–1903), backed by several of the largest Bohemian landowners, began organizing his own network of ombrometric (rain gauge) stations. Studnička was five years younger than Purkyně, the son of a schoolteacher, but he had achieved the kind of academic career that Purkyně could only dream of. By thirty-five he was a full professor of mathematics at Prague University. Purkyně could also have envied Studnička's success as a popular science writer. He was the author of numerous books ranging across the natural sciences and their history. Where Purkyně's style tended toward vigorous argumentation loaded with reams of evidence, Studnička wrote in a florid, politely antiquated style. Studnička liked to say that natural science could enhance rather than dispel the wonder of the natural world.[77]

Studnička believed that the scientific study of the Czech lands could nourish a healthy patriotism. Tellingly, he contrasted a patriotism rooted in the appreciation of the natural landscape with a cosmopolitanism arising from "slavishness and egoism." He also looked favorably on theories like those of Herder and Ratzel that attributed "aspects of national and individual character" to "geographical factors."[78] Purkyně, by contrast, took a more hard-headed attitude to what Czech speakers called *vlastivěda* (German, *Heimatkunde*, or homeland study). So, for instance, in contributing an article on an expedition to the Tatras for the third issue of *Živa* in 1853, Purkyně noted that Czechs and Poles had taken to lauding the unparalleled beauty of the Carpathians, while German tourists "declare this mountain range to be inaccessible, unattractive, without charm and without comforts." The Germans were not wrong, he admitted, "in that the most interesting spots do not always lie between two inns. . . . The Carpathians have their special charm and their comforts, also their inconveniences."[79] We can conclude that both men pursued natural science as a vehicle for national self-consciousness and cultural progress, but understood these ideals differently. In addition, from the early 1870s, Purkyně increasingly pursued this national goal by writing in a *Weltsprache* for an international audience.

Though Studnička believed Bohemia to be an unusually fertile land, he concluded that its climate was fast deteriorating. "Our country belongs," he wrote, "even in its roots to the lands blessed and favorably formed, and so nothing is needed but to try to maintain those conditions, and to improve them in those cases where it is desirable and possible to improve them; in no way, however, should they be worsened by inept human hands!" The vulnerability of nature to human destruction, intentional or otherwise, was a recurring theme in Studnička's writings. "How mightily," he wrote, "does weak man interfere with nature!" Extrapolating from effects he attributed to deforestation in North America, Studnička predicted a disastrous climatic shift in Europe's near future, which would rob Europeans of their global economic dominance, and turn their continent into "the least hospitable corner of the world."[80]

Although Studnička relied heavily on regional comparisons, such as that between Europe and North America, he assumed that climate could be analyzed province by province. He evaluated the data from his network without consideration of neighboring regions. This field of view is implicit in his somewhat awkward definition of "the climatology of Bohemia" as "the field of knowledge comprising the interpretation of phenomena occurring above the surface of the earth, insofar as they do not diverge from the borders of our *patria*."[81] This definition of climatology foreclosed the need to attend to the circulation of the atmosphere beyond national borders. By contrast, Purkyně urged scientists to track atmospheric phenomena across the entire surface of the earth.

Equally important, Purkyně and Studnička took opposing positions on the imperial centralization of meteorology and climatology. Studnička was critical of the entire centralizing program begun by Karl Kreil in the 1840s. He charged that Vienna's intrusions were to blame for a decline in the number of Bohemian weather stations in operation. By contrast, he credited several of Bohemia's large landowners, listing them by name, for supporting the proliferation of observing stations within his provincial network.[82] Meanwhile, Purkyně looked directly to Vienna for patronage.

"STRIKE WHILE THE IRON IS HOT"

Purkyně refrained from publishing on the forest-climate question until the early 1870s. In 1872, he was called on by the Bohemian Diet to investigate the devastating floods in western Bohemia, and he included in the resulting publication a brief comment denying the climatic influence typically attributed to forests.[83] The following year, at the World Exposition in Vienna, he exhibited a collection of large-format graphs displaying eight years of climatic measurements.[84] As he later recounted to Engelmann in Saint Louis: "My meteorological studies have taken me a dreadful time, so that everything else remained undone. I have put together the rain for all years from 1800–1807 for all observations in Germany according to differences . . . in order to prove, that the change in weather from month to month and year to year with less or more rain does not depend on local causes."[85] Repeatedly, Purkyně would complain to Engelmann that this work was leaving him no time for his true passion, plant physiology. But he could not give it up, "because it is a matter of deciding a question that plays a role in legislation."[86]

From his earliest publications in his father's wide-ranging journal *Živa*, Emanuel Purkyně had practiced writing about the natural sciences for a general educated audience. As he would soon learn, writing for an expert audience in German in the 1870s required very different skills than writing for a general audience in Czech in the 1850s. In the context of the Czech "awakening," it was not necessary to present original research in order to justify publication. It was enough if a text managed to extend the reach of the Czech language to describe a wider range of natural phenomena. It was for this reason that loose translations of foreign works were sometimes published as original research.[87] These articles often took a meandering, colloquial form, with plenty of digressions linking natural to cultural history. With the increasing specialization and professionalization of the sciences, however, scientists writing for an expert audience in German were expected to present original research and to do so as concisely as possible. As specialists had to wade through ever-growing numbers of scientific journals, novelty and brevity became the virtues by which scientific writing was judged.

Purkyně began to appeal to scientists in Vienna to get his research on the forest-climate question published. He started with Lorenz, earning goodwill by reviewing Lorenz's climatology textbook for a forestry journal. Soon he was asking his Viennese contacts to put in a good word for him at the agricultural ministry, since his situation at the forestry academy was becoming increasingly uncomfortable. And perhaps they might be able to find a scholarship for his son Ottokar.

How delighted this embattled botanist must have been to receive his first reply from Julius Hann in August of 1873. Purkyně had sent him a manuscript and requested that Hann communicate it to the Academy of Sciences. The letter opened with the assurance that Hann was "for the most part in agreement with him." To be sure, storms did not dump more rain on forested than on unforested land. Too little attention was paid to the fact that any modification of climate by forests would depend on geographical conditions. On the other hand, Purkyně's attempt at a mechanical explanation of the origin of storms made little sense to Hann. More important to Hann was the point that the forest-climate debate tended to neglect the role of forests in regulating water levels. Didn't forests allow the soil to soak up rainwater and so help to stave off flash floods? It was possible to read this initial letter as encouraging, but Hann's tone turned impatient over the following months. "Your article is welcome," he advised, "as long as it's short." Thereafter, Hann's letters to

Purkyně grew ever briefer. In 1877, when he declined to recommend an article of Purkyně's for publication, he pointed again to its insufficient "consolidation or concentration." For a journal like that of the Austrian Meteorological Society, which reported on "everything new in the field of meteorology," Hann advised Purkyně to condense his remarks to a single folio page.[88]

Purkyně turned next to Carl Jelinek, director of the ZAMG, in the hope that Jelinek could help him land a teaching position in Prague. Jelinek replied in January 1874 that in this matter he was "regrettably the least influential [*einflussloseste*] individual." Jelinek had been forced to give up his position at the ministry due to ill health, and in any case the ministry had little to do with the decisions of the Polytechnical Institute in Prague. In practice, hiring decisions were made independently by the Bohemian provincial council or Landesausschuss. Jelinek subsequently expressed interest in his writings, but reminded him of the obvious: "I cannot hide from you that you find yourself in the minority."[89]

Fortunately for Purkyně, he found far more cooperative patrons in Josef Roman Lorenz von Liburnau and Josef Wessely. From the start, both men assured Purkyně of their support. Lorenz agreed that the forest-climate question could only be settled by rigorous experimental research. The historical evidence was inconclusive, since one could never be sure of the conditions under which it had been produced. Wessely agreed that Purkyně's own survey of historical evidence demonstrated "how ridiculous it is to draw conclusions about a steady or changing climate from weather records for a few years."[90] Lorenz would later call Purkyně's measurements "far more valuable than parading around old chronicles." The one point on which Lorenz politely took issue (like Hann) was Purkyně's neglect of the forest's hydrological influence, a matter he would emphasize in *Wald, Klima, und Wasser*. He agreed with Purkyně that there was an urgent need for an experimental program, and he dangled before him the possibility that Purkyně might become the "sous-chef" for Bohemia, including Moravia and Silesia. "There are so many questions to answer in climatology and soil science that one could employ 10 Purkyně's, if one had them—and the money to do so!" Such an undertaking, however, looked unlikely at present, given the recent parliamentary deliberations on the ministry of agriculture's budget. More to the point, perhaps, Lorenz feared that Purkyně would face significant "prejudice" due to his "resolute" position on the forest-climate question. "I often occupy myself with the thought of finally creating a more suitable post for you," Lorenz consoled him. By a suitable position, he explained that he meant one that would allow Purkyně to support a family in Vienna, if not quite in a manner "befitting his class." Again, Lorenz stoked his hopes: "You'll see, forest meteorology will soon be in full swing over here." It was true, Lorenz confided, that a certain "Count M.'s" aversion to Purkyně would make the task no easier. Nonetheless, he hoped that the outcome would be "a new role" for the beleaguered naturalist.[91]

Lorenz confided further that he was fighting for their position in Vienna. It was a "seething, tumultuous" scene, with opponents like the "supposedly enlightened" Simony, author of *Protect the Forest!* The fight was no different than in Bohemia, except that it was verbal—"because I would have absolutely no time to enter into it in writing now."[92] There is no doubt that Lorenz was a busy man, but he was also savvy. His diplomatic skills had helped him rise from provincial origins to a powerful position in the imperial administration. Indeed, he found Purkyně's outspokenness alarming. One had to proceed with caution. Purkyně's critique of Ebermayer, for instance, might find a home in a prominent forestry journal, Lorenz reported after a conversation with its editor—but only if it were "very tame." Purkyně needed to learn the rules of the game: where, how, and when to publish.

Wessely, despite the colorful insults he hurled at his opponents in private, was no more of a risk-taker than Lorenz in public. As he told Purkyně in 1874, "Your polemic strikes me as apt and effective." But Purkyně would have to tread a fine line, "so that prominent people who aren't familiar with the facts aren't led to believe that the polemic is nothing but malice." It was out of the question for Wessely himself to take a public stand, given the state of his health. But he urged Purkyně to do so and to do it fast, since legislators were then at work on a revision to the forest law. "Strike while the iron is hot," Wessely advised him more than once.[93]

Lorenz agreed that timing was everything. He illustrated by means of his own career. In Fiume, he had been "too proud and too concerned with completeness" to publish, so that for years he was passed over for professorships in favor of men "who didn't have one tenth of my accomplishments." Hence his advice to Purkyně: "If you want to be recognized for your work soon, don't delay; rather, fire it off one part at a time." As he explained, "repeated smaller publications have more effect on the opinion of a broader public, concentrated larger works more on specialized experts." And the opinions of laypeople mattered "often enough" for the "fate of professionals." More to the point: "scholarly works" would have "absolutely no influence" over the minister of agriculture.[94]

And so Lorenz and Wessely laid out the rules of the publishing game. Do get the word out at the height of the debate, but avoid a polemical tone. Be concise, and keep details to the footnotes. Use the genre of the feuilleton to conceal your identity when necessary. Longer, "specialized" articles would never find a home in a newspaper. Some of the forestry journals would take such a piece, but "none of those professional journals has much of a circulation among the public." In general, writing for the press required more discretion. "One can only write entirely independently and according to one's own inclinations in brochures; there is always all manner of caution to be taken in journals."[95] Wessely was especially frank with his criticisms of Purkyně's manuscripts, though he begged Purkyně not to take it the wrong way. A "lighter," "less off-putting" style was appropriate for the wider audience Purkyně needed to reach. He was convinced that the writings of the Bohemian forester would "cause a sensation," if he would only follow Wessely's suggestions—which came at great length and in excruciating detail, even as he professed to be a hands-off editor. In the end, the nine installments of Purkyně's "On the Forest and Water Question" took up nearly three hundred pages of Wessely's journal.[96]

All this advice shows how wrong we would be to think of popular science writing as a simple, mechanical act of "translation." Wessely described it as a way of "conceptualizing" as much as a style of writing; it was a habit of mind that required years of training. The more one worked at it, the more it would be ingrained in one's "flesh and blood."[97]

Linked to Wessely's editorial advice were the guidelines he offered for the conduct of a scientific dispute. In advance of the 1877 forestry congress, Wessely wrote to Purkyně to let him know that "his topic" was on the agenda. He proposed a preliminary debate between Purkyně and one of his opponents in order to "light a spark" among the conference attendees. In the event, the spark caught fire, and soon it was blazing out of control in the forestry press.[98] In the aftermath, Wessely wrote that "the so-called forest-question debate, in which you are our *scientific* freedom fighter, has taken a very unfortunate turn. . . . I am disgusted by the manner in which our opponents are conducting the battle, as it neither serves their goal nor is it a noble approach." Yet Wessely was not prepared to devote the pages of his own journal to the dispute. In fact, he argued that "this battle cannot and should not be fought out *in detail in writing*." It was better to "let the opponents howl as they like, and occasionally launch a short but *significant* article, in which one sticks to the subject and if necessary mentions the opponents in a crushing but entirely *professional* way." Having sparked a confrontation, Wessely was concerned to maintain for himself and his journal an image of detached expertise.

Purkyně took all this advice graciously. He was optimistic that these well-placed mentors would help his ideas get a fair hearing. He wrote to his friend Engelmann in Saint Louis in 1878 to announce that he had attracted the attention of the agricultural ministry. He had even been invited to deliver a lecture on his research in Vienna, and it had gone well. He had received "flattering applause from great men, who, of course, understand nothing of the matter," and praise from "Dr. Hann," which was "very dear" to him. More importantly, he had received "the assurance of all the material resources for my studies," perhaps meaning funding for his station network. Best of all, from Purkyně's perspective, was that he could now turn back to other topics in forest botany. And Lorenz had even promised to win for these other studies "the greatest recognition and publicity," if they would "bring honor to Austria." Purkyně saw before him the career of a genuinely "Austrian" scientist.[99]

There is little doubt that his patrons in Vienna prized him highly and hoped to be of service. By a funny twist of fate, a letter has ended up among Purkyně's papers that was clearly not meant for his eyes, and which gives us a remarkably candid view of Josef Wessely's opinion of him. In May of 1878, after Purkyně's presentations in Vienna, Wessely wrote to an unknown correspondent, praising the Bohemian's research and his "goodwill." From Wessely's perspective, Purkyně's tragic flaw lay in his inability to compose his ideas on paper in a manner that a German-speaking public could digest. It was, Wessely suggested, a "truly Austrian affliction," this utter dependence of the man of science on the affirmation of the public.

Lorenz, meanwhile, felt a weight of responsibility toward Purkyně. "The trust that you have placed in me since we have become better acquainted distresses me in so far as I have still not been able to reward it with an effective step towards the desired change in your position." Lorenz hoped that the new research program in experimental forestry would provide such an opportunity. However, the friendly tone of their correspondence faltered in 1875. Purkyně had innocently offered to let Lorenz publish some of his ideas under Lorenz's own name. In a tone that could easily be read as either offended or jesting, Lorenz professed not to understand what Purkyně was suggesting; then he declared himself offended. Purkyně sent an apology, and Lorenz accepted. But a few months later, Lorenz revealed his intention to write something "more comprehensive" on the issue. "Were I indeed to publish such a thing before you," he told Purkyně, "I would appear a plagiarist."[100]

Yet publish first he did. Lorenz's popular book *Wald, Klima, und Wasser* appeared in 1878. Its arguments closely resembled those made by Purkyně in the manuscripts he had shared with Lorenz and in his publications in relatively obscure Bohemian periodicals. Bizarrely, Lorenz's book contains but one fleeting reference to his correspondent in Weisswasser. It comes on page 184, in the context of a discussion of the unreliability of historical evidence of climate change, and it records his first name as Emil, not Emanuel.

The correspondence between Lorenz and Purkyně ends soon after. Three years later, Purkyně suffered a fatal stroke at the age of fifty-one, a martyr to his vision of scientific truth.[101] For Purkyně, the fight over the forest-climate question was in part about winning a measure of autonomy from the state and the aristocracy—a form of independence that scientists in the German Empire and the United States already took for granted by the 1870s. His story shows, through his failure, what it took to climb to the perch of an imperial-royal scientist. Purkyně remained a mere provincial.

"THE PUBLIC GOOD"

Purkyně did find one other powerful if unlikely champion in Vienna. Georg von Schönerer began his political career as a liberal representative to parliament from Lower Austria. Only after the expansion of the voting franchise in 1882 did he develop his signature platform: anti-Catholic, German nationalist, and fiercely anti-Semitic. Today he is best known as the politician on whom Hitler modeled himself.[102] Who would have guessed that he was also a climate skeptic?

In 1876, von Schönerer took to the floor of parliament to denounce recent agitation in Bohemia on behalf of forest protection. He set himself up as the protector of small farmers and their right to develop their land as they saw fit. The theory that deforestation degraded the climate was, he charged, a "dogma" that had never been proven.[103] "The influence of the forest on rainfall and climate is, if present at all—which has not in the least been proven—so small indeed that laws that would constrain forest owners extensively are certainly not justifiable."[104] On whose scientific authority did this budding German nationalist rely? Ironically, a man whose name is indelibly associated with Czech nationalism.

Indeed, von Schönerer was so taken with Purkyně's research that he responded in a long article for the northern Bohemian paper, the *Leitmeritzer Zeitung*. Such an article was clearly a lightning rod in a region where a sustained drought had led to fierce agitation in favor of forest protection. The editors printed it with a note endorsing discussion of this important question "from various standpoints." The most interesting response they received came from a writer who identified himself as a "small farmer" and gave only his initials. He argued that desiccation could not be dismissed as a myth, and he appealed to the authority of collective memory.

> You must be dependent on the yield of a few small plots of land; you must have kept a lookout for a hearty rainfall day by day throughout the dry summer for

> several years, in vain; you must have seen how the crop gradually withers and the expected harvest shrivels to a minimum; you must have seen your favorite fruit trees mourning with wilting leaves . . . and only then, over the long winter, when there is little work in the barns and little to bring to market, will you come to the conclusion, that it has grown much drier than it used to be.

Then, suddenly, the writer's tone changed.

> And yet, automatically, the question also looms: will it continue this way, or even grow worse? What is the cause? Is there no remedy? I am neither a scientific hero nor a Methusalem who could speak from a century of experience, but only an entirely ordinary small farmer. I want to entrust the solution to these questions to experts, in the hope that those who, with their trove of knowledge, have so often lent a hand to agriculture, will also be of assistance to redress this calamity, if it is at all possible.

Here the writer delimited the sphere of his knowledge and of his uncertainty. He was sure that the climate had grown drier in recent years, but he could not be sure of the longer-term trend, nor of the cause of this change. In this exchange, we can see how the process of scaling played out as a negotiation between collective memory and expert knowledge.

It was on this fertile ground that Lorenz's 1878 *Wald, Klima, und Wasser* fell and flourished. Thereafter, a proposed revision to the imperial forest law, drafted in 1878, failed to be brought up for debate in the Reichsrat. When the floods of 1881 renewed the controversy, Emperor Franz Josef appointed a commission to consider proposals to strengthen the imperial forest law. Among the fifteen commissioners were five university experts (including Eduard Suess), seven large landowners, one attorney, and two civil servants. Eight of the fifteen held aristocratic titles, and only four made their homes in the flood-stricken lands, while nine came from primarily Slavic regions of the Monarchy—all undoubtedly results of the influence of the conservative primate minister, Count Eduard von Taaffe. The commissioners were critical of proposed changes to the law. They objected to the assumption that measures introduced for France could be uncritically adopted for Austria. They were also critical of the proposals for failing to combine forest management with infrastructural protections against floods into a single law.[105] In the subsequent debate in parliament, one speaker in particular signaled the changing tide. This was a Herr Neumayr, age fifty-two, a farmer from Salzburg. Neumayr denied that the proposed law protected the interests of small farmers. He attacked it as undemocratic, since it allowed for afforestation of privately owned land without the permission of the owner, based solely on the advice of experts. At minimum, he urged, *local* experts with genuinely practical experience should be consulted.[106] This analysis was in keeping with the political critique of forest protection articulated by Purkyně and later taken up by socialists like Otto Bauer and Walter Schiff. These critics all charged that strict protective measures served the powerful interests of large landowners, expert foresters, and the railways (the last of which supported afforestation to prevent damaging floods and landslides). They argued that forest regulation could best serve "the public good" by allowing for multiple uses. And they warned against allowing reforestation to become the quick fix for problems that grew out of social inequality, not only environmental neglect.[107]

No further protective legislation was passed at the imperial level. Indeed, the Forest Law of 1852 was not supplanted in Austria until 1975. Instead, in keeping with Lorenz's analysis, the issue of reforestation was addressed administratively at lower levels of government. In 1884 and 1885, for instance, Tyrol and Carniola each passed laws requiring afforestation as a means of flood protection. After much negotiation in the crown land diets, the cost of these measures was divided among the imperial state, the crown land, and the local inhabitants, with the locals bearing no more than 5 percent of the burden.[108] This result represented a compromise between conservation and development, and between provincial autonomy and imperial oversight.

CONCLUSION

The effect of these negotiations was to rescale the forest-climate question, transforming it from a matter of imperial legislation into an issue for local action and international research. In order to provide a framework for research, the imperial ministry of agriculture approved a "General Organizational and Operational Plan for Experimental Forestry Stations" in 1886. The plan stipulated:

> Forest-meteorological observations. Of the utmost importance is the principle that those observations should be undertaken or continued that fill holes left by the treatment of this subject in other countries. Among these are two questions in particular:
>
> a) "How does the moisture of the air within and over the forest canopy behave in comparison to the moisture of the air of unforested surfaces at the same elevation?" The answer to this problem will furnish the most important

material for deciding the question of the influence of the forest on the humidity of the atmosphere.

b) "How does the forest affect the climate of its surroundings and how far does this effect reach?" Here, of course, is to be included not only the moisture of the air as in a), but also the temperature, the frequency and amount of rainfall, and the strength of the winds.[109]

This proposal stood at the vanguard of a rising international trend: European and North American governments were just then beginning to assume responsibility for the investigation of the environmental impacts of human activities, as inquiries into river pollution and urban smoke pollution attest.[110] Lorenz, the author of the proposal, argued that Austria-Hungary was uniquely positioned for experimental studies of the forest-climate question, due to the variety of climatic conditions within its borders.[111] Three experimental stations were established: one in Lower Austria, one in eastern Galicia near the Russian border, and one in the Carpathian foothills. This research confirmed that the climate inside a forest was moister and its temperature range less extreme than on unforested land nearby, but it failed to find a climatic effect beyond the forest's border. On the basis of this research, Lorenz's collaborator Franz Eckert, director of the forestry school in Aggsbach, Lower Austria, argued that the forest-climate question had, at last, been solved. The climatic effect of the forest had previously been "greatly overestimated," and forest protection had been guided by "motives of a non-climatological nature." Forests could not influence the surrounding climate directly by means of radiation or conduction, but only indirectly and dynamically—that is, by "the mediation of air currents." It was necessary to distinguish in this respect "between local air currents coming from the surface vegetation, and the general air currents that mediate this influence."[112] In this analysis, Eckert was following the schema of dynamic climatology and the scaling practices of imperial-royal science.

From a comparative perspective, it is not surprising that Austria failed to pass a stricter imperial forest law. Studies of the German Empire and the United States suggest that centralized state environmental regulation was not a viable option elsewhere in this period. In the case of smoke pollution, for instance, a Prussian abatement law of 1848 was rarely enforced (much like the Austrian forest law passed four years later). Most policy makers of the nineteenth century seem to have assumed that smoke abatement was the responsibility of cities, not states.[113] Likewise, in the case of river pollution (whether household or industrial), regulation was left up to municipalities. These, not surprisingly, were happy to overlook waste that collected downstream from them. City planners explained that rivers themselves could remove the threat of pollution through a natural process of self-cleansing (*Selbstreinigung*). Rarely was it asked how far downstream of the source of pollution one had to be to benefit from this effect.[114]

In fact, one of the first studies to investigate the spatial dimensions of river pollution was carried out by hygienists at the University of Vienna. Among them was Ernst Brezina, a collaborator of Wilhelm Schmidt's on studies of urban climatology. In 1903, Brezina borrowed a canoe from the imperial Danube Regulation Commission and sailed downstream from Vienna toward Pressburg/Pozsony/Bratislava, observing and taking samples of the water as he went. Between kilometer 10 and 20, the sewage of the Danube canal had mixed with the fresh water of the river. By kilometer 31, the river showed "traces" of pollution, but "one cannot speak of contaminated water in the usual sense there, not even under unfavorable circumstances."[115] In this case, as in the forest-climate debate, Habsburg scientists pressed a question of scale that seems to have remained unexamined elsewhere.

Indeed, Habsburg scientists, foresters, statesmen, and farmers reformulated the forest-climate question as a problem of scaling: a comparison between the economic benefits of easing restrictions on forest use and the climatic benefits of tightening them. This was a paradigmatic dilemma of early environmental politics, the type of trade-off between preservation and modernization that Harriet Ritvo documents in her study of the controversy over the construction of the Thirlmere Reservoir in the 1870s. The difference in the case of the forest-climate question was the degree of uncertainty attached to the environmentalist argument. No reliable quantitative measures of the regional impact of deforestation existed. Indeed, even today, the question of the regional (as opposed to global) climatic impact of deforestation continues to provoke debate. Today the question turns on the role of feedback mechanisms. Positive feedbacks mean that even partial deforestation could push a region into a "permanently drier climate regime," turning parts of the Amazon, for instance, into dry savannah. On the other hand, feedbacks can also operate to restore the original equilibrium: in some cases, deforestation on a smaller scale has been shown to lead to more precipitation and more plant growth.[116] Like Lorenz, Purkyně, and their collaborators, scientists today stress the need for analysis at multiple spatial and temporal scales. Otherwise, it may be impossible to distinguish spatial and temporal *fluctuations* of a local climate from progressive climate change.[117]

What did humans know in the nineteenth century of their potential to transform the earth's climate? Fabien Locher and Jean-Baptiste Fressoz are right to question the assumption, implicit in the idea of the Anthropocene, that awareness of anthropogenic climate change dawned all at once in the late twentieth century.[118] Before we attribute knowledge to past societies, however, we need to consider whether existing evidence met their own standards of inference. As this chapter has shown, many of the leading nineteenth-century experts in climatology judged the data incomplete and equivocal. Certainly, there was no evidence that deforestation had a climatic effect that was irreversible or global. Indeed, there could not have been, because available series of consistent instrumental measurements covered at most a few decades, and because the infrastructure necessary to "make global data," in Paul Edwards's terms, was only in the earliest stages of construction.[119] Before judging these nineteenth-century scientists for their failure to speak out against anthropogenic climate change, we should consider the dilemma that confronted them. Advocate for tighter forest protection on the basis of shaky evidence? Or allow rural populations a chance at economic development, an opportunity to even out the vast social inequality that was feudalism's legacy to east-central Europe? It is the historian's responsibility to convey the complexity of the choices they faced.

By reconstructing this debate and its legal resolution, we have been able to follow a process of social scaling. The resolution of the forest-climate debate represented a novel conceptualization of climate as a matter of the "public good." As we have seen, Austrian law contained a contradiction: it assumed that the removal of woodlands degraded the surrounding atmosphere, even as it treated the atmosphere as a boundless and therefore unregulated resource. Dynamic climatology offered a way out of this confusion. On one hand, dynamic climatology showed climate to be a planetary system, uninterpretable at the local scale without reference to the general circulation. On the other hand, dynamic climatology defined the degree to which climate *was* a local phenomenon, according to the degree of mixing of the air.

Thus the *global* climate was not a material fact waiting to be discovered. To see this, it helps to recognize that the "global" of late nineteenth-century climatology was not that of twenty-first-century climate science. The global data of the nineteenth century was two-dimensional instead of three, confined to the earth's surface; and it was constructed by pen-and-paper interpolation and theorization from first principles, not from satellite images and computer simulations. Yet the history of the forest-climate question shows that the global scale of climate in this nineteenth-century sense was not just a technical achievement; it was equally the *social* product of negotiation between expert opinion and common sense. What's more, the global climate was not the sole result of this process. Just as crucial, as the second half of this chapter has confirmed, was the *local* scale of climate that emerged from this debate.

To conclude, we might return to Freud's analogy between the draining of swamps and the taming of the unconscious. The nineteenth-century experience of environmental transformations like swamp drainage and deforestation may well have had a tinge of the uncanny. A parching drought or a devastating flood brought Freud's contemporaries face to face with forces that were not altogether natural, in that natural science could neither predict nor explain them. To those who believed that they were civilized by virtue of having mastered brute nature, such uncontrollable forces were not unlike visitations from the underworld. As Amitav Ghosh has written of anthropogenic climate change today, "uncanny" comes closer than any other word to "expressing the strangeness of what is unfolding around us."[120]

CHAPTER 10

The Floral Archive

The historian has rarely lived through the events of past times that he describes. He has not seen them with his own eyes; rather, he describes them on the basis of the documents at hand, whether these are the yellowed leaves of old codices and parchments, or the brown fossil leaves in the coal shales, or the green leaves of the living plant world.

Anton Kerner von Marilaun, 1879[1]

Among all the sophisticated new tools introduced in the nineteenth century for scaling the climate—for understanding the weather of the here and now in relation to large-scale, long-term processes—none proved quite so effective as nature's own climatic indicators: plants. Along with rocks and fossils, plants were objects that made climate change visible and palpable. But unlike rocks and fossils, whose transformations were invisible on the time scale of a human life, living plants functioned to bridge the temporal scales of human history and geohistory. Flowers and fruits were climatic indicators that spoke to nearly all the senses at once. Not only did they speak of climatic relationships that spanned vast gaps of space and time; they did so in ways that nonscientists could easily grasp. Today, botany rarely figures in genealogies of our knowledge of the earth's climatic history, which focus on discoveries in geology and atmospheric chemistry. Yet the study of plants had a unique potential in the late nineteenth century to connect concern over the forest-climate question to the scientific project of reconstructing the variability of climate in the geological past.[2]

Nineteenth-century scientists had great respect for plants as climatic measuring instruments and as clues to the health benefits of a local climate. To be sure, plants weren't much use for taking readings of individual meteorological elements like temperature or air pressure. Rather, vegetation measured the combination of factors most relevant to that most immediate of purposes, life. Researchers noted that no measuring instrument yet invented "was affected by these influences in quite the same way as plants." In the 1780s, the Bohemian naturalist Thaddäus/Tadeáš Haenke meticulously recorded the locations of plants in the mountains between Bohemia and Galicia. Haenke's example was an inspiration to Alexander von Humboldt, who reviewed Haenke's book in 1791, praising his method of classifying plants according to vegetation type and location over Linnaeus's binomial system. It was here that Humboldt made the crucial observation that Haenke had provided the world with "a vegetable thermometer."[3] Thirty years later, Humboldt began publishing exquisite cross-sectional maps that correlated vegetation type with altitude. He likened the changes in vegetation as one ascends a mountain to those encountered when traveling from a temperate to a polar region. These maps made famous Haenke's method for using plants to delineate climatic zones. It soon became a commonplace among naturalists that "everywhere the plant world is a reflection of the local climate."[4]

And yet a plant's relationship to the local climate was too complex for the analogy to a measuring instrument to hold strictly. As the forest-climate question implied, it was necessary to consider not only the influence of the climate on the plant, but also the influence of the plant on the climate. In addition, generalizing about these influences was tricky. The enterprise of phenology was challenged by the difficulty of pinpointing the moment at which any given vegetative change occurred. Seasonal responses varied across small spaces and even from plant to plant. Those willing to grant subjectivity to vegetation even termed these responses "subjective," arguing that "plants as individuals are not all of the same disposition."[5]

Thus, by the middle of the nineteenth century, naturalists commonly recognized plants as indicators of the climate of the land in which they grew. What was novel was the idea of using plants to gauge the climates of the past. At the time, geologists were still a couple of decades short of reaching a consensus that a major shift in climate had occurred before the advent of recorded history. Some, though not all, were convinced that there had been a period approximately twenty thousand years earlier when much of the now temperate regions of the earth were covered in thick sheets of ice. Only recently had some naturalists come to understand elements of familiar landscapes as evidence of this Ice Age. Sediments known as "drift" were reinterpreted as glacial deposits, while the seemingly randomly placed rocks known as "erratic boulders" were said to have been transported by the former ice sheets. The Habsburg geological survey, however, ruled out the Ice Age theory, as did most British geologists, following Lyell.[6] It took a good twenty years for the theory's proponents to win their case. The 1870s saw the completion of a global survey of the evidence, including maps of the world's remaining glaciers and measurements of sea levels.

In the 1850s, then, the study of past climate was still a contentious field, riddled with problems of interpretation. It was not easy to reconcile the apparent stability of the earth's climate in the present with a theory positing such a radical shift in climate in the relatively recent past. Some of this work was accomplished not by physicists or geologists but by botanists. It was in part the achievement of a network of researchers, professional and amateur, that spanned the Habsburg Monarchy and beyond.

THE IMPERIAL-ROYAL BOTANIST

The network was spearheaded by the visionary naturalist Anton Kerner von Marilaun (1831–98). According to Kerner's biographer, the Jewish botanical artist and writer Ernst Moritz Kronfeld, it was Kerner's lifelong intention to write a complete flora of Austria-Hungary; all his other research was preparation for that undertaking. He taught scientists and nonscientists alike how to view the Habsburg territory through the lens of plant geography. For this he came to be known in Austria as "our Humboldt."[7] He also created the first floristic map of the Monarchy and authored the chapter on botany for the "natural-scientific overview" volume of the *Kronprinzenwerk*. *The Plant Life of the Danube Lands* (1863) inspired a new generation of Habsburg botanists, while its elegant prose was excerpted in school readers.[8] What's more, the terminology Kerner introduced to classify central Europe's vegetation remained in place long after his death, and it was often applied to plants throughout the European continent.

Kerner is occasionally cited today as an originator of the concept of the plant association and thus a founder of the field of plant "sociology," a cornerstone of modern ecology. In his own day, Kerner's achievements in the study of plant breeding brought him the admiration of Charles Darwin, who cited Kerner's attention to "various small, and apparently quite unimportant, details of structure," which constituted "most wonderful adaptations for various purposes."[9] However, Kerner also deserves to be recognized for his contributions to climatology. Notably, his son, Fritz Kerner von Marilaun, who collaborated closely with his father, went on to become one of the first experts in the new field of paleoclimatology.[10] Today, as climate modelers search for ways to make their predictions vivid to a general audience, Kerner should also be remembered for his talent for the communication of science in visual and literary form. He built on a growing demand for popular botanical literature at the time, evident in German-language magazines like *Gartenlaube*, *Bonplandia*, and *Flora*. Kerner skillfully employed artistic images and poetic text. He also took an ethnographic interest in the role of the plant world in the folk cultures

FIGURE 33. Portrait of Anton Kerner von Marilaun (1831–98), by Julius Viktor Berger.

of central Europe. Through painting and lyrical prose, he taught the public to see a vegetative landscape as an indicator of a changing climate.

Kerner was born in the Wachau Valley of Lower Austria, roughly halfway between the plains of Hungary and the Austrian Alps (see figure 34). His father was a high-level administrator for the powerful Schönborn family, and the Kerner family lived in the Schönborn castle in Mautern. The setting was rich in well-preserved art historical treasures; indeed, the Krems district would provide the subject for one of the first art topographical surveys to be published by the imperial monument commission in Vienna.[11] Anton's brother Josef shared his love of the natural world, though Josef pursued a career in law and botanized only as a hobby.[12] In one of his earliest publications, Anton described the landscape of the Wachau as "a romantic forest valley, in which the most diverse flowering plants can be found densely packed in a narrow space."[13] This fascination with the diversity of the vegetation of the Wachau Valley would shape Kerner's early botanical research, as he followed the trail of familiar plants west to the heights of the Alps, east to the Hungarian steppe,

FIGURE 34. Watercolor of the Wachau Valley, by Anton Kerner, ca. 1852.

and south to the Adriatic coast. Throughout his life, he retained a sense of himself as originating in a land straddling these sharply contrasting environments.[14]

Anton, following his father's wishes, enrolled as a student at the medical faculty of the University of Vienna, the preeminent medical school in Europe at the time. There he studied the latest scientific medicine with luminaries like the physiologist Ernst Brücke and the pathologist Karl von Rokitansky. Alongside his medical courses he enrolled in Franz Unger's class on "the history of plants," which must have had a powerful effect on him. By 1851 Kerner was reporting on his own botanical research to the newly formed Zoological-Botanical Society in Vienna.[15] Already in this earliest research, conducted in the immediate aftermath of the 1848 revolutions, Kerner had begun to infuse his botany with patriotic overtones. Having discovered in the Wachau a type of fern previously known only in Bohemia and Moravia and one native to the Italian Alps, Kerner took these as "evidence . . . of how rich and inexhaustible the flora of our beautiful fatherland is."[16]

Meanwhile, Kerner's commitment to a career in medicine was shaken by the cholera outbreak of 1855, in the face of which physicians seemed helpless.[17] His teacher Unger had previously made the switch from medicine to botany, and now Kerner was ready to do the same. In 1855 he passed the examinations to become a Realschule teacher in chemistry and natural history. His first posting was to an imperial high school in Ofen (Hungarian, Buda; today part of the city of Budapest). In 1860 he accepted a call to the University of Innsbruck and the directorship of the botanical garden there. Two years later he married Maria Ebner von Rofenstein, the wellborn sister of a medical student at Innsbruck. They moved to Vienna in 1878 so that Kerner could assume the directorship of the imperial botanical garden. Emperor Franz Josef honored him with the rank of Hofrat (court counselor) and ennobled him as Knight of the Order of the Iron Crown and, two years before his death, awarded him the Decoration of Honor for Art and Science.[18] Kerner's career thus led him from Vienna to Hungary, to Tyrol, and back to Vienna. It can be seen as a continuous effort to produce a botanical history of the Habsburg lands. Yet it was marked by a series of reorientations, as he tried and failed to fit the plant world to the model of patriotic imperial histories of migration and cultural exchange.

THE VANISHING LIFE OF THE PUSZTA

In Hungary, Kerner was alert to the environmental transformations under way. Rapid deforestation under Ottoman occupation had left the region prone to both floods and droughts, and population growth further strained the land and its resources.[19] Kerner noted, for instance, that places with "birch" in their name no longer contained any birches. "In a half century the romantic life of the puszta will have vanished, along with the vegetation that originally covered the ground of the steppe."[20] In particular, he was drawn into public debate over the climatic effects of swamp drainage in Hungary, over a decade before the forest-climate question swept through Austria. In a series of articles for the *Wiener Zeitung* in 1859, Kerner described the history of land reclamation in Hungary, including what he labeled the premier hydrological project of European history: the regulation of the Theiss in the 1840s to drain three hundred square miles of wetlands.[21] With his medical training, Kerner brought to this topic a concern with the atmospheric conditions conducive to organic growth. Like most naturalists of his day, he attributed a moderating influence to swamps similar to that of forests. By collecting and evaporating moisture, swamps could temper the sudden swings between hot and cold characteristic of a steppe environment.

His contribution was to insert a new perspective into this debate, the perspective of plant geography. At first, as he admitted, his specialty must have seemed remote from the public's concerns. Indeed, Kerner had the impression that the majority of Austrian foresters had nothing but disdain for natural science, an opinion that Emanuel Purkyně might have seconded.[22] What plant geography offered in this case was the tool of "local vegetation lines" (*örtliche Vegetationslinien*). Addressing the Zoological-Botanical Society in Vienna in 1859, Kerner had to define the meaning of this term.[23] In contrast to the "principal vegetation line," which marked the region in which a given plant could be found at all, the *local* vegetation line marked the internal boundary between a smaller area in which a given plant was common and an area in which it was rare. As Kerner later explained, he thought of local vegetation lines as analogous to meteorological isolines; both were techniques for "making vivid" the distribution of climatic conditions.[24]

As he tried to make sense of the effects of modernization in Hungary, analogies to other parts of the Monarchy came to mind. Kerner likened the boundaries of forest growth in the Hungarian plains to the upper limit to forests in the Alps. Drawing on newly available meteorological observations from the ZAMG's network, he pointed out that these two sets of boundaries resulted from the same effect: the limited duration of the growing season in each climate. Tall trees require at least a three-month growing season. In the Alps, these limits were set by nighttime frosts in the spring and fall, whereas in Hungary they resulted from the shortage of water at the height of the summer drought.[25] The point was that forests could only grow in certain environments, and only there would it be possible to *improve* the climate by planting trees. Thus Kerner insisted that haphazard afforestation would bring no benefits.

Implicit in this judgment was the recognition that vegetation and climate exist in a reciprocal relationship. Already in the 1860s, he warned that damage from deforestation, drainage, and other human interventions (*Bodenumgestaltungen*) was irreversible; where it had occurred, the dream of reforestation was futile. Such efforts would benefit the empire only as signs of the danger of wanton deforestation.[26]

Essential to this argument were the analogies Kerner drew among Austria's varied "forestless" landscapes.[27] Implicitly recalling Humboldt's analogy between the climatic effects of increasing altitude and increasing latitude, Kerner arranged seven types of landscapes found in Austria-Hungary in a progression from glacial peaks to "barren desert." All had in common that they were used primarily for pasture. Then, drawing on his personal measurements over eight years, he presented a preliminary table of the variation in the upper tree line across the Monarchy, showing that its height was lower in more northern regions and those closer to sea level.[28]

By these means, Kerner was pressing a new point of view on readers of the popular German-language periodicals in which he published. First, he was training them to see ecological analogies between the most disparate environments of the empire. Namely, the environmental challenges of the high mountains paralleled those of the steppe. What's more, the environmental comparison had social implications. The plight of the inhabitant of the Hungarian steppe was not so different from that of the Alpine peasant. If the second deserved to reap the benefits of agricultural and technological development, so did the first. But in both cases, development would have to proceed responsibly. Kerner urged measures to offset the ill consequences for the natural environment. This meant, above all, better irrigation in Hungary and reforestation in the Alps.

In order to mitigate the climatic consequences of swamp drainage in Hungary, he recommended the construction of a "system of irrigation canals, running through the entire plain, and large artificial water reservoirs, which would collect the flood waters and save them for later dry periods," plus "watering stations, which would see to the maximal distribution of the occasional rainfall over long time periods."[29] Kerner acknowledged that irrigation works in Hungary would require "great sacrifices, much money and much time," and "the investment of large amounts of capital with a delayed return." But the work would be "of lasting value." Urging his readers to consider the benefit to future generations, he quoted the German poet Emanuel Geibel: "All we need, we have been blessed to receive from our fathers; it falls to us to prepare the ground for those who follow."[30]

Implicitly, his transregional and multigenerational perspective suggested that these tasks were in part the responsibility of the imperial government in Vienna and not only a local matter. An undertaking on such a large scale would have to originate with the imperial government. Kerner was arguing that the Habsburg state had a duty to preserve the climate of the Hungarian plain for the sake of its future citizens. Austria-Hungary was a "state of contrasts," of "oppositions between east and west, north and south," and "the contest among contrasting elements has therefore become nearly unavoidable on Austria's territory." Yet Kerner insisted that Austria had nothing to fear from this competition. On the contrary, he wrote:

> Even if a spark occasionally flashes from the friction between these oppositions, it is a wholly beneficial flame, which enlivens our nerves and keeps our powers fresh and alive; it will never be allowed to erupt into a consuming flame. Austria is a natural historical necessity. It is simultaneously a bulwark and the mediating link between the eastern barbaric, homogeneous continental steppe landscape and the western coastal landscapes, which, thanks to their highly varied ground, have been fortunate to develop a rich cultural life.[31]

Here the young Kerner fell back on a facile opposition between the "barbaric" east and the "cultured" west, along with a crude environmental determinism. Yet his botanical research was already opening his eyes to a far more nuanced geography of difference. Note his emphasis on the productivity of the encounter between East and West on Habsburg territory, which he likened to the electrical (nervous) force generated by friction between unlike materials. He too sought to naturalize this state's diversity in terms of the dynamism of local contrasts.

If the mood and tone that Kerner cultivated in his popular writing in these years seem to anticipate those of the *Kronprinzenwerk*, the similarity is not coincidental. In the 1870s, as Crown Prince Rudolf was planning that series, he turned to Kerner for help. Rudolf intended to write the natural historical sections on the Danube floodplains and Vienna Woods himself. But he was better versed in ornithology than botany, and so he wrote to Kerner to request his expert input. From that interaction came Rudolf's invitation to Kerner to author the botanical chapter for the natural historical overview, "Austria-Hungary's Plant World," one of five chapters in that volume. It was an assignment befitting Kerner's ambitions. Through his studies of the flora of the Monarchy, he was attempting to construct a vision of continuity across time and space that could ground his appeals for the preservation of nature's own monuments.

Although the concept of a "monument of nature" (*Naturdenkmal*) was not yet in common use (chapter 2), it was implicit in Kerner's writings. Already, with the intrusion of modern farming and the railway, "the culture of the present has usurped the world of the Hungarian *puszta*." Kerner suggested that it was the duty of the imperial-royal scientist to advance modernization and, at the same time, to salvage what was left of traditional natural-cultural landscapes, to "preserve in image and word these last vestiges of what is authentic." These remains were still visible, he attested, but "we must make haste if we want to record and pass on to posterity the images that are still displayed there."[32]

Kerner's project shared much in common with the contemporaneous movement for architectural preservation launched in Vienna by Rudolf von Eitelberger in the early 1850s. Much as Kerner was motivated by concern for the effects of human activities on plant life, so did von Eitelberger seek to preserve Austria's art historical monuments in the face of urban and industrial development. Both men were trying to teach the public to look at landscapes throughout the Monarchy with a new appreciation for the marks of history. Whether these took the form of the weathered stones of a medieval church or the weathered bark of a stand of old-growth trees, their aesthetic effect

FIGURE 35. A moor in the Burgenland region of Hungary, painted by Anton Kerner, ca. 1855–60.

should be similar. Their surfaces made visible the passage of time, the cycles of growth and decay.[33] The preservation movement linked such relics, natural or man-made, to others strewn across the Habsburg lands, conjuring visions of connectedness across space and time.

PLANTS FIND THEIR VOICE

It was a spring morning in 1855, six years after Austria's defeat of the Hungarian revolutionaries, when Anton Kerner first set foot onto the puszta, the Hungarian steppe.[34] Kerner felt "powerfully drawn" (*mächtig ergiffen*) to this landscape, to its "seemingly endless flat sea of green with its countless fields for grazing, its isolated cottages and wells, with its marshes and slow-moving streams always wearing the same garb."[35] (See figure 35 and plate 4.) During his five years in Buda/Ofen, he returned again and again to study this fast-changing landscape before its defining features vanished forever. He mapped geological strata, compiled phenological observations, and gathered ethnographic information on local agricultural practices and lifestyles.[36]

At first, however, Kerner found little sympathy for the local inhabitants. He wrote home that he and his companions "were sincerely happy when we found ourselves once again among Germans, with German manners and German as their native tongue."[37] His early descriptions rehearsed stereotypes common among commentators on Europe's "East." He lamented the "mismanagement" (*Unwirtschaft*) of the land and the superstitious character of the Wallachians.[38] His local guides showed no love for nature: "Indeed, not once did they find it worth the trouble to gaze into the distance. With the exception of the nearest path that he is to follow, the Romanian is indifferent to the whole terrain."[39] To Kerner, it seemed obvious that the locals were deficient in aesthetic sensibility.

If the steppe was barren of artistic culture, Kerner suspected that this was because it was (and, as he gradually came to believe, always had been) barren of trees. Kerner took for granted that forests were a cradle of cultural creativity. In this, he was in good company. By this time in the nineteenth century, Germans, Poles, Russians, and Lithuanians had all invented myths of origin that traced their culture's roots to one or another "primeval" forest.[40] The gaze of the steppe dweller, by contrast, struck Kerner as narrow and instrumental. Not once did the locals lift their eyes to look off into the distance. He faulted them for their insensibility to what Riegl would call *Stimmung*, the "mood" created by viewing a landscape from afar.

A lonely twenty-four-year-old, Kerner turned to poetry. In "Meine Blumen," he wrote of the Alpine flowers that he had brought with him from Vienna and planted on his windowsill, attributing to them his own sense of displacement and observing their transformations with a physician's eye. In "Saxifrage," he likened the plant's "wan, sallow leaves" to the poet's own face, "grown pallid and wan." Neither had acclimated to the "hot puszta." This homesickness needs to be read both as a Romantic trope and a medical diagnosis, part of a long tradition of thinking about the psychic and physiological dependence of organisms on their native climate. The poet promised that man and plant would both recover "when we return home, leaves green and cheeks red." In "Gentiana" (another Alpine flower also to be found in low-lying meadows of Europe and Asia), the poet imagines a drop of water on the plant's leaf to be a tear cried for its "sisters, who stand together on the mountains of the distant homeland." He begs the flower to stop its tears, or else he too will cry "like a child" out of sympathy for the gentiana, his "dearest friend": "Dearest to me of all, far and wide / Of all my friends here in this strange land / We are bound together by our sweet, shared home."[41] The metaphor of the homesick plant came naturally to Kerner, aware as he was of the strong affinity of certain plants for certain places.

As Kerner would later remark, plants play various roles in poetry. Beyond serving as symbols or evoking the "characteristic mood" of a landscape, they can function anthropomorphically, to voice human feelings.[42] This was the role that flowers played in the poems that the young Kerner composed in Hungary. They expressed the poet's own homesick longings.

Imagining himself in conversation with flowers, Kerner began to consider how the voice of a plant might actually sound. Thus we have one poem from his time in Hungary in which a plant "speaks" in a more literal sense. These verses concern a linden tree, a species common in Lower Austria but rare in Hungary. The poet listens to the sound of the wind in the linden's leaves and hears voices from afar: "sweet tales from the homeland, that lies so far away,—so long. And above me rustling I hear a dear, familiar song.—Oh, sweet tones, pause a while, why must you rush along!"[43] With few familiar human voices to keep him company, Kerner was learning to listen to nature in new ways.

Indeed, the imaginative exercise of giving plants their voice gradually reoriented Kerner's perceptions of the steppe. Initially, he had found the local culture to be insensible to the beauty of the natural world, its art and architecture void of references to plants. But exercising his poetic ear attuned him to the *aural* culture of the steppe, its poetry and folk songs. He collected samples of these verses and even tried his hand at translating them. And here, at last, he found botanical motifs. The steppe had only appeared to him as an aesthetic wasteland, because he hadn't known what to look for. Or rather, he hadn't known what to listen for.

By the end of his five years in Hungary, he had reached a new interpretation of the culture of the steppe. In 1862, he presented his conclusions in *Die Gartenlaube*, the German-speaking world's most popular illustrated family magazine. "What plant forms might the imagination attach itself to," he asked, in the absence of tall trees and enduring greenery? The author had scrutinized the visual arts and architecture for answers to this question, but in vain. The answers lay instead in music and poetry. In Hungarian folk songs, plants found their voice.

Music translated the "elegiac" sounds of the Hungarian steppe. The cymbal imitated "the murmur and lisping of a thicket moving in the autumn wind"; the fiddle mimicked the song of a warbler in the reeds. Kerner asked his readers to picture the fishermen on "the reedy banks of the Theiss," who "look out all day long over the water, and all that strikes their ears is the melancholy rustling of the cane, and the plaintive, mournful song of the aquatic birds that make their homes in the reeds." The song of these fishermen "harmonizes perfectly with the surrounding natural world." In Kerner's judgment, Hungarian lyrics likewise expressed a reverence for the natural world no less "exalted" than the regard of the Swiss for their Alps. They recreated the *Totaleindruck* or "holistic impression" of the steppe landscape (what Humboldt had called *Physiognomie*). They also celebrated particular elements of that natural world. Among these, naturally, were the roses, tulips, lilacs, and other ornamental flowers that had been brought to Hungary from the east. What interested Kerner more was the poetic attention lavished on native plants, like the tribulus and evening primrose (the latter actually native to North America), both hardy flowering plants adapted to the soil and dry climate of the puszta. More surprising to western European readers would have been the popularity in Hungarian poetry of a lowly species of grass *Arvaleáanyhaj* (*Stipa pennata*, *Waisenmädchenhaar* or "orphan girl's hair"), so called because of its "whispy, feathery white awn." It was a species that could be found in isolated spots throughout southern Europe, west to the Rhine and north to southern Sweden.[44] But Kerner insisted that it was only "at home" in steppe regions, namely, in southern Russia and Hungary, where it was "authentically characteristic of the vegetation," essential to the "physiognomy of the landscape." Kerner likened the role of *Stipa pennata* in Hungarian poetry to that of the *Alpenrose* (*Rhododendron ferrugineum*, snow-rose) in the lyrics of the Alpine world. As an example, he offered his own German translation of the opening stanza of one Hungarian folk song. Rendered into English, the verse might read, "I festooned my hat with orphan-girl's-hair / I chose an orphan girl to be my wife / The first I plucked on the wide steppe / and in the village I found my mate for life." The illustration that accompanied the article, based on one of Kerner's watercolors, showed *Stipa pennata* as part of a typical *Pflanzengruppe* or plant community of the puszta, its tendrils blowing in the wind very much like long, loose hair (see figure 36).[45]

By the end of his stay in Hungary, Kerner had rejected his earlier characterization of eastern Europe's landscapes as monotonous. "The inhabitant of the Alföld, who loves his homeland exactly as the Swiss loves his Alps, knows how to reclaim [*abgewinnen*] inspiration from the plain in abundance, which he exalts in his poetry and develops into images and metaphor. A mountain miner's son, upon first stepping onto the desolate, unvarying plain, is seized by a feeling of loneliness and isolation. But the man of the puszta feels himself elevated and elated by the view."[46] He now saw floral diversity all through Europe, matched by human cultures to appreciate it.

THE PLANT LIFE OF THE DANUBE LANDS

In *Das Pflanzenleben der Donauländer* (1863), Kerner recreated his encounter with the Hungarian steppe in the register of allegory. The book as a whole was skewed toward regions of the Monarchy that Kerner knew well, yet which had received relatively little scientific attention. Eleven chapters dealt with Hungary; four with the Carpathians along the Hungarian-Transylvanian border; four with southern Bohemia, Moravia, and Lower Austria; and six with the Alps, specifically northern Tyrol. He opened the section on Hungary by

FIGURE 36. *Stipa pennata*, watercolor by Anton Kerner, ca. 1860. This archival watercolor captures better than his published engraving the sheen that gives the grass an almost liquid look.

emphasizing how strange the steppe must appear to a first-time traveler from "western Europe," who would "feel himself transplanted to an entirely new world." Kerner set out to impart this sense of displacement by literary means. He described the illusions to which one fell prey on the great plain. A shimmer on the horizon, for instance, could falsely lure the observer by creating the appearance of distant hills (see figure 37). A meadow might appear to be a forest, a farmhouse, a castle. Kerner dutifully provided an explanation: the hot soil heated low-lying air unevenly, causing wavelike motions that distorted distant objects. No sooner had he cleared that up, however, than he moved on to further illusions and distortions. He recounted an encounter with *betyars*, the Hungarian term for highwaymen, only to reveal that it was a dream, in which the *betyars* quickly transformed into water sprites. Even in broad daylight and apart from the distorting effects of the atmosphere, the steppe could be highly disorienting. "With the uniformity of the surrounding world, and the absence

FIGURE 37. *Délibáb (fata morgana) on the Alföld*, by Paul Vago, 1891.

of all reference points, it seemed impossible to me for one to orient oneself here. And I marveled in astonishment at my guide's knowledge of the area."[47]

As we have seen in earlier chapters, experiences of being lost were viscerally important to the work of scaling. Accounts of ballooning or being lost in a snowstorm or a forest signaled the need to reorient oneself in space and time, to look for new measures of distance and proximity. For the visitor from "western Europe," the steppe demanded such a reorientation, for distance was not what it appeared. Kerner drove home this point by quoting a hymn to the landscape of the plains by the Hungarian nationalist poet Sándor Petőfi, who died in the revolutionary warfare of 1848–49. In lines written in 1844 and famous to Hungarians by the 1850s, Petőfi had attested that the celebrated mountain landscape of the Carpathians left him unmoved compared to the vision of the puszta, his native land. In a recent English translation:

What are you to me, land of the grim Carpathians,
For all your romantic wild-pine forest? . . .
My home and my world are there,
In the Alföld, flat as the sea,
From its prison my soul soars like an eagle,
When the infinity of the plains I see.[48]

Kerner's lyrical German translation chose to emphasize the *Unendlichkeit* (infinity) of the steppe by placing it at the end of the line. His point was that the view across the plains commands just as wide a perspective as the view from a mountaintop. The effect is not necessarily sublime; perhaps it is better described as the experience of reverence that Riegl would term *Stimmung*, though he would associate it instead with alpine and ocean panoramas. When looking for a vantage point from which to paint an overview of the "fatherland," German-Austrian writers tended to choose a view from a peak in the Lower Austrian hills or from the tower of Saint Stephen's. Kerner used Petőfi's poem to make the point that there was more than one perspective from which to gain a comprehensive view of the Monarchy—a point that we have already seen made by Stifter and Kreil in these years. On the other hand, Kerner went on to close this preface with a far more predictable overview: an outlook from a hill outside Vienna, taking in the "near and far" of the Habsburgs' dominions. Nonetheless, by beginning with Petőfi, Kerner had reminded the reader that the imperial traveler need not set out from Vienna in order to acquire a proper sense of scale. As Petőfi's made clear in "Az Alföld," his love for the steppe had also prepared him to appreciate the "proud mountains," even if they did not speak to his "heart."

GOOD SPECIES AND BAD

Kerner's years in Hungary also proved crucial for his understanding of the diversity of plant life. Much as Joseph Hooker and Alfred Russell Wallace were doing in Britain, Kerner began to question the concept of a "typical" species.[49] When he had taken up the study of botany as a youth, he had been taught to collect what were termed "good species," meaning plants that were clearly distinguishable from each other according to an authoritative list of traits. The goal was to find ideal-type specimens for each species, while all other samples could be left behind in the field. When Kerner arrived in Hungary, however, he found himself at a loss. "For a long time I couldn't get my bearings. Nearly all the plants had a somewhat altered appearance and nearly all diverged from those that I had come to know in my western homeland as the typical 'good' species." He went on to describe this predicament in the language of scaling: "So I saw that the Hungarian flora, measured against the ruler I had brought with me from Vienna [*mit meinem von Wien mitgebrachten Massstabe gemessen*], were in fact composed primarily of 'bad' species." "So I had landed in quite a bad society," he joked, poking fun both at traditional botanists and at stereotypes of backward Hungarians.[50]

Kerner also began to notice that the same name was often used for different species in different regions. The tree referred to as *Schwarzföhre* (black pine), for instance, was really three different species of tree, one found in Transylvania and the Banat; another in the Vienna basin and parts of Galicia, Croatia, Dalmatia, and Bosnia; and the third in the border region between Bosnia and Montenegro.[51] Similarly, the undersides of the leaves of the flowering plant known as primula had but the lightest covering of fluff in Lower Austria, a velvety gray layer in Hungary, and a white felt in Transylvania. And so Kerner began to hold on to "bad species," instead of throwing them away.

In this way, his travels eventually supplied him with specimens that he could arrange in a morphological series, such that plants next to each other were self-evidently of the same species, while the ones at each end were utterly unlike—so unlike, in fact, that Kerner never would have guessed they were related, if he hadn't found the specimens in between.

These experiences convinced Kerner that "the great majority of our classifications are merely artificial. . . . In most cases there could be no speaking of sharp divides between types traditionally regarded as species; and, in general, there were no species at all in the sense in which they were customarily understood." Kerner still believed that the classification of species served important purposes, but the definition of species would have to be modified. The new definition would need to take account of geographic distribution and not only morphology.

Here we have a striking illustration of how the structure and ideology of the Habsburg Empire set different incentives for the sciences than did overseas empires. Britain's chief botanist at this time, Joseph Hooker, insisted that his metropolitan perspective allowed him to see commonality where collectors in the colonies saw difference. He was the preeminent lumper.[52] Kerner, on the other hand, was primarily a splitter. He repeatedly took it upon himself to show that specimens grouped as members of the same species should be recognized as distinct, because they were found in geographically separate areas. Kerner himself argued that he might never have had these insights if he had worked in a state other than Austria. To illustrate, he offered a parable about a botanist from one or another country of western Europe, a man who "trusted authority," whom he referred to as Simplicissimus. Simplicissimus decides to travel to Austria, "the land of contrasts, which encompasses such completely different vegetation regions," in order to study the unique characteristics of different floral regions. To his great disappointment, however, all he finds are "bad species," useless versions of those already listed in his books. "'An odd flora,' Simplicissimus thinks once again, 'in which so many typical plants are only bad species or even worse than bad species.'"[53] Kerner himself had been a Simplicissimus until his stay in Hungary. It was there, he implied, that he had truly learned to see and appreciate the variety of the natural world.

RELICS OF ANOTHER AGE

At the end of his 1863 survey of the "plant life of the Danube lands," Kerner proposed to illustrate what he called, borrowing a term from linguistics, the "genetic [i.e., historical] relationship" among the plant formations described in earlier chapters.[54] The site that would make this relationship visible was a riverbank in the Alps. This bank was distinguished by its tumultuous geological history: avalanches and floods had repeatedly churned the layers of earth and created new surfaces for plants to settle. Kerner's description of this process may be the earliest account of what ecologists later came to call succession: the process by which one plant formation prepares the ground for its successors.[55] Not only did he posit this sequence of events; he went on to estimate the time it must have taken. To do so, he used the azaleas that bloomed briefly in early spring as a tool of temporal scaling. He compared "the extraordinarily large mass of humus" in which the azaleas grew to "the size of the plant and the short length of its annual growing season." In this way, he estimated that this 1.5 foot layer of humus would have taken at least a thousand years to form. The azalea groups were the capstones of this process, and Kerner concluded that they would endure and spread unless disturbed by man or natural disasters.[56]

But where would they spread? How did climatic factors demarcate the geographical range of Alpine vegetation? In lower-lying regions with higher humidity (perhaps a shady spot where snowfall was late to melt, or near the spray from a waterfall), one could find plant formations similar to those at much-higher altitudes. In some such cases, one could assume that seeds or whole plants had been transported down the mountainside by streams. However, when Kerner came upon isolated stands of Alpine flora farther from the high peaks, he found it harder to imagine how they had arrived. These groups appeared to be in the throes of "extinction," and he noted that much of the Alpine vegetation was "retreating . . . towards the moister region higher in the Alps." All this, he had no doubt, was "connected to the reduction in moisture that has followed in historical times from the clear-cutting of forests, from the draining of swamps and countless other incursions on the primordial state of our lands." Kerner was highlighting a robust form of empirical evidence for the destructive impact of human activity on the plant world: vegetation lines.

Emphasizing the reciprocal influence of man, plants, and climate, he speculated that the climate of the northern and eastern Alps must have been more "coastal" in an earlier epoch. Freshwater lakes would have been more plentiful, and the "Alpine" plants would have covered far more of the low-lying land. With the desiccation of the climate and the disappearance of many lakes, the vegetation too had been transformed.

From that point of view, the riverbank he had described, with its tree stumps and snow-roses, was a surviving microcosm of the Alpine landscape in the "diluvial" age. Here was "the most authentic image" of the vegetation of this region in an epoch before men's eyes could appreciate its "splendor and majesty." "The imagination takes pleasure in forming for itself an image of this long past age. We find ourselves transported to a cool, misty mountain land."[57] By means of unassuming plants, Kerner transported his reader deep into a past that no human had ever seen. Yet this feat of temporal scaling was only a trial run, tested on one small patch of a sprawling empire. Kerner's ambition was to tell a history of all of Austria-Hungary, and to do so with flowers.

"OUR HUMBOLDT"

Kerner found it ironic that botanists so often lavished description "in image and word" on the flora of distant lands, while the plants in their own backyards were left waiting for their chronicler.[58] In this vein he quoted the end of Grillparzer's "Entzauberung" (1823): "and my India lies in Moravia." This line, which would later appear in the Moravian volume of the *Kronprinzenwerk*, could well have stood as a motto for Habsburg science in Kerner's day. It was in this spirit of discovery that Kerner set out to explore regions of the Monarchy still virtually unknown to the scientific community, such as the eastern Carpathians, the Lower Austrian *Waldviertel*, and the southern high Alps. He also inspired other Habsburg researchers to explore the Balkans, including Bosnia, Albania, Bulgaria, and Montenegro.[59]

What was the purpose of such surveys? The Zoological-Botanical Society promoted their practical value; they would answer "all the questions raised by a rational agriculture as it strives to meet the high economic demands of our day."[60] Botanical surveys were thus linked to the emerging field of economic geography and the rational basis it provided for Habsburg governance.

At the same time, Habsburg plant geography was infused with the *aesthetic* ideals of the new *gesamtstaatlich* or "whole-state" research programs. As we have seen, it was in the 1850s that self-proclaimed "Austrian" schools of research formed as explicit challenges to the research programs of the nationalist movements. Against arguments for the distinctiveness and antiquity of national cultures, these "whole-state" disciplines collected evidence toward a history of cultural exchange and ethnic mixing in the Habsburg lands. This was, initially, the model that Kerner adopted for his vegetative history of Austria.

To this end, his illustrations aimed to capture local details as vividly as possible. An illustrator with whom Kerner worked explained their modus operandi: "We both held to the principle that for the representation of the *paysage intime*, which is particularly relevant to the botanist, it is necessary to adjust the standpoint of the viewer appropriately, as well as the image's borders or else its contents."[61] The *paysage intime* was the nineteenth-century French ideal of a naturalistic, unromanticized rural landscape. Applied to Kerner's work, the term referred to the unusual close-up point of view of many of his illustrations, in which the smallest plants take center stage (see figure 35 and plate 4). Indeed, Kerner argued that the smallest plants were often the most significant for the representation of a landscape because of their proliferation.[62] His overview of Austria-Hungary's plant life would emerge from countless intimate landscapes.

Kerner explicitly likened his historical inquiry to that of a "curious" reader of a political map, who might wonder how the present "political and national borders" had come to be.[63] Like von Czoernig in ethnography and von Eitelberger in art history, Kerner set out in the 1850s to explain the diversity of vegetation across the empire in terms of a history of mixing and exchange. The plants, however, seemed to tell a different story.

BOTANICAL MIGRANTS

In 1860, as Kerner left Hungary behind for his new post at the University of Innsbruck, he was already formulating an ambitious research program, one that would add to plant geography the axis of time. Kerner once defined plant geography as the discipline concerned with the spatial limits to the distribution of species, and with the local conditions of soil and climate that explained those boundaries. But his very next sentence made an extraordinary leap—from a geography to a history of plants: "Likewise the displacement of the vegetation lines, the advance of a given species in one or another direction, the retreat and extinction of others in historical times are brought under observation, and thus stimulus given to a chronicle of plant migration."[64] This move from spatiality to temporality was anything but obvious to Kerner's contemporaries. Oddly, historians have overlooked Kerner's distinctiveness in this respect, routinely discussing him alongside contemporaries such as August

Grisebach, a more prominent plant geographer who denied the possibility of climatic change.[65] Clearly, we need to understand how and why Kerner broke with that conviction.

Plant geographers of Kerner's day were quite reflective about the spatial scale on which they chose to work, even if they gave little thought to the temporal dimensions of their research. Supported by an army of amateur researchers and low-paid artisans, they were producing successively finer-grained maps of vegetation. In a process that Nils Güttler terms "zooming in," these new representational practices transformed the Humboldtian plant geography of the early nineteenth century into the plant ecology of the late nineteenth century. Magnification made it possible to interpret spatial patterns in the distribution of plants in terms of geographic conditions more specific than the gross generalizations of the Humboldtian era (such as "tropical" or "alpine"). "Zooming in" thus enabled the ecological point of view: a precise understanding of the environmental conditions necessary for life.[66]

Kerner, however, was the first to demonstrate another crucial consequence of "zooming in": it paved the way to a historical perspective on plant life. Questions about plant migration in response to shifting environmental conditions could only be answered once vegetation had been mapped precisely enough for distributional anomalies like oddly shaped tree lines to come into view. In Kerner's terms, a tree line "oscillates" with temporal variations in climate.

Kerner cited his teacher Franz Unger and the British naturalist Edward Forbes as the first historians of the plant world. It was in reflecting on their work that he began to formulate his own research program. Unger's turn to geohistory had come in the 1830s, likely as a result of an opportunity to inspect plant fossils unearthed from a coal mine.[67] He had realized that the present distribution of plants could not be explained only by present-day forces. Unger was an evolutionist, but he did not believe that transmutation was the result of the direct action of the environment. Rather, he favored the idealist theory of the *Bildungstrieb*, a universal drive that created new species out of old, propelling the earth from one stage of organic life to the next, as if the plant kingdom were a single, developing organism. He argued that the earth's climate had gradually cooled over the course of the planet's history, and this explained the greater frequency of plants with "tropical" characteristics in the deep past.[68] Unger likened the study of botanical prehistory to the study of plant geography, but in effect he abandoned the spatial axis of analysis when he took up the temporal one. That is, he described climatic conditions and characteristic vegetation in the past without consideration of geographical variation. Also, unlike Kerner, Unger considered the earth's climate in his own day to be essentially a stable factor, aside from certain cosmic fluctuations and the degrading effects of certain human activities. Thus natural fluctuations and human interventions could cause a temporary advance or retreat of a plant's range, but not a permanent change. The future course of evolution would be determined, he believed, not by environmental conditions but by nature's innate striving for perfection.[69]

In 1851, Unger published *The Primeval World in Its Different Periods of Formation*. In it he conjured lost worlds with evocative texts and lavish illustrations. His desire, he said, was to "express only a possibility," yet to "awaken in men of liberal education a taste for the contemplation of these long-banished periods, and cause them to look upon the Present as the result of this great Past."[70] There is no space here to recount the shock with which Unger's hypotheses were met in a cultural climate still reeling from the revolutions of 1848.[71] More relevant is the inspiration that Kerner took from them. Although Kerner broke with Unger's idealism, he shared his mentor's motivation to synthesize empirical observations into a vivid, if speculative, and popularly accessible account of the history of the plant world.

Just as inspiring to Kerner was the work of Edward Forbes (1815–54), a widely respected geologist and naturalist from the Isle of Man. Forbes had attempted to explain the same phenomenon that perplexed Kerner's Simplicissimus: the existence of similar but not identical species, growing under analogous conditions yet at great removes from each other, such as "Scandinavian" plants in the highlands of the British Isles or "Spanish" plants in western Ireland.[72] Forbes attributed the distribution of flora in Britain to migrations of plants from elsewhere in Europe. He did not suppose that their seeds had arrived borne by currents of air or sea. Rather, following Lyell, Forbes proposed that Britain had once been linked to the European continent by a system of land bridges, formed during a period of continental uplift. Foreign plants and animals had settled on the British Isles during this epoch. As the land subsided, the changing conditions had led to the "retirement" of older species and the "introduction of such as were fitted for the more temperate climate."[73] Thus Forbes adhered to the belief that species had been divinely created not once and for all but in different locations and at different times in earth's history.

Forbes died at the height of his career, aged thirty-nine. The following decade, his work came under attack by the Göttingen botanist August Grisebach. Grisebach believed, like Forbes, in multiple centers of creation, but he drew from this tenet quite a different conclusion. If each species had been designed for its particular environment, there was no need to turn to theories of migration and environmental change in order to explain the spatial distribution of plants. Grisebach did not stand alone, and it may be that the prevalence of theories of multiple creation in this period explains the lack of interest on the part of many European naturalists in tracking and explaining changes in the spatial distribution of species over time.[74]

Grisebach went so far as to accuse Forbes of falsifying his evidence, and Anton Kerner stepped up to defend him. It is from this defense that the epigraph to this chapter is drawn. Kerner argued that Forbes, in piecing together a history of vegetation, had simply acted as any historian would. In defending Forbes, Kerner ended up describing his own research program, which combined historical with experimental methods. He would evaluate the hypothesis of evolution by collecting historical and geological evidence of the prior distribution of plants and by experimentally investigating their mechanisms of dispersal.

A NETWORK BLOOMS

The first step was to build a network of observers, for only on the basis of close cooperation would it be possible to produce a uniform map of plant distribution. Karl Fritsch's phenological network provided a foundation, as did Kerner's personal contacts. While conducting a botanical survey of Carniola, he struck up a friendship with Josef Lorenz (chapter 9), later the president of the Austrian Meteorological Society.[75] He also initiated exchanges with lesser-known naturalists throughout central and southern Europe. One correspondent in Serbia replied in 1865 that he "gladly accepted" Kerner's "friendly invitation to enter into a botanical exchange." Kerner's popular writings also contributed to the expansion of his network, as is evident from letters written to him by fans who then became collectors.[76]

In 1871 Kerner convinced the editor of the *Österreichische Botanische Zeitschrift* to add a new column to the journal: "Chronicle of Plant Migrations." As Kerner explained, he often received plant specimens from regions where the plants had "only recently begun to grow." Often a collector would attach a note about the history of the plant's distribution, but few thought such details to be worth publishing. Individually, observations like these were of little value, but they could be tremendously informative if collected and "interpreted from a synthetic perspective."[77] Kerner offered an example from his own research. The previous summer he had received a specimen of *Rudbeckia laciniata*, a relative of daisies and sunflowers, from an estate owner in Upper Austria. His correspondent remarked on the recent proliferation of these flowers along the banks of the Aist, where they had been cultivated in gardens attached to the mills and ironworks. Kerner decided to investigate.

This involved consulting "numerous" rare books from the seventeenth and early eighteenth centuries, as well as dozens of nineteenth-century articles. He managed to trace the flower from its native home in North America, through its introduction to Europe in the seventeenth century and its cultivation in gardens in Paris and the Low Countries, and finally to its "migration" out of gardens and along riverbanks. In this way, the species had become "naturalized" (*eingebürgert*) in Europe. What surprised him was that he found no references to *Rudbeckia laciniata* in the wild until the middle of the nineteenth century—and then suddenly such references abounded. Where and when the plant had first escaped the confines of gardens, Kerner could not say. But it became clear that the species had turned wild only over the past two to three decades and only in favorable locations.

Kerner's plant geographical network was thus subtly but significantly different from earlier ones. To the task of tracking variations in vegetation across space, he added that of tracking variations across time. He was after stories of how certain plants and plant groups "migrate from place to place and conquer a new field for themselves." Of interest were the "climatic barriers" they encountered and could not "overcome," and thus the emergence of "new vegetation lines and upper limits." Equally of interest were the plants that "do not participate in this movement," those that might instead be "driven back," and might now be found "gradually dying out."[78] Kerner thus set out to collect the data for an imperial history of migration, evolution, and extinction in the plant world.

This project gained new momentum when Kerner was named professor of botany at the University of Vienna and director of the university's botanical garden, founded by Maria Theresa in 1754. Kerner set the garden on a new model. Convinced that modern botanical gardens must serve both experimental research and public instruction, he divided the garden into spaces with different purposes. New greenhouses accommodated experiments, and a portion of the garden was cordoned off from the public to serve research and university instruction. The public area was divided in three, according to different organizational schemes: medicinal and economic uses, systematics, and geographic origins. This last area was the most innovative part of the design and was widely imitated elsewhere.[79] Here the visitor could explore typical plant groups from around the world, including displays of species from Japan, China, Iran, the Himalayas, Siberia, the cape of southern Africa, Australia, and Mexico. This area also introduced visitors to the geographical divisions that Kerner had developed to characterize the flora of Austria-Hungary.

At the same time, Kerner began work on a comprehensive herbarium representing the full range of vegetation in Austria-Hungary. Each specimen was labeled according to identifying characteristics, alternative names, and data on the geographic conditions of the location in which it had been collected. Kerner shared the editorial work with twenty-nine other botanists from across the Monarchy, and they could count on regular contributions from approximately 150 members of his network.[80] In 1881 Kerner published the first installment of the *Flora exsiccata Austro-Hungarica*, which was sent to gardens around the world in exchange for their own catalogs.

Kerner's transformation of the botanical garden and his construction of a collecting network were geared to public outreach as well as research. As he told the liberal daily *Die Neue Freie Presse* in 1886, the *Flora exsiccata* was intended "to enliven and awaken interest for floristics in the widest circles of the fatherland by including all those who take an interest in it and by means of the widest possible distribution of the results of the collective work." The *NFP*'s report stressed that the network transcended divides of nation and class. In easing national tensions, the paper suggested, Kerner's project had succeeded far better than the conservative administration of Minister von Taaffe:

> In order for this many-headed organism to function, for its energies not to be diffused and for information to arrive appropriately selected and on time, a well-designed and unified organization is naturally necessary. . . . The collaborators are distributed over all parts of the Monarchy. The most diverse nations, confessions, and classes are represented among them. Here the German is joined on the peaceful field of scientific activity in harmony with the Czech, the Pole, the Ruthenian, the Magyar, Croatian, and Italian. And what still eludes Count Taaffe in the political realm, the reconciliation of the peoples of this monarchy, has been accomplished here in miniature. Just how diverse the social positions of the collaborators are can be shown by a survey of them from this angle. Naturally, academics and clergymen make up the largest group. To the first belong 27 collaborators, from university professors down to village schoolteachers; the latter includes 13 collaborators, from the cardinal down to a humble curate in the Tyrolean Alps or the pastor of a lonely valley in Transylvania. Alongside these are: 5 physicians, 8 civil servants of the most various types, 3 pharmacists, 3 students, and 5 private citizens.[81]

In a period when other metropolitan naturalists downplayed the role of "local" collectors,[82] Kerner and his admirers celebrated the collaborative character of Habsburg plant geography. His history of botanical migration was a project custom made for the ideology of the supranational state.

Kerner's network connected observers far removed in space; less obviously, it also connected those observers to past and future generations. It aimed to recover observations hidden in historical sources and preserve new observations for posterity. Kerner recognized that natural history needed intellectual continuity, and he prioritized this goal even in his personal life. Just as he showed deep respect for his scientific forerunners, he took care to groom his successors. His son, daughter, and son-in-law all assisted him with the writing of his late chef d'oeuvre, *Das Pflanzenleben*. His son, Fritz Kerner von Marilaun, would carry on his father's climatological research. Likewise, Kerner's son-in-law and former student, Richard Wettstein, would replace Kerner as professor of botany at Vienna, paying frequent tribute to his memory.

Whether spanning across nations or generations, scientific cooperation seemed to Kerner to require a certain ethical effort. In this respect, he echoed the botanist Eduard Fenzl, first president of the Imperial-Royal Zoological-Botanical Society, who had urged his colleagues to master their self-regard and appreciate the seemingly "narrow" interests of their fellow Habsburg researchers. Becoming an imperial-royal scientist meant recalibrating the significance of one's local expertise in relation to a wider geographical field and a longer history of knowledge. Kerner insisted that any contribution to botany could be valuable, whether from a professional or an amateur, from a collector across the world or one nearby. The key was to learn to think on a new scale:

> In our age, which clings to the principle of the division of labor, it has almost become a rule that each researcher progresses only along a single very narrow path. However, narrowness too often has hubris as its consequence. Thus the paths that others simultaneously tread are frequently arrogantly undervalued. In much the same way, with overconfidence in the infallibility of the scientific results of the present, little value is attached to the work of earlier times.[83]

What Kerner described here was a social process of scaling. In order to protect against "narrowness" (*Einseitigkeit*) and "hubris" (*Selbstüberhebung*), the researcher would engage with a network of respected collaborators, recalibrating his estimation of what was distant from him either in either space or time.

FOUNDLINGS OF THE PLANT KINGDOM

In his youthful travels through Hungary and Carniola in the 1850s and 1860s, Kerner had paid particular attention to what he came to call "foundlings"

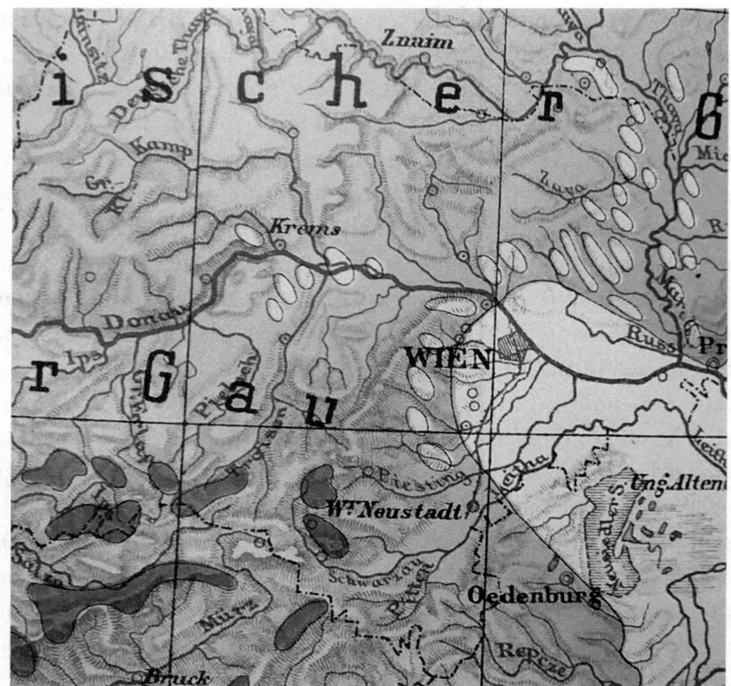

FIGURE 38. Detail of the "Floral Map of Austria-Hungary" (1888), by Anton Kerner von Marilaun, showing the Wachau Valley and the border between Baltic flora and Pontic flora, with Alpine flora at higher elevations.

(*Findlinge*). These were islands of plant associations far from their main population centers, such as high-altitude flowers like Alpine roses and edelweiss on the low coastal mountains of Carniola, or Mediterranean plants like fig trees, tulips, and peonies in Hungary. At that time, the historical explanation for such anomalies seemed self-evident. These vegetative islands must have resulted from the dispersion of seeds through human migration and trade, assisted by the action of wind, water, and migrating animals.

Consider, for instance, Kerner's explanation of the distribution of flowering plants in his native region of the Wachau Valley (see figure 38). In the geographical terms that Kerner introduced in the 1860s but later replaced, the Wachau represented the intersection of three distinct geographical domains: *herzynisch, alpin*, and *pannonisch*. Hercynia referred to forested areas of central Europe and Pannonia to east-central and southeastern Europe; his use of ancient toponyms handily avoided any reference to the political borders of his own day. The Wachau was, for this reason, a particularly interesting field site:

> When crossing the valley the image of the vegetation in many cool shady spots reminds one of the flora of a valley of the foothills of the Alps. Further along, a plant group of the hercynian region recalls to us the dry, sandy forests of pine and spruce of the Bohemian-Moravian hills. And a few steps further perhaps we find "the hair of orphan girls" blowing from a terrace of loess or encounter a citizen of the pannonic floral realm in the shrubbery surrounding some white and turkey oaks, reminding us of the far-off Hungarian plains. And so we come to the conclusion that we find ourselves on a very remarkable patch of ground with respect to plant geography, on which three great central European plant regions converge.[84]

How had this remarkable variety come about? Kerner turned in part to historical evidence to find out. It seemed that the pannonic plants had shifted to the west in the course of human history. Since the local rivers all flowed east, water could not have carried "pannonic" seeds westward from Hungary. Human trade must have been responsible. Thinking ethnographically, Kerner noted that one pannonic species had long ago come to be used in the Wachau region as a signpost for a "forbidden path," as if recalling its foreign origin.

And yet human trade alone could not explain the present distribution. Kerner found that other eastern species had arrived in the Wachau more recently. No corresponding eastward shift of western species had been observed, making it unlikely that these newcomers were the result of livelier traffic between "East and West." In manuscript notes from the 1860s, we see Kerner's observations of plant distribution in the Wachau. He was coming to suspect the influence of climate change. Taking the climatic influence of forests as a given, he inferred that deforestation was pushing the climate of the Wachau in the direction of "continentality," toward lower humidity and greater extremes of heat and cold.[85] Kerner understood himself to be tracking a shifting border between Europe's "East and West." At this point, he was still relying on a crude cultural geography as his template for botanical geography. He suspected that this creeping line would eventually come to rest at the foot of the Alps: "and just as this landscape once appeared as the border between the nations of west and east, now it forms the western wall to eastern Europe's vegetation."

Well into the 1860s, Kerner thought of plants, like human colonists, as actively taking advantage of favorable climates and soils. Like Darwin, he emphasized "the individual organism's capacity to migrate."[86] So, for instance, the Mediterranean plants growing in Hungary appeared to Kerner as a living memorial to the Ottoman conquest. When Moritz Wagner put forward similar ideas in 1868, Kerner was among his more receptive readers. Wagner argued that migration and isolation were the key factors enabling speciation. Without a physical barrier to prevent interbreeding, hybrids would revert to the characteristics of the original species in future generations.[87] Although Wagner was quickly shouted down by the gatekeepers of evolutionary theory, Kerner explored his ideas in an 1871 article, "Can Hybrids Become Species?"[88] By this time, however, Kerner had embarked on a remarkable series of experiments to put the role of migration in speciation to the test.

Soon after his arrival in Innsbruck, he began to note morphological differences between plants growing in the valley and in the mountains. This suggested to him a way to verify the Lamarckian hypothesis. He replanted common flora of the valley in his high-altitude garden, then observed whether the new environment produced heritable changes in the transplants. The results were negative: most of the plants from lower-lying areas failed to reproduce at all in the experimental gardens, but those that did showed no signs of the inheritance of acquired characteristics. As late as 1866, Kerner was still advancing Lamarckian views in publication, but by 1869 he had come out in favor of natural selection.[89] In "The Dependence of the Plant Form on Climate and Soil," Kerner presented his observations on the spatial distribution of related plant species. He had long wondered why the same name was applied to what appeared to be different plants in different regions. Previously, he had assumed that species transplanted to new environments were directly modified by local conditions and the new traits passed on to offspring. Having abandoned the assumption that transmutation resulted from the direct action of the environment, he now suggested that speciation occurs when a variation allows organisms to "fill a hole" near the periphery of the range of the original species.

However, his real interest lay not in the mechanism of heredity but in the relationship between populations of living things and the changing environments to which they had to adapt. Thus Darwin's theory mattered to Kerner above all for its interpretation of sexual reproduction. By reproducing sexually, plants avoided the extremes of "species chaos" and "monotony." In other words, sex allowed a species to generate enough variations to weather environmental fluctuations, while keeping the range of variation narrow enough to allow something identifiable as a species to stabilize. As Kerner put it in 1869, it is a remarkable fact of nature that species are both variable and constant, and this implies no contradiction. Crucially, "the ability to transform into new species is not unlimited," while "the constancy is only temporary." Two aspects of this formulation are noteworthy. First, in the aggregate, the remarkable characteristic of living things was their resiliency in the face of environmental change. Second, the fact that the "constancy" of species is "merely temporary" was a claim about the scale of observation: plant species appear fixed only on the time scale of human experience. Plants had taught Kerner to measure time according to a scale other than the human.[90]

THE DYNAMICS OF PLANT MIGRATION

The "migration" of plants had rarely been studied experimentally when Kerner took up the problem in the 1860s. Naturalists tended to assume that the present distribution of plants had resulted from the transport of seeds by humans or animals on the move, by running water, or by air currents. In *The Origin of Species*, Darwin had attested to the capacity of seeds to germinate after such trials as being submerged in seawater, swallowed by a bird, or embedded in a beak, claw, excrement, or floating log. His evidence suggested that plant seeds were hardy long-distance travelers. He was therefore inclined to explain present-day distributions in terms of migration and to reject explanations that appealed to geographical changes such as the appearance and disappearance of land bridges, as Forbes had proposed in 1846.[91]

Kerner too had long thought of plants as avid migrants, yet in the late 1860s, he began to doubt this assumption. Like Darwin, Kerner had convinced himself experimentally that many types of seeds could survive passage through a bird's digestive system with their fertility intact. However, consideration of the timing of seasonal migrations and of the preferred foods of migratory birds made this an unlikely explanation for the presence of southern plant varieties in the high Alps. Indeed, none of the fauna found in the high Alps were known to live in distant regions. Nor could human traffic explain the presence of such foreign plants in uninhabited areas.[92] Meanwhile, by the 1870s, the younger de Candolle had concluded that most species of seeds could not in fact survive water transport.[93] With these possibilities eliminated, it seemed logical to attribute plant migration to the action of winds. Kerner took up the challenge of testing that hypothesis.

Doing so meant taking up the perspective of dynamic climatology as it was then being developed at the ZAMG. Kerner was an enthusiastic participant in the growth of atmospheric science in Austria in these years. Three of his research papers appeared in the *Journal of the Austrian Society for Meteorology*, and he founded a weather station of the imperial observing network near his summer home in Trins. His study of the variation in soil temperature with exposure was cited approvingly by Julius Hann in his textbook on climatology. Like the best of the dynamic climatologists, Kerner insisted on the importance of observing atmospheric phenomena continuously as they unfolded in time. This meant striking out at the critical moment, rather than relying on the scheduled measurements of a permanent observing station. In the opinion of the physical geographer Carl von Sonklar, who dealt extensively with questions of climate, Kerner's work was remarkable for "the thoroughness of the research . . . and even more the all-around highly developed physical understanding of the author."[94] Kerner even designed an instrument to solve a problem at the heart of the forest-climate debate: how to standardize the measurement of dew. Decades earlier, Karl Fritsch, director of the ZAMG's phenological network, had complained to Kerner that his dew measurements were "looking like a failure." Nowhere could he find reliable observations, and "still less am I able to say how the amount of dew depends on variations in exposure and air currents."[95] Kerner's drosometer filled this need. It used a thin leaf of aluminum to collect dew, while the weight of the dew was measured by the displacement of water within a thin pipe, which was calibrated both by absolute weight and by its equivalent as a height of precipitation. One forestry journal judged that this device had finally succeeded in producing "relatively inter-comparable data."[96]

Kerner began his investigation of the large-scale mobility of plants with an observation of botanical life on the smallest of scales. Imagine yourself in the high Alps in the middle of a bright sunny day. Were you to take a rest in the partial shade of an overhang, you might notice that the air, lit from the side, is teeming with tiny seeds. Each of these seeds is equipped with miniscule hairs or wings for gliding. Kerner attempted to estimate their number by catching them in his hands. He found that an average of 280 seeds, some weighing as little as 0.02 milligrams, passed through an area of roughly one square meter per minute. The question that mattered to Kerner—the question critical to the history of Alpine plants—was just how far such seeds could travel, once they had hitched a ride on a mountain breeze.[97]

This was, first, a matter of empirical aerodynamics. Using a fan to produce a controlled horizontal air current, he tested how far seeds and fruits of different types could glide. By far the most flight-worthy were those equipped with what Kerner described as feathery parachutes; these did indeed seem capable of covering vast distances and ascending to "extraordinary heights." Kerner admired the "marvelous structure of the appendage," which optimized the surface area of the seed in contact with the air relative to its weight. Just as significant was the fact that this parachute was made of hairs rather than a continuous membrane, achieving the same surface area with less weight. It would seem that such seeds could indeed be carried far and wide by modest winds. But was there evidence of such journeys?

For this purpose, Kerner judged, moraines provided "natural testing grounds." Within a few years of a glacier's retreat, plants would begin to grow here and there in the sand between rocks; after a decade or so, those pioneers would have produced enough humus for a second generation of settlers to move in.[98] Therefore, where flowering plants could be found growing amid the till left behind by a retreating glacier, it seemed safe to conclude that they had arrived as seeds relatively recently. Still, there were no surprises among the species Kerner found growing on the moraine, nor buried frozen in the glacier; all were well represented in the surrounding valley. Nonetheless, Kerner considered the possibility that seeds from afar might arrive with "equatorial air currents." A southern wind might carry seeds as "a gift and a memento from its warm homeland far away."[99] Notice that Kerner was still following Dove in thinking of warm central European winds like sirocco and foehn as blowing from tropical regions. And yet, remarkably, Kerner next proceeded to apply his own thermodynamic considerations. In effect, he turned a problem of botanical geography into one of dynamic climatology.

Kerner noted that for a seed to be borne and remain aloft, it must remain dry. A seed in moist air would drop quickly to the ground, as Kerner's experiments demonstrated. But this situation could never obtain in reality, he argued. It was ruled out, "always and necessarily," by the laws of thermodynamics. Any air current bearing a seed aloft would expand and cool as it rose, and its relative humidity would increase accordingly. The seed would be lifted to a certain vertical limit, and then it would be able to rise no further. That limit would vary with the type of seed, the season, the time of day, and the topography. Kerner estimated its maximum at five hundred to six hundred meters above the highest peaks, taking this to be the upper limit of a vertical air current, presumably on the basis of experiments with kites. What's more, Kerner observed that on days with warm sun and low humidity—conditions conducive to generating rising air currents and keeping a seed dry—the horizontal wind component tended to be weak. So no matter how high the seed rose, it would land a relatively short distance away, easily confined by the towering peaks of the Alps.[100]

Kerner's reasoning rested on painstaking observations of atmospheric conditions up and down the valley walls (see figure 39). From the 1850s into the 1880s, he filled dozens of notebooks with measurements of meteorological elements, observations of clouds, and diagrams of wind directions. He even invented his own method for detecting winds too light to turn a wind vane. It used nothing more than a wax candle attached to a pole and matches. Once extinguished, the candle's smoke would mark a trail indicating the wind's direction.

On the basis of this dynamic analysis, Kerner ruled out the possibility of the "transport of airborne seeds across broad expanses of land and sea," at

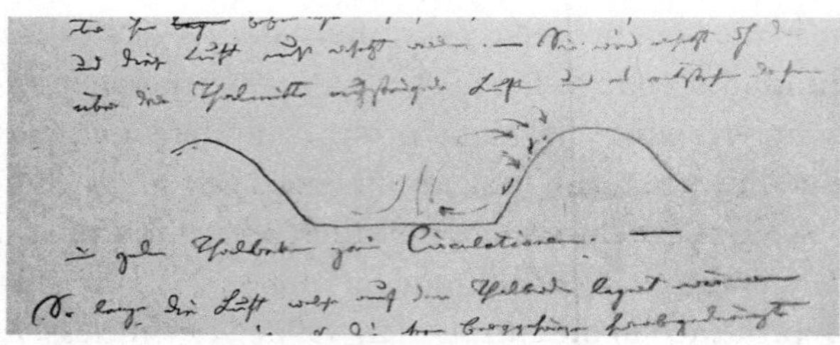

FIGURE 39. Example of Kerner's unpublished notes on atmospheric dynamics, showing the direction of airflow in a valley.

least in the Alps.[101] Recent studies using dispersal and seed-bank analysis have confirmed this finding.[102] This conclusion had important consequences for the curious phenomenon of "foundlings," particularly the isolated occurrence in the eastern Alps of plant communities ordinarily found far to the south. Kerner had now rejected the idea that these seeds had been transported by humans, animals, water, or air. There remained little doubt in his mind that such outliers were instead "forgotten outposts of a former population that had previously lived throughout the region." He continued: "the fact that such colonies of southern plants appear in several places in the eastern Alps leads to the conclusion that a warmer climate reigned in the region of the eastern Alps after the last Ice Age . . . but that later, due to a change in the climatic conditions, those species were restricted to more southern areas and only remained at isolated, climatically highly favorable locations to the north."[103] In this way, Kerner's painstaking observations of winds in mountain valleys, combined with his early grasp of atmospheric thermodynamics, convinced him that the mobility of plants in mountainous areas was tightly constrained. For this reason, the present distribution of Alpine plants was a crucial clue to the climate of the past. With plants to guide him, Kerner had succeeded in turning the instruments of dynamic climatology—thermometer, barometer, hygrometer, anemometer, candle, and kite—into tools of temporal scaling.

A HISTORY OF AUSTRIA-HUNGARY IN FLOWERS

Kerner published less frequently in the last decade of his life—not for lack of energy, but because he was preoccupied with work on his popular magnum opus, *Das Pflanzenleben*, which appeared in two volumes in 1888–90, with a revised edition between 1896 and 1898, the year of his death. In the 1880s, he also made two final contributions to the floral history of Austria-Hungary. First, building on his botanical overview for the *Kronprinzenwerk*, he constructed a "Floral Map of Austria-Hungary" for the new *Physikalische-Statistische Handatlas* (chapter 5). This was a visual synthesis of a lifetime of research into the diversity of Habsburg vegetation. It portrayed the empire as the site of the convergence of four floral realms (see plate 5). His final contribution, presented to the Austrian Academy of Sciences in 1888, took this map into the fourth dimension.[104]

This exercise in temporal scaling began with the familiar problem of interpreting the occurrence of "foundlings" in the Austrian Alps, at the boundary between the regions of Pontic and Mediterranean flora on his botanical map. By 1888, Kerner had amassed a vast trove of empirical information to back his claim that floral foundlings were signs of radical and relatively recent shifts in climate. He had acquired more detailed evidence of present-day plant distributions in Austria-Hungary and other mountainous and steppe environments in Eurasia, including Scandinavia, southern Russia, and the Himalayas. These observations showed that Alpine flora grew well beyond the Alps: east into the Carpathians, Caucasus, Altai, and Himalayas, north into Scandinavia, even south to the Dinaric Alps, the Pyrenees, and the Balkans. How could this wide dispersal be explained? Certainly not in a Humboldtian framework. Kerner reminded his readers that Humboldt had posited the identity of Alpine and Arctic flora, assuming a neat correspondence between latitudinal and altitudinal variation. Detailed surveys showed this was not the case. No, Kerner insisted, the question was how specifically Alpine varieties had spread so far and wide, and only a geohistorical explanation would suffice. Fortunately, he had new evidence to draw on: plant and animal fossils, as well as recent geological discoveries such as that of Ice Age high-altitude bridges (anticlines) between the Alps and Carpathians.

From there, he opened out to furnish a sweeping vision of the Habsburg lands from the Ice Ages to his own day and into the future. The evidence suggested that the present distribution of vegetation was due to an extended period of warm, dry summers following the last Ice Age. It was then that the Scandinavian and Alpine glaciers had begun their retreat from the plains of central Europe. Seeking cooler conditions, some Arctic varieties would have gradually migrated to higher elevations to the south, while some Alpine types would have found refuge to the north. Migration in this sense meant an exceedingly slow shift of the distributional limits, not a colonizing expedition. The presence of Alpine flowers in the Balkans was harder to explain, since there was no geological evidence of glaciation there. Kerner suggested that these species were in fact natives of mid-altitude regions that had moved upland in the Alps as the climate warmed and the glaciers receded. Like most of his contemporaries, Kerner mostly left aside the question of the *cause* of such a "profound" climatic shift. He did, however, venture a hypothesis to account for the regional change that most interested him: the separation of Pontic from Mediterranean flora. Echoing a theory put forward by Eduard Brückner at this time, he suggested that it was due to geological "changes in the configuration of the continent" around the Black and Caspian Seas, changes that had given the eastern and southern parts of Eurasia distinctly different climates.[105]

In this way, Kerner arrived at a general theory of the response of vegetation to climate change in mountainous regions. In periods of cooling, plants requiring a longer growing season would die out in the valley, making room for vegetation from higher altitudes. Conversely, in a period of warming and glacial retreat, plants from the valley would move up the mountainside. They would arrive in stages, each community preparing the soil for the next. In the case of the eastern Alps, these settlers would have come from the south and east, that is, from the regions of Mediterranean and Pontic flora.

Those "homesick" plants had proved their value as scaling tools. They afforded a vision of spatial and temporal continuity: from the eastern Alps north to Scandinavia, east to the Eurasian steppe, south to the Balkans, and from the last Ice Age into the present. "The present cannot be sharply marked off from the warm period following the last great advance of the glaciers. The transition was a very gradual one."[106]

Looking ahead, Kerner saw no reason to believe that climatic conditions would remain as they were at present. "If for once the constraints that restrain plant migration at the boundary of high-altitude flora fall away or expand or contract, then even these stable flora will again be in movement and flux."[107] Kerner spoke of the plant world as a *Lager* or "storehouse"—not in the traditional sense of a stock of resources for the satisfaction of human wants, but in a new sense: that of a system equipped to adapt to a changing climate. He observed that mountains were endowed with such a wide variety of plant types, whether sprouting or dormant, that they would always house suitable colonists for lower altitudes, no matter how radically the valley should cool.[108] In much the same way, plant geneticists today speak of ecosystems like forests or coral reefs as "storehouses" of biological diversity, transforming an anthropocentric metaphor into a biocentric one.

WITH THE BLESSING OF THE LAND

Kerner was an eloquent advocate for the role of imagination in scientific research. Even if it often led to error when allowed to "soar unanchored by empirical observations," imagination (*Phantasie*) was an essential "aid to research."[109] Indeed, Kerner's experiments in scaling depended on his capacity to imagine a temporality other than that of human life.

When Kerner was in his forties and had attained the rank of Hofrat, he and his wife decided to have a summer cottage built near the Trins moraine in the Gschnitz Valley in Tyrol. This is where they spent all the subsequent summers of their lives. Kerner's son Fritz and son-in-law Richard Wettstein built their own cottages nearby, and Kerner frequently hosted visiting international scientists here. In this way, Trins became one of the most thoroughly studied geobotanical field sites. Thanks to this historical tradition, it still serves today as a benchmark for the climatic history of the eastern Alps.

In choosing this site, Kerner was already thinking of its historical significance. For it was here, in 1867, that he had discovered the *Primula auricula*, the Alpine flower that Clusius had hunted in vain three centuries earlier. Kerner's insights into the changing climate of the Alps might have given him a new appreciation for this find. Clusius's years in Vienna (1573–88) coincided with the start of a period of pronounced cooling in the Alps (1570–1630). Perhaps when Clusius came looking in the 1570s, the seeds of the auricula were already frozen in a layer of ice. Kerner, on the other hand, was working in the warming phase that followed the end of the Little Ice Age. The moraine near Kerner's summer home was deposited after the retreat of the Trins glacier some fifteen thousand years ago, when much of the rest of the Alps was already ice free, and it may be for that reason that it was home to plants not found elsewhere. The glacier advanced again during the Little Ice Age, reaching its recent maximum extent circa 1850. Evidence suggests that the hillside where Kerner made his discovery had remained glaciated longer than surrounding peaks.[110] According to recent research, it is possible for seeds of many Alpine plant species to survive in a frozen state for long periods of time and remain germinable.[111] Arriving in 1867, Kerner might well have found the auricula blooming in the moraine for the first time in three centuries.

To mark the occasion of the completion of the summer cottage in 1874, Kerner wrote a festive play. It was in the style of the Renaissance pageants in which courtiers had been cast as features of the natural world. In this case, the roles consisted entirely of rocks and plants. The play opens with the character Flint reminiscing about the Ice Age, remarking on how much the landscape has changed since then. Suddenly he hears a hammering and sees the glint of an ax on the moraine. He turns to Larch (an evergreen) and Charlock (a flowering wild mustard) to ask who is "disturbing the peace of the valley?" Larch informs him that a human couple arrived "two moons ago." Charlock urges

Flint to let loose an avalanche and flood to scare these intruders off. Instead, the characters begin to consider what manner of humans these are. As they speculate on the couple's intentions, we see how the work of climatological research might appear from the perspective of the plant and mineral world. "With rod and rule they began to measure, note, and sketch the contours of the hill." Then the humans "stretched themselves out comfortably on the ground," where the grasses eavesdrop on their conversation. In the end, the characters conclude that the couple mean no harm. The stones and flowers bless Kerner and his wife, wishing that they may take pleasure in the surroundings "until their hair is the color of the mountain faces that encircle their beloved cottage."[112]

This glimpse into Kerner's personal life suggests how deeply his experiments in scaling were informed by his poetic imagination. By giving voice to plants, once again, Kerner managed to juxtapose his personal experience of aging with the time scale of geological change. The dialogue among Kerner's vegetal neighbors beautifully illustrates Suess's dictum that nature must be measured by man but not according to man. In this case, the result is a nonhuman perspective on the measurement process itself. Not only does it convey the ephemerality and fragility of human life from the point of view of geological forces and geological time. It also underscores how partial a view of the nonhuman world is captured by human measuring instruments.

CONCLUSION

Kerner's achievement was, above all, to grasp that radical climatic shifts might be a part of the earth's future as well as its past, and to offer his contemporaries, scientist and nonscientist alike, the tools to imagine how familiar living things might react to a more or less radical shift in climate. Openly acknowledging the uncertainties inherent in his method, he nonetheless presented a possible future that was consistent with all known data. It was an imaginative act—that Kerner freely admitted—but it was a vision grounded in vast empirical research and careful calculation.

Essential to Kerner's botanical history were his practices of spatial and temporal scaling. Plants were tools for situating unfamiliar regions and distant epochs with respect to a nascent map of the Monarchy. Exotic vegetation became a metric of distance, while familiar plants were signs of cultural proximity via migration and exchange. At the same time, Kerner mined archives, libraries, herbaria, and botanical gardens to situate his observations in relation to earlier scholarship. "Each of our theories has its history . . . and only narrow-mindedness and learned foolishness [*Gelehrtendünkel*] can view and promote the [scientific] laws of the present as infallible and immutable."[113] His finely honed historical sensibility reflected in part his self-consciousness as heir to an Austrian tradition of natural history, from the courts of Maximilian and Rudolf to the reincarnation of the Vienna Botanical Garden under his own direction.

And yet Kerner's capacity to discard old theories delineates the scope of the imperial-royal scientist's autonomy from Habsburg ideology. Although he began his vegetative history of Austria on the model of other patriotic histories, his meticulous empirical research eventually convinced him that this model had to be modified to account for the precise distribution of the empire's diversity of plant life. The hypothesis of migration had to be qualified by the evidence of dynamic climatology. Imperial history had to be rewritten as climate history.

Kerner recognized that these conclusions were of little interest to most botanists of his day, focused as they were on the cellular scale. But he felt sure that times would change: "Even in science, fashions change. . . . And so there will surely come an age that will better recognize the value" of these observations. "For the recording of data that make it possible to follow the gradual changes in vegetation, they will owe us the greatest thanks."[114]

CHAPTER 11

Landscapes of Desire

In public, the imperial-royal scientist appeared confident in his ability to mediate between competing systems of measurement and valuation. In private, however, this work was fraught with anxiety. When the young hero of Stifter's *Nachsommer* announces to his mentor, Risach, that he has learned the lesson of the relativity of scale, the wise naturalist suggests that the lesson is not so easily learned: "When there is a surfeit of wishes and desires in us, then we listen only to these and cannot grasp the innocence of things outside us. Sadly, we call them important when they are objects of our passions, and unimportant when they bear no relation to those passions, although the opposite may often be true." The hero reflects: "At that time I didn't yet understand the full meaning of these words. I was still too young and often listened only to my own inner voice, not to the things around me."[1] Stifter's prose, so far removed from the ecstatic sublime of the Romantics, reads as an effort to teach the meaning of Risach's words. What might strike a modern reader as belabored descriptions can be read as a lesson in the patient, sober, selfless observation of the material world, the antidote to the narrator's "surfeit of wishes and desires."

The persona of the imperial-royal scientist represented at once a moral and a physical ideal. Morally, it implied the subordination of local concerns to the goals of "whole-state" research. Physically, it meant the acquisition, mile by mile, of bodily knowledge of the diversity of the Austrian lands. Pursuit of this ideal plunged many scientists into a tense struggle. In their travels around the Habsburg lands, they were frequently beset by longings they did not know how to satisfy, whether tempted by the exotic or lured by nostalgia for home. Charged with distinguishing between things that were small yet significant, and those that were truly petty, these scientists became tormented by their liability to be distracted by irrelevant details or merely personal concerns. The private side of scaling was the effort to mediate these conflicts.

The archival documents that give access to such private experiences are of a kind that many historians of science would skip right past. They are legible only with the greatest effort and might sooner be classified as poetry than science. Yet I will argue that any genuinely historical understanding of dynamic climatology must confront the emotional turmoil that these scientists experienced as they reoriented their sense of near and far.

The theme of this final chapter, then, is desire. It is about men who experienced themselves as ruled by desire. In their hands, climatology took on a role more commonly assigned to religion and acquired something of its ethical force. The science of climate became a tool for gauging the significance of things apart from the fickle longings of body and soul. Here was a science that would allow its practitioners to disentangle personal inclinations from universal laws, local contingencies from general patterns. It would enable them to right their own sense of proportion, so as to see clearly the significance of those "small things" that truly mattered, while overlooking those that were indeed petty. In this respect, the work of scaling was, in part, an attempt to stabilize a masculine sexual identity that could support the social relations of imperial-royal science.

THE PRIVATE WORK OF SCALING

In the spring of 1856 the eighteen-year-old Julius Hann began to record "little notes about experiences, intellectual and personal interests in the form of a diary." They were nothing more, he said, than "little nature sketches," notes on his readings and studies, and descriptions of "the changing character of the landscape according to weather and season, which always left a deep impression on me." They seemed to him in many parts "entirely meaningless and petty." He later considered destroying them, but he did not, because he had come to value them as records of his personal development. Invoking a spatial metaphor that recurs throughout his diaries, he wrote that personal memories could free the mind at times when "our vision and efforts are lost in the minutia of a shallow, flat environment."

Retrospectively, Hann cast these diaries as a chronicle of his struggle to adapt to the social world of science in the imperial capital, to rise to his professional "calling." "It took repeated, painful struggles and experiences in order to do away for the most part with the habits and training of this solitary and

FIGURE 40. Photograph of Julius Hann (1839–1921).

ostensibly self-sufficient way of life, and to be fit for truly, uniquely enriching and uplifting relationships with other people." And so Hann decided to preserve these books, and he justified this act with reference to the relativity of scale:

> Until recently I was of the opinion that it was best to throw my past away, once I had the opportunity to measure [*messen*] it in its simplicity and triviality against richer, finer lives. But that would have been a mistake. Each of us can do no better to fulfill his duty, and for his own good, than to foster and develop in good measure [*maßvoll*] what Nature gave him as his own. . . . And if it appears quite small when compared to richer, stronger, more highly endowed natures, it has its value in its place; it is one of the driving forces in the rich unfolding of life.

What appeared small and merely personal might hold value when viewed from an appropriate distance.[2]

Hann's feelings of weakness may have reflected his social status. His father, Josef Hann, the son of a craftsman, had risen to the position of caretaker of Schloss Starhemberg, a sixteenth-century palace near Linz where Julius spent his first nine years.[3] This early setting remained vivid to him: he could recall the quality of air and light, and even recounted an attempt to draw the estate from memory.[4] "I often have dreams of that time, spent entirely in the lap of nature in the solitude of the estate I called home, in fields of flowers, and which now, in memory, is but one unbroken springtime."[5] Josef lost his post in the course of the revolutionary year of 1848, as a consequence of the reform of the land-tenure system. He then received an imperial appointment as a magistrate in nearby Kremsmünster. But he died two years later, in 1852, when Julius was thirteen. The family spent the following year in Linz, where Julius attended his first year of gymnasium. His instructor in natural history was Dominik Columbus, the first person to serve as meteorological observer in Linz at the station newly founded by Karl Kreil, himself a native of Upper Austria. But Julius did not remain in Linz for long. His mother had been left without a pension, and in order to support herself she moved with her children to Kremsmünster, where she opened a boarding house for students of the monastery's gymnasium. The boarding house flourished, becoming a mainstay of student life into the twentieth century. It was maintained by Julius's beloved sister Anna after his mother's death in 1873. Meanwhile, Julius had the good fortune to find his way to the Kremsmünster gymnasium. There he proved to be an unusually earnest, eager, and attentive student.[6]

Hann had mixed feelings about monastic life. As he wrote in 1861, after eight years of studies: "Corpus Christi.—These days interacting with the novices (some already priests) I clearly sensed the reverse side of monastery life: rather awful gloominess, spiritual apathy—undoubtedly a consequence of the miserable, uniform, intellectual and bodily lethargy-inducing life of strict isolation without stimulation and ventilation of the mind, grown musty. Five years of this!" Although he was not attracted to the seclusion of a monk's life, he was drawn to the monastery's Mathematical Tower. He befriended a priest by the name of Gabriel Strasser, who was then an adjunct to the director of the observatory and a teacher of physics and mathematics at the gymnasium. Strasser was the son of a miller, but his intellectual abilities had been recognized early on by a village schoolteacher, making it possible for him to study at Kremsmünster. From 1873 until his death in 1882, he would serve as the observatory's director, and in that period he published extensively (in the yearbook of the ZAMG and elsewhere) on meteorological observations made at the abbey.[7] Hann admired Strasser and envied aspects of his career: "And so I still thought back then, if this could be endured, then I could probably be quite happy as a priest. Particularly alluring to me was the prospect of a position like that of Father Gabriel, which seemed to me from just about every angle to fulfill all of my desires. I thought of Gabriel, equipped with aesthetic feeling and a great depth of feeling and imagination—how could he not be happy there! Oh that you would surely be, I thought!"

Hann was indeed a deeply spiritual if unorthodox youth. He addressed his prayers not to God but to the natural world. In doing so, he echoed the Benedictine trope of listening to the divine word with a "wide" heart. Thus in his diary he apostrophized his natural surroundings: "Welcome me again into your circle, spring and summer days! May you make my eyes more radiant, my spirit wider, my heart larger and stronger, that it will cast away all the weight of sullen worries that rob one of time and mirth."[8] In a similar vein, he addressed the atmosphere: "You fortifying, invigorating, purifying forest air! Flow through me, cleanse me, purge from all that is dull, slack, ailing."[9]

It was in these years that Hann first read the great prose works that would provide models for describing both his emotional life and the natural world. He quoted long passages from Goethe, including the *Italian Journey* and *The Sorrows of Young Werther*, which found echoes in his private accounts of his inner states. His style also borrowed from Alexander von Humboldt, whose *Cosmos* he read with fervent excitement in these years, and whose death, reported in the press, struck him with a "wondrous, shuddering feeling."[10] He was inspired, too, by the ideal of scientific landscape painting proposed by the Saxon physician and artist Gustav Carus. Carus foresaw a great future for landscape art, christening the new genre of *Erdleben* (earth life, by analogy to *Stilleben* or still life), which included above all paintings of atmospheric phenomena. Carus, a Romantic, understood landscape painting as a process of uncovering the mystical aspects of nature. "How infinitely varied and delicate are atmospheric phenomena! Whatever finds a resonance in the human breast; processes of lightening and darkening, evolving and dissolving, building and destroying: all this displays itself to our senses in the delicate forms of the clouds."[11] In his diary, Hann quoted Carus on the potential for science and art to motivate each other: "From awareness comes *knowledge*, or science; and from skill comes art. In science, man feels himself in God; in *art*, he feels God in himself." In this spirit, Hann's diary moved effortlessly between aesthetic sketches and scientific observations. Poetic description might shade into the scientific observation of clouds and their classification, and from there into a synoptic vision of a storm.[12]

Hann's diaries chart his deepening engagement with the science of weather and climate. In later entries, his reports of meteorological phenomena grow increasingly detailed and precise. Starting in 1861, he records his own thermometer readings, along with telegraphic reports of meteorological conditions elsewhere in Austria and beyond. His memory for weather events was remarkably keen.[13] Upon reading a paper published in 1862 about a storm that struck Vienna in 1854, he was able to recall the details of the storm's effects in Kremsmünster: "I vividly recalled the day. It was a Friday, with a long Latin class to come in the afternoon, as I headed home from school around 10. A strong east wind was driving the clouds and frequently beat the branches of a tall thin pear tree against the window of our living room. Just a couple of hours later the storm swept in from the southwest."[14] From these recollections and the time of the storm's arrival in Vienna, he even made a rough calculation of its velocity. On another occasion, upon acquiring a treatise on meteorology by Heinrich Dove, Hann remarked: "With what joy does one receive such a book, if one is prepared inwardly and thirsts for instruction." Just over a month later, he was able to observe for himself phenomena that Dove had described.[15] He wrote repeatedly of his engagement with the new science of meteorology as "the fulfillment of my wishes." Hann's passion for meteorology and climatology reflected the broad aesthetic and ethical aspirations associated with cosmic physics by a leading Habsburg physicist like Baumgartner, who had spoken of training "an eye that reaches out across the earth."[16] From such grand aspirations sprang the intensity of Hann's excitement for this science.

In the years chronicled by the diaries, Hann alternated between periods of self-doubt and moments of ecstatic optimism. He often expressed these contrasts through spatial metaphors. Confidence was represented by an image of a wide horizon, and despair by narrow constraints. He expressed his determination to rise above the *Kleinigkeiten*, the petty concerns, that distracted him from his studies and his pleasure in the natural world—in his words, "the constantly agonizing disproportion [*Missverhältnis*] between my weakness and the high aims towards which my powerfully building enthusiasm pointed me." In a similar spirit, Humboldt had written in *Cosmos* of the "contrasts between what is morally infinite and our own narrowness, which we aspire to escape."[17]

Cosmic physics in the tradition of Baumgartner confronted the individual with the undeniable fact of his own smallness, "allowing one to feel more deeply one's own narrowness [*Beschränktheit*]." Gone was youth's illusion of omnipotence, its blindness to its own constraints. In this way, cosmic physics could right a young man's sense of scale: "Eternally constant nature! Thus you mock our inner storms, our joys and our pains!"[18] How harmless the "storms"

FIGURE 41. *View of the Falkenmauer from near Kremsmünster*, by Adalbert Stifter, ca. 1825.

of one's daily life appeared when measured against those of the heavens. Hann thus set himself the goal of developing a more realistic sense of proportion.

LONGING

A central theme of Hann's diaries is the conflict between his longing for foreign landscapes and his persistent *Heimweh* or homesickness. On homesickness, Hann cited the German geographer Carl Ritter, who held that the impression made by one's native landscape was "so deep" that a separation from this environment brought on a form of sickness, "in which a person's whole being may dissolve into longing." Ritter explained *Heimweh* as an "uncanny" remnant of a primitive stage of human life that persistently haunted modern societies that had otherwise freed themselves from the bonds of nature.[19] When Hann wrote of *Heimweh*, he implied both a general nostalgia for youth, typically symbolized by a beautiful boy, and a longing for a particular place: the abbey at Kremsmünster.

Hann had formed such close ties with his teachers and classmates at Kremsmünster that parting from them between terms caused him great pain. He expressed special affection for the younger students—"nice, lively lads"—whom he occasionally instructed. Of one young student, he remarked: "The refreshing directness and unaffected spiritedness of the handsome boy moved me strongly. Here bubbled a well of fresh vigor with its natural force intact."[20] A reunion with a group of boys provoked a similar outpouring: "Greeted all the dear fellows, the nice chaps, whom I always find so cheerfully stimulating. A marvelous charm lies in the freshness and innocence of boyhood, a charm that we feel all the more vividly and longingly, the more the expansion of our knowledge confines us for good to a small sphere, and then again are bound to feel ourselves outwardly confined by conventions." Hann singled out one companion in particular, a boy he referred to as "der kleine Alex."[21] Upon Alex's departure from Kremsmünster, he recorded:

> The days that followed were dreary and bleak! The parting, the separation from everything I loved so much, everything that had become so intimately, so completely a part of my inner life, plunged me into misery—which, with my sensitive nature, and in the absence of other stimuli, crushed me nearly to a stupor. All my attempts to master myself [*mich zu ermannen*] failed, none of my studies could hold my attention, nothing could tear me from the tantalizing [*quälend*] memories.[22]

Hann experienced these longings, along with his loneliness in the absence friends and family, as a crippling weakness; he felt "unmanned." He went on to upbraid himself for his timidity and solitary ways. Apparently, the homoerotic bonds that he had formed at Kremsmünster threatened to derail his pursuit of a career in the imperial capital.

In these years, Hann fantasized frequently about traveling. He often read travel stories by writers like James Fenimore Cooper and Charles Sealsfield, admiring the latter's "splendid, vivid descriptions of nature and society in North America." His diary from these years contains a list of books on scientific exploration that he hoped to purchase, from Ross's account of Antarctica, to Darwin on South America, to a collection of narratives of round-the-world journeys.[23] He writes often of his "longing" for certain foreign places. "My readings have once again inspired a powerful longing to see the seaside and filled me with day-dreams and imagined scenes."[24] The landscapes he envisioned, often quoted from literature, combined clichés of the Mediterranean with wild fantasies: a purple moon, the roar of steamships on the Mississippi, the golden light of Vesuvius radiating over the bay of Naples. One summer in Kremsmünster, he wrote that the quality of the light "awakened my desire to travel and a deep longing for the wonders of distant southern lands, especially for the richly colored barren expanse of the blue sea!" And then he copied out the 1838 poem "Sehnsucht" (Longing) by Emanuel Geibel, composed as a farewell to his friends in Berlin before embarking on a long-awaited trip to Athens. It begins with the traveler's ambivalence:

> I look into my heart and I look into the world,
> Until a burning tear falls from my eye,
> Though the distance glows with golden light,
> Still the North holds me fast—I will not arrive.
> O the bonds are so tight and the world so wide,
> And time goes so quickly by![25]

Through poetry, Hann was finding a way to articulate the conflict he felt between his longing to travel and his craving for home. In his familiar Upper Austrian hills, in the popular lakeside resort of Gmunden, he expressed disdain for the superficial gaze of the tourist: "This place is teeming with tourists and life here disorients my sun-bedazzled heart, used as it is to solitude and familiar surroundings. The eye blinded and seduced by the gleam of the surroundings forgets that infinitely more lies in the depths of one's own soul [*Gemüt*]."[26] Nonetheless, Hann believed that travel could also benefit the soul. Shortly after parting from his gymnasium friends in August of 1860, in the depths of loneliness, Hann was able to distract himself with his first rail journey. In Salzburg he seems to have thoroughly enjoyed the role of tourist. Shortly after, he took an overnight trip to Wels, Upper Austria's second largest town. He described this emphatically as an expansion of his horizons: "This little excursion had the benefit of teaching me to recognize that there is much that is beautiful and desirable beyond my former horizons."[27]

Hann's diaries thus record his struggles to master his sensitive character, in order to make his way into the circles of imperial-royal science—a quest that began with his enrollment at the University of Vienna in the fall of 1860. The city did not agree with him. He felt insecure in academic circles, awed by his professors. The amusements of student life made him uncomfortable, even physically ill. "At home this evening," he reported. "Lifeless, in an ill humor, and physically unbalanced from indisposition. And outside bleak weather.—I'd taken part in some vacuous, inane form of fun, which always leaves me quite morose."[28] Hann particularly hated laboratory work, always comparing it unfavorably to field research in the open air. He spent Christmas conducting chemical experiments with charcoal powder, succeeding only in transforming "plenty of time and money into filth and stench." Meanwhile, the university was rife with "electrical-machine fever." Students built the machines themselves out of wood, glass, and metal, as the cuts on their hands attested.

> Ceaseless labor with Leyden jars and batteries, until the golden spring peered so brilliantly through the window that it drove us irresistibly outdoors into the newly blooming natural world, and all the papers and cartons and jars and disks are left in peace. Then thoughts and feelings that were cooped up until now in the narrow room and nailed to all the tools turn outwards again; again they follow the clouds and winds, towards the pure blue of the sky and throw themselves like lovers [*sich liebend*] on the blooming earth.[29]

Spring meant liberation, but waiting for it was agonizing. His laboratory work gave him the disturbing impression that "all the roots of my life had been cut and ripped out of their soil."[30]

In December, Hann returned to Kremsmünster for the Christmas break, which he spent happily studying the role of rising air currents in the formation of storms. "The familiar, friendly landscape suddenly showed me a thousand things that were new and beautiful and spoke to my heart."[31] In lines that Hann later copied into his diary, the Moravian poet Hieronymus Lorm explained the significance of *Heimatsgefühl*, or affection for the homeland. Lorm observed that this affection would be regrettable if it came at the expense of an engagement with the wider world. "It would be a profound shame, that even an active individual can know and enjoy such a vanishingly small part of the infinite variety of creation, if it were not the case that even the smallest part is a microcosm of the whole."[32] Learning to see home as a microcosm of a wider world was a key step in the process of scaling.

Gradually, as Hann began to feel at ease with his new colleagues in Vienna, his confidence returned. Coming home to his rented room in Vienna at two in the morning after an uncommonly pleasant evening, he looked out his window and saw the summer sun already beginning to brighten the eastern sky. Glimpsing a distant rainstorm, he felt the "cool, fresh, plant-scented air blowing from the east." These impressions "mingled" with joyful memories of early morning hours in the mountains of his homeland. The effect proved unsettling: "Experienced a painful morning, painful evening! The time is not yet ripe! When will the day come when I can finally be at peace: blue skies, sunshine again at last!"[33] The expression "the time is not ripe" (*das Maß ist noch nicht voll*) points to Hann's sense that his *Heimweh* was a state of disproportion between past and present. Another evening, having taken great pleasure in a concert at a friend's home, it felt to him that "every little worry, the vanities and petty cares of life, they all vanished for me, since I paid no attention to them and they had no effect on me." The small, the local, the personal were not without significance, but one had to learn how to read their meaning in the context of a larger whole. He turned to the science of the atmosphere to orient himself in a wide world that was just beginning to reveal its horizons.

One spring evening, as the sun was starting to set, Hann wandered out of the "torment" of the laboratory, feeling "irritated," and made his way through the "dusty, foul-smelling streets" to the Prater, where "the evening sun was spreading golden sparkling light over the majestic trees." He moved quickly through spaces that were "teeming with people" and "full of life, noise, and music," and soon found himself in a forest. The setting was "exquisite," and the air was fresh and piquant from the new greenery, with a gentle breeze from the east. Hann wandered "as if in a dream" from one charming scene to the next, "intoxicated by the sultry, misty air," visited by "alluring, wistful" visions of his homeland in springtime, "half fantasy, half memory."[34] The Prater became a frequent destination for him, a space in which he could reorient himself. He began to grow fond of life in Vienna—of his classmates and professors, and of the cultural offerings in the capital.

In early October 1861, Hann was preparing to leave Kremsmünster for his second year of university. He spent his days in the forest, readying himself for the transition to Vienna, and paid particular attention to the relationship between nature on the small scale and the large.

> Took pleasure in the colorful fusion [*Farbenschmelz*] of the near and the far—out in the forest the freshest, most singular life of nature on the small scale—listened to the mutual stimulation of plants and animals by the water that pools between entangled, moss-covered roots of fir trees. . . . Adieu, beloved valleys and peaks, you have comforted me once again! You have taught me to extract from every milieu the side that allows me to let adversity wash over me—everywhere, to preserve the inner freedom that stands above external circumstances with mastery and control.

Hann's apostrophe is to be taken seriously. The moss, fir trees, and mountain peaks of this landscape functioned as tools of scaling. They taught him to perceive the interrelations of "the near and the far," which here "fuse" into a single impression.

Now Hann seemed ready to venture further. He spent a day traveling the Danube by steamer, with the narrow, winding river creating "new landscapes" at every turn. "A tremendous desire to travel took hold of me."[35] Many of his travel fantasies came and went, but the dream of the Italian coast held constant. In 1857, a stretch of the Südbahn railway had been completed that connected Vienna to the Adriatic. Hann learned in 1862 of the "Triest pleasure train." It was, at last, an opportunity to see Austria's own seaside, and Hann seized it "with joyful excitement." On the seventh of June, he set off from Vienna's Südbahnhof. The route led along the Bruck River through Styria, past Ljubljana in Carniola, across the karst plains, and along the Istrian coast. He found the views breathtaking, although the coffee was "undrinkable" and the water "horrid."[36] Upon arriving in Triest after a twenty-four-hour journey, he was overcome by emotion. "It was too much all at once, so overwhelmingly powerful was the view. A moment of undiluted joy, of bliss, as is accorded to one who sees his most ardent, secret wish suddenly fulfilled." From the train station, Hann and his traveling companion headed for the harbor and hiked up to the town of Muggia for the views over the sea. After checking into their hotel and eating a late lunch, they returned to the harbor and boarded a ship to tour the coast. Hann recorded the look of the clouds, the unusual quality of the light, the sublime immensity of the sea, and, repeatedly, the great breadth, the "freedom" of the blue horizon. His experience of enchantment extended from the scenery to the company in which he found himself: "There was plenty of merriment on the ships, and people even danced to the music that accompanied us." Hann decided to devote his "full attention" to the "wonders of the sky and sea" and the "exhilarating life" around him. The "cute, tanned sons of the Italian captain tussled and chatted" next to him, and the gentle sounds of their language moved him deeply. "I almost envied the youths, for whom the sea would be life's vital principle."[37] After his initial sense of disorientation, Hann found his bearings in this new environment; he was able to interpret its light and clouds and to identify with its inhabitants.

CRISIS

In the fall of 1862, Hann accepted a position teaching physics and mathematics at a secondary school in Vienna that was a short walk from the university. At the ZAMG, he had access to the growing archive of geophysical data from across the Habsburg territory. There he found colleagues who shared not only his enthusiasm for climatology, but also his determination to organize the new science and give it a public outlet. In 1865 the Austrian Meteorological Society was founded, and in 1866 the first issue of its journal appeared, with Hann as one of two editors. That year also saw the publication of his article on the origin of the foehn wind, which would make his reputation as a pioneer of dynamic climatology.

At this point in his diaries (1866), his entries become more erratic, and his handwriting less legible. Often they consist of notes on distant weather events, copied from newspapers or scientific publications. This is clearly the journal of an energetic young researcher with international connections and a wide frame of reference, plunged deep into his investigations. Already, Hann was beginning to correlate weather patterns across great distances—commenting, for instance, on observations of polar lights near Riga and in various North American cities, or drawing an analogy between seasonal weather patterns in Siberia and in western Europe. It is here that he begins to outline a popular work on cosmic physics, which was to address the "earth as a whole, in relation to the movement and illumination of the cosmos,"[38] evidently the germ of his *Allgemeine Erdkunde* of 1872, which went through five editions. It is here too that he begins to compile information on the climate of Greenland. Weather observations from Greenland had been recorded since the eighteenth century by Moravian missionaries and published in German journals.[39] But this data had never been scrutinized. Hann was interested in patterns of anomalous weather in Greenland and their timing in relation to anomalies in central Europe. His younger colleagues Exner and Defant would follow up this line of research. Hann was thus the first to begin to pose the kinds of statistical questions that would eventually lead to the identification of the correlation pattern now known as the North Atlantic Oscillation.[40] Hann would go on to publish on this topic in 1890 and again in 1904. By this time, other researchers had begun their studies of such "teleconnections." In a published article of 1906, Hann reflected on this shift in scale:

> It is very gratifying to observe that meteorological research is striving once again to gain a wider horizon. For too long it has studied and followed exclusively the small atmospheric disturbances that pass over Europe in the form of barometric depressions. . . . If one examines the relative space that Europe occupies on a globe, one must immediately realize that meteorological progress will not be served by precise studies only of the local phenomena that play out in this confined space.[41]

Hann was suggesting that European climatology of the nineteenth century had suffered from its provinciality. What was necessary was a new sense of proportion, a recognition that the weather of the European continent was but a small piece of a much-larger puzzle—for the atmosphere of distant regions might well affect the climatic patterns experienced in Europe. Hann observed that the British Empire would seem to have been custom made for this kind of research. Yet the British were so narrowly focused on storm warnings that they did not consider the climatological significance of their data. What they failed to realize, according to Hann, was that the study of atmospheric phenomena in the southern regions of their empire might help to explain their own weather patterns back home in the British Isles. Climatology offered the tools necessary to attain a proper sense of proportion.

As Hann progressed in his education and came closer to seeing himself in the role of imperial-royal scientist, he reflected on the qualities necessary to serve the state well. Was he cut out for such a position of leadership? "The state demands complete individuals," he observed, "not scholars, artists, industrialists."[42] And yet he expressed equal disdain for those who "fall into the sin of those remarkable personalities who believe themselves suited for any office—even though, from one administration to the next, they have visited no other classroom than the ministers' cabinets."[43] Compounding his self-doubts were his ongoing attacks of *Heimweh*. He still hoped to learn how to make himself at home in the world, to attain the "knowledge to be able to live anywhere, cheerfully at work and in regular contact with people." Yet "the heart wishes for a Heimat, an isolated point amidst endless wanderings."[44] He sought a center, an origin: what Husserl would call the "null point."

In the early summer of 1869, he began to express more intense feelings of homesickness and an almost desperate nostalgia for the scenes of his youth. Then comes the entry of 9 August, in which he confessed a profound uncertainty about his future: "A period of sadness—the decision is made, I will not pay the price of my life for an uncertain future—very likely vanity," and he refers to himself as one who has been set back by fate. "Take me back to you, forests; your peace, your mercy to a tormented, wounded heart, as in the days of youth . . . I return to you, a penitent son, return to the depths of your joyful peace." We are left to wonder what the setback might have been that had made his future so uncertain. Two weeks later he was at Kremsmünster, reading poetry, hiking, and reuniting with old friends. His entries for the fall of 1869 and spring of 1870 are sparse. Then, in the summer of 1870, after a couple of fleeting sketches of hikes in the Vienna hills, he scrawled: "25. July.—Contrast—received the full impression—firm decision to break with this life." There follows the annotation, "see 9. Aug. 1869!"—the day on which he had renounced his "uncertain future" in Vienna. The following page of the diary has been cut out. The next entry is from 24 August, reporting that he has arrived in Kremsmünster. No further reference is made to his "decision."[45]

Could he have been contemplating leaving his academic career to follow a fantasy described earlier in his diaries—that of living as a monk and naturalist at Kremsmünster? There are no further clues.

Seven years later, at age thirty-eight, he became the director of the ZAMG and a full member of the Vienna Academy of Sciences. The following year he married Louise Weinmayr, a daughter of a district court official in Upper Austria and a granddaughter of Karl Kreil, and the couple went on to have four children.[46] There is no sign that he regretted these choices. Yet it was his crisis of 1869–70 that gave shape to the narrative that Hann retrospectively intended his diaries to tell. It was indeed a story of the making of an imperial scientist, fraught with self-doubts and with conflicts between desire and duty—a quest in which Hann looked to the tools of climatology to recalibrate his sense of the relative significance of things.

INTERLUDE: HISTORICIZING DESIRE

Hann was not as quick as Stifter to dismiss desire as a distraction from the study of the natural world. Hann rejoiced in the vitality that he associated with desire, and he even seems to have accepted homoeroticism as appropriate within an educational setting—as long as it could be displaced onto the natural world or otherwise serve as a stimulus to the quest for knowledge.

In the span of the generation that separated Hann from Heinz Ficker (1881–1957), both the geography and the connotations of desire changed significantly. In the 1860s Hann fought his nostalgia for his Alpine Heimat by trying to cathect the domesticated wilderness of the Prater and the karst landscapes along the tracks of the Südbahn. In this way, he developed a sense of himself as an imperial-royal scientist and gained confidence in his capacity for scaling. While Hann's fantasies roamed south to the Mediterranean, Ficker's would wander east, to the Muslim world. In 1878, the Habsburg army occupied Bosnia, bringing 499,000 Muslims under Austro-Hungarian rule. Historians have argued that the sympathies expressed by Catholic Austrians for Muslims thereafter functioned to legitimate this move, by suggesting that Austrians and Turks were bound by "shared experiences."[47] Like Hann's imaginative identification with the Italian-speaking boys in Trieste, Ficker's fascination with Muslims both in Croatia and central Asia should be understood in relation to an ideology of empire.

Ways of speaking about desire were also shifting by the late nineteenth century. What had been a Romantic concept was increasingly framed by the discourse of medical pathology. Viewed through a Darwinian lens, deviations from normative heterosexuality required neurophysiological explanation. It was in this period that European medical experts began to attribute a powerful influence on human sexual function to climate. They cited climate as a key factor in sexual maturation, often prescribing climatic cures for those suffering sexual dysfunction as a result of nervous disorders.[48] "Even amongst the civilized peoples of Europe," Sigmund Freud wrote in 1905, "climate and race exercise the most powerful influence on the prevalence of inversion and upon the attitude adopted towards it."[49] While historians have focused on racial theories of sexuality, many of Freud's contemporaries emphasized the effect of climate even to the exclusion of race. Sir Richard Burton famously argued that pederasty was "popular and endemic" within what he called the Sotadic Zone. This was a region encompassing most of the inhabited world—except northern Europe, Russia, and southern Africa, where the inhabitants were "physically incapable of performing the operation and look upon it with the liveliest disgust." Burton stressed that this phenomenon was "geographical and climatic, not racial. . . . I suspect a mixed physical temperament [i.e., mixed between masculine and feminine] effected by the manifold subtle influences massed together in the word climate."[50] This conviction found echo in Ellsworth Huntington's influential theory of the global "distribution of human energy on the basis of climate."[51]

Within Austria-Hungary, the effect of climate on sexuality was a central theme of the controversial novels of the Galician writer Leopold von Sacher-Masoch. His fiction is thick with erotic descriptions of the landscapes of the northeastern periphery of the Habsburg lands. Consider his most notorious novel, *Venus in Furs*. The story unfolds at a "little Carpathian health resort," where the protagonist "lies in his window," taking the air. He lives according to a "system" remarkable in part for the role of climatological self-observation by means of "thermometer, barometer, aerometer, [and] hydrometer." Outside, the mountain landscape is said to "tremble" and "undulate"; it seduces and intoxicates. At one point, he rides out on a donkey, intending "to numb my desire, my yearning, with the magnificent scenery of the Carpathians." He returns, however, "tired, hungry, thirsty, and more in love than ever." His love object, whose hair is described as "electric," attributes her own sexual power to the atmospheric electricity generated by her fur pelts: "It is a physical stimulus which sets you tingling, and no one can wholly escape it. Science has recently shown a certain relationship between electricity and warmth; at any rate, their effects upon the human organism are related. The torrid zone produces more passionate characters, a heated atmosphere stimulation. Likewise with electricity."[52] Given this set of associations, it should

FIGURE 42. Heinrich von Ficker (1881–1957), ca. 1920.

not be surprising to find that late nineteenth-century Habsburg climatologists took an interest in the effects of atmospheric conditions on their own excitability.

CLIMATOLOGICAL MODERNISM

Just as Goethe's Romantic fiction provides a key to unlock Hann's journal, so does modernist poetry help us to decipher the personal diary of Hann's younger colleague Heinz Ficker (figure 42). To read Ficker's journal of his 1913 expedition to central Asia, one must know something of the literary circle in which he moved back home in Innsbruck. For Ficker was not a product of Kremsmünster, like Hann and Kreil, nor of Bohemia, like Purkyně and Suess. He had been born in Bavaria, but moved as a teenager to the Tyrolean capital, where his father, Julius Ficker, became professor of history. In Innsbruck, Heinrich's brother, Ludwig Ficker, became the center of a famous circle of literary modernists. This was a movement fascinated by a realm of experience that was inaccessible to logic and empirical science.[53]

In 1906 Heinz Ficker completed his dissertation on foehn at Innsbruck; three years later, his brother Ludwig began work on a new Tyrolean literary journal titled *Der Föhn*. A coincidence, but a revealing one. *Der Föhn*, the journal, was a celebration of the Tyrolean *Heimat* or homeland. As its editors decreed, its goal was to give a "complete picture of Tyrolean intellectual life." It was packed with hymns to Alpine nature and reports on Tyrolean folk culture. Soon after its founding, however, Ludwig Ficker broke with *Der Föhn*'s editors over their narrowly provincial vision for the journal. It was then that he conceived the project of founding his own journal, an enterprise that took shape in conversations with the poet Carl Dallago. As their letters record, Ficker and Dallago struggled to reconcile the value they placed on their Tyrolean roots with their far more cosmopolitan ambitions. This tension reflected a growing conflict between conservatives and liberals in Tyrol, which became acute during the 1909 commemoration of the one hundredth anniversary of the Tyrolean rebellion against the Napoleonic occupation. Two versions of Tyrolean regional identity clashed at this juncture: one politically conservative, chauvinistically German, and anti-Semitic; the other progressive and cosmopolitan.[54] For Ludwig Ficker, the result was a new literary venue in which, as he explained to another friend, the Tyrolean aspect would appear only in the title, "otherwise nowhere; only in the topical article with which I intend to conclude each issue will local problems be tackled, but only in so far as they lend themselves to perspectives of more general significance."[55] The new journal would thus make it possible, as Dallago put it, to "go from my Heimat out into the public."[56] Drawing out the political implications of this venture, one Jewish poet wrote Ludwig to confirm that the new journal would be an appropriate place for him, a Zionist, to publish.[57]

Ludwig Ficker's new journal would prove one of the most influential in the history of German modernism. The title that Ficker chose as the only mark of its Tyrolean roots was *Der Brenner*, after the famous Alpine pass that connected northern and southern Tyrol. In Austria, the Brenner pass symbolized a vital link between Germanic and Mediterranean cultures. But it also evoked the nationalist tensions between German and Italian speakers south of the Alps. It was a site rife with meaning for the historian Julius Ficker, father of Ludwig and Heinz, who had defended the integrity of the Austrian Empire against the project of "small-German" unification and in the face of Italian nationalism. His study of Italy concluded that it would fare best under Austrian control, which offered unity and protection while nurturing diversity. The Brenner, both the pass and the journal, were thus more than a road out of provincial isolation—they symbolized the world-historical mission of the Habsburg Empire. The Brenner played an equally central role in Heinz Ficker's volume on Tyrol for the *Climatography of Austria*, in which the pass evoked simultaneously opposition and unity between contrasting climates.[58]

He too was staking Tyrol's claim to a kind of provincial cosmopolitanism, as a bridge between the cultures of northern and southern Europe.

Like the literary modernists around Ludwig Ficker, Heinz Ficker experimented with symbolism in his landscape descriptions. In an article on foehn for a popular mountaineering journal, he conjured "a cold winter's night in a mountain valley." Even in the silent, wind-still valley, disconcerting signs of the foehn's approach could be heard and seen—the rustling of fir needles in the forests high above and the uncanny quality of light: "full of a wild, sickly fervor is this play of colors, a sinister weather sign."[59] As it happens, a strikingly similar vision of foehn's arrival appeared in one of the first and most famous poems to be published in *Der Brenner*, which introduced the world to one of the great German modernists of the twentieth century, Georg Trakl. "Suburb in Föhn" (1912) moves from a description of the oppressive, noxious atmosphere during foehn to an evocation of quite a different world—an orientalist vision of eastern romance and indulgence.

> The foehn winds tinge meager shrubs more colorfully
> And the redness slowly creeps through the flood . . .
> . . . From clouds gleaming avenues surface,
> Fulfilled with beautiful chariots, daring riders.
> Then one also sees a boat failing on cliffs
> And sometimes rose-colored mosques.[60]

As both Ficker and Trakl emphasize, foehn's approach, accompanied by falling pressure and stagnant air, accentuates the oppressive isolation of their provincial home. Yet the wind arrives with such a play of light and color that Heinz Ficker described foehn as a "master painter."[61] For Trakl, the wind evokes the mysteries of foreign lands, whence perhaps it blows. Dramatized in this way, foehn becomes a fitting metaphor for the confluence of the provincial and the cosmopolitan that Tyrol's young liberals were then working to promote.

ALAI, ALAI!

Ficker would continue his literary experiments during his explorations of the Russian east. He was still a student of geology in Innsbruck when he first traveled to the southern periphery of the Russian Empire in 1903. The conditions of this journey were spartan, and Ficker traveled not as an imperial scientist but as a mountaineer, determined to "conquer" the region's peaks. He and his sister Cenzi joined a party of central Europeans led by Willi Rickmer Rickmers, the man who introduced skiing to Tyrol. Cenzi scaled a 4,700-meter peak and so impressed the local aristocrat that he offered her the mountain itself, "as a gift."[62] Cenzi was soon known as the foremost female alpinist of her era. She returned east with Rickmers in 1907, recording meteorological observations that her brother Heinz would rely on for a study published the following year. In 1913 both siblings would follow Rickmers to Turkestan.

In a journal labeled *Diary of the Pamir Expedition, 1913*, Ficker left behind a dozen pages of never-published sketches with titles like "The Most Beautiful Hour" and "A Distant Evening." On the surface, these are descriptions of landscapes and the moods they evoked, along with telling details of what Rickmers called the "piquant little amusements" that the party indulged in. Read more closely, these sketches are canny descriptions of what happened when Ficker's exoticizing fantasies collided with his on-the-ground experiences. Evident throughout the journal is Ficker's imaginative identification with many of the locals he encountered, above all with those he recognized as victims of oppression. So, for instance, one entry opens with an orientalist vision of "the goal, the limit of our journey. There lay Afghanistan—shuttered and full of secrets." A local chief points out the border, bids them good night, then retreats to what Ficker realizes must be the "dirty hole" in which he keeps his harem. When the door opens silently, Ficker glimpses the women within: "the whites of the eyes of two girls. Then a low, strange noise, that escaped hauntingly from the mouths of the women, and a strained murmur from the man." "A Thousand and One Nights!" Ficker exclaims, mocking his own naïveté. What he has seen is unclear, but the very ambiguity of it and the surrounding silence—"the mountains stood silently and guarded the last, defiant sickle of the expiring moon"—are unsettling. The situation raises a question, one that Ficker poses explicitly elsewhere in the journal: who in this wide world bears responsibility to respond to this apparent injustice?[63]

There was, of course, a much-older tradition of European laments over the mistreatment of women in Muslim societies, which was often racist and hypocritical. But Ficker's observations were free of moralism and remarkably self-aware. In a village near Samarkand, his companion, a coarse Russian, leads him in search of liquor and whores. "Not many years ago people who sold Schnapps and wine were still tortured. 'A dreadful culture!' my companion whispered. Then he laughed. 'But we'll get some yet!'" Soon this acquaintance goes off with some prostitutes, and Ficker learns that the women earn fifteen kopecks per client. "Just think," another Russian remarks. "If they do it a hundred times, they've only earned 15 rubles!" As they walk out into the night, Ficker reflects to himself that once these women too had been infants, and a mother had "stood by the cradle and prayed for her child." The second Russian expresses pity both for the prostitutes and for the local men: "Isn't it horrible: we've conquered this land and now we supply it with Schnapps and whores!" Ficker's next line is crossed out, and the sketch ends simply: "I inquired no farther."[64] Silences of one sort or another typically close the sketches, acknowledging the complexity of the questions they raise.

Another entry describes Duschanbe, "the summer resort of the Ottoman Empire." Here Ficker's party experiences "refreshing nights, humid mornings, then shimmering heat all day with a fiery hot wind and blowing dust." One evening, musicians start to play for them, and a girl begins to dance—"A pale, delicate, contented, and probably depraved girl." As Ficker watches her move, he begins to wonder who exactly is singing. "We don't understand it. We see it only in the gleaming eyes of the men who swarm around the girl. . . . 'This could make one crazy,' a companion near me says and wipes the sweat from his forehead. I know what he means. Yes, we all know it: this is no girl. . . . A dancing boy, a boy in girls' clothing! How beautiful the little rascal is!" Ficker then admits that he had known what to expect and had laughed at the prospect. Indeed, rumors had long circulated among Europeans of cross-dressed dancers in the Muslim world.[65] But these Europeans find they cannot tear themselves away. Ficker despairs at this sudden surge of desire. "How beautifully everything has gone until now. No stirrings of desire, no buoyant longing. And all of a sudden it's this push and pull [*hin und her*]. As if the senses were made to order [*Wie hergestellt sind die Sinne.*]" He and his companions prepare to depart, but the thought won't leave him: "And if there are no women, why not boys?" They've "had enough"; they feed their dogs and turn to go. Ficker thinks: "God knows—what would happen if one had to remain longer in such a land!"[66] The dreamlike quality of this sequence, where reality blurs with fantasy and desire surges unexpectedly, is reminiscent of Sacher-Masoch's fictions of the imperial periphery. In the tradition of orientalism, Ficker attributes a seductive quality to the landscape itself, connecting this sudden wave of desire to the "shimmering" heat and torrid wind. The question of what he would do were he to remain in Turkestan follows from this writing style. Like Sacher-Masoch, Ficker adopted an imaginative engagement with the oppressed, which easily shaded into the thought experiment of "going native."

At a deeper level, then, the journal tracks the subtle effects of Ficker's encounters in Turkestan on his perception of himself. This is evident on his return journey, as he passes through Muslim communities within the Habsburg borders, in Croatia. There he encounters a beautiful girl of about fourteen, at work driving an ox in a mill, as she is brutally beaten by the miller. "The child suffered in silence, and her dark eyes shone as if with mercy." "Appalled," Ficker makes some gestures about the rough treatment. "The men laughed outright and gesticulated and took pleasure in my anger. I was shocked. How did this all come to us [*Wie kam uns alles*]?" Was this not a country where children were treated "more royally than elsewhere," where women and children were spared hard work? In this way, the question of moral responsibility came home with him.

Ficker was forced to admit that it was not in Russian Asia but in Habsburg Croatia that he had witnessed the most disturbing brutality. "When I think of that moment, sweat still forms on my brow. Why didn't I feel like hitting the miller in the face with my whip? Why could I tolerate a human being tortured before me for hours? It was not fear. Perhaps the feeling that I could do nothing to change the girl's fate." Ficker felt he would forever be haunted by the suffering eyes of this young girl, "these eyes that tell me that, in that moment, I was not a man." Once again, his masculinity seems to hang in the balance. Perhaps in answer to this self-accusation, he looked back over his voyage and arrived at a reckoning. He had not seduced any women. "Sleeping with them is only an adventure for me when it is a testament to a long courtship." Having secured his masculine identity in this way, he went on to suggest that the value of these recorded experiences was to "give perhaps a better idea of the constitution of a traveler in distant lands than a broad description of the foreign land itself."[67] Indeed, the theme of these sketches is the question of how profoundly travel might change the traveler.

TYROL TO TURKESTAN

With these experiences in mind, let us turn to the scientific conclusions that Ficker drew from his firsthand research in central Asia. In his 1908 paper on the climate of Turkestan, published before his expedition and based solely on station data, he had argued that Russian Turkestan was a "dying," "forsaken" land. This mountainous, arid region owed its fertility to runoff from melting glaciers, but these glaciers were shrinking, and, as a result, the region faced irreversible desiccation. Its present fertility was due solely to artificial irrigation, an unsustainable practice: "For the more water that I artificially divert to a region today, the more I deprive another region and deliver it up to destruction."[68] In his studies published after his return from central Asia, he mustered further evidence of desiccation in this region, from the shrinking of forests to the retreat of glaciers. He was more convinced than ever that Turkestan was a dying land. Now, however, Ficker saw Turkestan's problems in a wider frame: "In no case however will what man destroys in this region be replaced by nature; indeed, how quickly under such conditions whole regions can become permanently void of trees is clearly shown by the Karst region and by many stretches of the southern Alps."[69] Thus Ficker did not stop at cautioning Russians against overexploitation of their natural resources; he also hinted at the need for conservation in regions of Austria. Instead of a contrast between enlightened west and uncivilized east, he saw two regions suffering a related fate, linked by a global climate system.

Ficker's journal can thus be read as a record of the emotional work of scaling. As his published reports make clear, observations of the natural environment could not be separated from observations of human customs, since natural conditions were so sensitive to human intervention. Likewise, the imaginative effort to relate the Alps to the Pamirs seems to have been inextricable from the emotional effort to empathize with the inhabitants of Turkestan. Scaling was at once a bodily, kinesthetic process and a moral act. In order to replace the absolute distinction between near and far with a measure of degrees of separation, firsthand observation alone is never enough; at some point, it becomes essential to trust in foreigners. In the capacity to imagine communities beyond one's own lies the origin of the mathematical concept of infinity, according to Husserl's student Ludwig Landgrebe. In short, the empathy that Ficker exercised in his journal was an essential part of the process by which he scaled from what he knew of the Tyrolean Alps, to what he learned of the high Pamirs, to what he extrapolated for the earth system as a whole.

EPILOGUE

Ficker's journal also reveals a keen intuition of historical transition. From 1908 to 1919, an implicit theme of his publications on the desiccation of Turkestan was the decline of civilizations and the fall of empires. Looking back on their journey of 1913 from the perspective of the late 1920s, Rickmers would write of the "transition from the voyage of discovery to the voyage of processing, from the scout to the mobile observatory, from the hussar to the armored car of science."[70] For Rickmers, this was the moment when the instrument replaced the eye and the statistician replaced the storyteller. Such a contrast could only emerge in retrospect, yet already in 1913 Rickmers had hinted that the role of the explorer was in jeopardy. He lamented that a large expedition with a strict "division of labor" left little room for personal liberty. In his view, it was Ficker who represented freedom from such constraints. While the expedition leader and the topographer were bound to fixed points on the ground, the geologist was "a restless spirit," requiring "freedom" for his work, able to "poke around in every corner and taste the joy of discovery."[71]

Ironically, Ficker's next and last trip to the Russian east came as a prisoner of war. This time, his research would be restricted to calculations based on published data. In this way, calculation acquired a taint of imprisonment, while firsthand observation was all the more firmly associated with a lost age of freedom. The opportunity to present his "Investigations on the Meteorological Conditions of the Pamir Region" did not come until after the war, in 1919, the year of the founding of the League of Nations—a moment that, in retrospect, heralds the systematic internationalization of science and the end of the age of individual exploration. Against this background, Ficker's conviction that the meteorological travel narrative would remain "highly popular in the future" may seem incongruous. Of course, what he failed to foresee was the rise of computers and satellites, with their unprecedented power to synthesize masses of data, produce vivid images, and ultimately support global climate models. In the computer age, the globalization of atmospheric science would mean the abstraction of meteorological knowledge from the people and places that produced it. Bjerknes's "polar front," a concept that began its rise to fame in 1919, is a case in point. Ficker and Bjerknes both identified a hemispheric discontinuity between polar and tropical air and recognized its role in generating cyclones. But Ficker focused on the climatological—one might even say ecological—implications of this finding. Seeking the origin of cold waves in Eurasia was, for Ficker, part of a broader quest to understand the connections between fragile climates like those of Tyrol and Turkestan.[72] Bjerknes, by contrast, hailed the polar front more narrowly and instrumentally as the key to storm prediction.

With the triumph of Bjerknes's interpretation in the 1920s came a loss of meaning. Gone was the incentive to draw out the implications of this phenomenon at a regional, even local scale. The age of weather prediction was dawning, demanding that meteorological observations be reduced to neat numbers that calculators (human or mechanical) could manipulate. Meteorologists and physical geographers would soon part ways.[73] Ficker's report belonged to a different age, brimming as it was with observations and digressions that resisted abstraction and quantification. His climate science remained a geographical enterprise in the holistic style of a continental empire. Hence his decision to pack in as much as possible—instrument curves, fine print, footnotes, and all.

Ficker's journal from the Pamir expedition acknowledged the ambiguous line between personal experience and scientific observation. One entry describes a distant view of the Peter the Great mountain range, noting that the czar was "a proud man, who is only supposed to have moved proud mountains." It continues: "Science is something beautiful, especially when one ~~like us~~ can do it on horseback. Before we reached the middle range from Almalik, amid blooming meadows, high in the saddle, we forgot science." Ficker's reference to Peter the Great is tinged with irony, as the mountains dwarf the spatial and temporal scale of human life. As it happens, Ficker struck a minor blow at Peter's legacy when he determined that the range "Peter the Great" was not one but two mountain chains and proposed to apply Peter's name only to the western one. For the eastern chain he proposed the name Catherine the Great—an appropriate choice, given that his companions included two female mountaineers.[74] Indeed, Ficker's conflicted experience of sexual desire in Turkestan links the theme of imperial decline and fading heroism to that of imperiled masculinity. Against this background, the passage works reflexively to underline the loss of meaning entailed in shifting from a personal to a scientific register. By "forgetting" science, Ficker could experience the mountain and the moment in their full, sensual complexity. In it one might hear an echo of Carl Dallago's reference in *Der Brenner* to "science, where it is authentic; but there it is no longer merely science."[75] Or one might be reminded of the more famous words of another member of this literary circle: "What we cannot speak about we must pass over in silence." This is the final line of Ludwig Wittgenstein's *Tractatus Logico-Philosophicus* (1921), another exploration of the limits of science, which was originally intended for publication in *Der Brenner*.[76]

Of course, the historian of science might be advised to stop reading at the point where a scientist "forgot science." But the question of what comes next, of what lies "beyond" science, is too important.

> Science led us here.
>
> But it was not at the beginning.
>
> Was it not in truth the longing for these unknown, mysterious mountains that brought us here, 6000 km away? . . .
>
> We pined for mountain beauty . . .

Would reading further constitute an indiscretion? Should we think twice about spying on the scientist as he privately swoons before a mountain he describes as "powerful, commanding," as a "Tamerlane of mountains"?

> There is something more beautiful, far more beautiful, as great science: great mountains.

> In the shade of a tree stood a young girl, as if made of stone, with large eyes, lightly clothed. In the snow her cheeks glowed and her brown, gentle eyes looked around helplessly. Like a grouse chick the beautiful child stood before me and lifted her gaze and lowered it again. The girl on the Almalik! My heart warms when I think of that moment. A wish—still it seems to me there is no desire in me—my buoyant, sinful desire.[77]

Perhaps it is time to replace the manuscript in its folder and return to serious research. But what if this sensual, speechless moment does tell us something about Ficker's "science"? What lay "beyond science," in Ficker's terms, is reminiscent of what Husserl called the "prescientific" or the "natural world." For Husserl, to recover this dimension of experience was not to renounce science. The goal was instead to recapture the experiences that gave a scientific idea its original meaning in its original context. For Husserl, this was the solution to the crisis of meaning in European culture of the early twentieth century. In pressing the question of the relationship between science and desire, Ficker may have had a similar intuition. "Now what was science, what were the mountains?"

CONCLUSION

After Empire

In 1949, the geographer Hugo Hassinger, born in Vienna in 1877, was nearing the end of his life in a defeated, discredited, foreign-occupied Austria, roughly one-tenth the size of the Austro-Hungarian Empire. Hassinger had been allowed to continue in a leadership position at the University of Vienna after 1945 because he had never joined the Nazi Party. Still, doubts remained about his politics. During the war, as director of the "Southeastern Europe Research Community," Hassinger had produced scholarship designed to legitimate Hitler's conquests in that region.[1] He had even plotted the resettlement of ethnic Germans in the conquered lands. His publications had been shot through with the language of "German living space." Now, in Allied-occupied central Europe, this vocabulary was untouchable. So it was that in the twilight of his long career, at the dawn of the Cold War, Hassinger was returning to the rhetoric of imperial-royal science.

In *Austria's Nature and Destiny, Rooted in Its Geographic Situation*, Hassinger made a case for this ailing country's significance in the postwar world order. He argued that Austria was more than a "border zone" between ideologically opposed regions. It was the protector of the supranational tradition of the Habsburgs and of the very idea of Europe. The Habsburg Monarchy had been "European," he argued, not only in its thinking, but even in its landscapes. He pointed in particular to the "diversity of climatic phenomena," manifest in the "climatic contrast between the icy region of the high Alps and the Mediterranean climate; between the forests and Alpine meadows dripping with moisture and brushed by the oceanic westerly winds, and the dry pannonic climate zone with poor irrigation and lakes without outlets."[2] This "rapid variation within a small space of the energy associated with relief, soil type and regional climate" results in the most varied "forms of economic activity." As in the images propagated a century earlier by Minister von Bruck, Hassinger attributed Austria's vitality to the dynamism of neighboring contrasts, to the "energy" associated with "rapid variation" in space. However, he borrowed the imagery of atmospheric warfare more common in the tradition of Dove and the Bergen school than in Habsburg climatology. Central Europe was the region "where continental and oceanic forces struggle against each other," and "precisely here lies the deeper significance of Central Europe's existence." It was the "*battleground of oceanic and continental forces in the atmosphere and on earth*, where plants, animals, and people and their cultures stand locked in conflict with one another." And yet he insisted that these opponents "also depend on each other to seek a compromise [*Ausgleich*], such that Central Europe can also be regarded as *a European space of equilibration* [*Ausgleichsraum*]."[3] Like von Bruck and his nineteenth-century contemporaries, Hassinger tied the fate of central Europe to a pregnant linguistic slippage between the physical tendency of neighboring extremes to "balance" each other out and the diplomatic tendency of nations to "compromise."

How naturally this vision flowed from the pen of a man who, just four years earlier, had hoped to subject this entire region to Nazi rule! A man who had been planning the expulsion of non-German populations from South Tyrol, the Burgenland, and the Inner Carpathians was now paying tribute to the marvelous diversity of these lands. How could this be?

CLIMATE AND THE FATE OF CENTRAL EUROPE

The Habsburg trope of unity in diversity had met resistance from the start, and the skepticism only intensified with the outbreak of war in 1914. Two years in, the economist Ludwig von Mises lashed out against the idea that there was anything natural about the Monarchy's unification of contrasts. Arguing that barriers to free trade are always irrational, von Mises ridiculed the notion that a state could foster political unity by imposing protective tariffs at its borders. His prime targets were two popular proposals for the postwar European order: Fredrich Naumann's economic conception of *Mitteleuropa* (1915) and Karl Renner's federative socialism. Renner had described Galicia as "organically bound" to Austria—despite its mountainous border—by the grain, timber, petroleum, and spirits that the former supplied to the latter, as well as by the iron, textiles, and paper that flowed from the latter to the former. Von Mises quipped that one might just as well describe Austria as "organically bound" to

Great Britain and its colonies, in view of the sugar, tea, and cocoa in Austrian pantries. Ultimately, von Mises argued that any future protective union would impede at least one member from the development of industry. In this sense, it would constitute not a state of "economic community" but one of "economic war." And yet, even as he rejected these organic metaphors, von Mises adopted the physicalist language of equilibration. In his analysis of the flows of labor and capital, he wrote of the "balancing out" (*Ausgleichung*) of contrasts of wages and profits.[4]

The Hungarian historian Oszkár Jászi likewise scorned the ideology of natural diversity in his bitter 1929 history, *The Dissolution of the Habsburg Monarchy*. Despite their political differences, Jászi, like von Mises, saw this rhetoric as a cover for the economic exploitation of one population by another (though he emphasized divisions of class rather than nation). Here is his apt caricature of the old arguments for the Austro-Hungarian free-trade zone: "Behold the Habsburg monarchy gives the privilege and opportunity to many peoples and countries different from each other in natural conditions, in language, in culture, in economic development, to trade with each other without the obstacle of custom barriers and, therefore, to complete each other in the most harmonious way. . . . How advantageous and progressive this free trade is!"[5] Jászi recognized that this argument had not lost its power in the 1920s. That much was evident from the significance of free trade for the Pan-Europe movement. Notably, however, despite his patronizing tone, Jászi accepted the premise that natural diversity was inherently valuable. "The more two or several economic territories can offer to each other," he affirmed, "the more they complete each other, the more advantages free trade promises to them, the more disadvantage if tariff walls separate them." Remarkably, among the reasons he cited for the failure of the Habsburg customs union was its relative climatic uniformity. "The preponderant part of the monarchy furnished the articles generally produced in the temperate zone, and in this respect the differences within the monarchy were not great or decisive."[6] That Jászi would engage in debate over the degree of climatic diversity in the former Habsburg lands suggests that the question of central Europe's destiny continued to be framed partly in the naturalistic terms introduced in the 1850s.

In fact, Austria-Hungary's claim to be a "natural unit" was a matter of intense dispute during the Great War. Among geographers in imperial Germany, it became common even before 1918 to describe Austria as an "unnatural" state doomed to disintegrate along national lines.[7] On the other hand, Austrian geographers continued to defend the "geographical foundations" of the Habsburg state in terms inherited from the "whole-state" discourse of the

1850s. Norbert Krebs, for instance, wrote in 1913 that Austria-Hungary's natural diversity made it "self-evident that an active exchange of varied products supports cultural development and that the possession of these products makes the state more autonomous."[8] Robert Sieger, writing in 1915, insisted that the physical diversity of the Habsburg territory inevitably produced "the strongest cultural contrasts" and was, therefore, a sound foundation for economic self-sufficiency.[9]

After the Monarchy's collapse in 1918, climate remained an organizing principle in discussions of the fate of the former Habsburg lands. Now, however, climate was more often construed as a deterministic force than an adaptable resource. One of the most radical visions for the future came from the philosopher and statesman Richard Nikolaus Coudenhove-Kalergi, son of an Austrian diplomat and his Japanese wife. Coudenhove-Kalergi's "Pan-Europe" movement, launched in 1923, envisioned a unified Europe as technocratic utopia. Technology, in Coudenhove-Kalergi's terms, was a European invention, a "spiritual" endeavor to harness nature for human ends. In Europe, he argued, the social question was at base a matter of climate. Europeans lived in a state of "climatic unfreedom," at the mercy of their cold winters and short growing season. By means of technology, Europeans had already succeeded in transforming "northern primeval forests and swamps into cultural paradises."[10] The solution to the social question, therefore, was not political but technological. Coudenhove-Kalergi has been cited as an inspiration to the simultaneously technocratic and Romantic ideals of the European Union. It has gone unremarked, however, that his eclectic commitments were held together by the idea that climate was the ultimate obstacle to freedom in Europe.[11]

Climate was equally central to the future of central Europe as imagined in influential works by the Austrian geographer Erwin Hanslik. Hanslik had studied with Penck at the University of Vienna. In 1915 he founded the Vienna-based Institute for Cultural Research, dedicated to reinventing the Austrian Idea for a new age.[12] Hanslik's formative research concerned the "German-Slavic language border." As he explained in the introduction to his dissertation, the origin of this project lay in his personal experience of the contrasts between German and Polish cultures, while growing up as the son of a factory worker in Galicia in the 1880s and 1890s. As Jeremy King and other historians have made clear, it was in the years of Hanslik's youth that nationalist agitators began to compete to identify individuals and communities as speakers of one national language or another, even if the individuals declared themselves to be comfortably multilingual.[13] As Hanslik explained, he hoped that his research might "have a moderating influence on the language war" by providing "scientific insight into the nature of the language border and the nature of Austria-Hungary."[14] That is, he hoped to settle the question of linguistic identity once and for all by means of a line on a map.

Hanslik, a working-class Galician by birth, seems to have coveted the mantle of imperial-royal science. In that tradition, he rested his claim to scientific authority on his personal experience of the Monarchy's natural diversity, as "someone who for many years has been in the habit of combing through the western and eastern mountains alternately," and whose mind could distill, "among all the details," a true "natural border." Hanslik's analysis was ostensibly modeled on the climatological and plant-geographical studies of Hann and Kerner von Marilaun, from whom he also drew most of his data. He borrowed their methods for tracking small spatial variations in climate and vegetation in order to define natural regions. What he did not adopt was their recognition that "climatic borders" were contingent and subjective, dependent on the choice of statistical procedures and modes of botanical classification. In Hanslik's hands, the methods of Habsburg climatology became a crude means of dividing Europe in two. From the evidence of meteorology and botany, he claimed, there "leaps into view" the great geographic division of Europe between west and east, "oceanic" and "continental." The eastern and southeastern portions of the Monarchy constituted not an "equilibrium climate" (*Ausgleichsklima*) but a "border climate" (*Grenzklima*), an "alteration of mastery [*Herrschaft*] between western and eastern climates." In the manner of climatography, he rested this claim on bodily experience: "Whoever lives in these lands, learns this well enough by observing himself. He comes to orient himself accordingly for the sake of his health."[15] From this simplification, Hanslik made a conceptual leap not to be found anywhere in the writings of Hann or Kerner. Namely, he associated climatic zones with national spaces. "The great Germanic-Romanic nations of western Europe are limited to the oceanic region, the Slavic ones to the continental and transition region. Thus the horizontal classification of Europe and the climatic one are especially closely related." This quickly brought him to the remarkable conclusion that "the language border between Germans and Slavs is not a random line that derives simply from historical processes." Rather, "it is a line already drawn by nature."[16] In this spatial boundary Hanslik also saw a temporal divide, a *Kulturstufe*, a contrast between two stages of cultural development.

Hanslik pressed this image of European geography on a broad audience through the publications of his Institute for Cultural Research, which illustrated his border between west and east using bold, modern graphics.[17] He also advanced it in an essay on the Carpathians for the 1915 illustrated

FIGURE 43. The photograph that introduced Hanslik's essay on the Carpathians in *Mein Österreich, Mein Heimatland* (1915).

compendium *Mein Österreich, Mein Heimatland.*[18] This two-volume work might have looked, at first glance, like an updated and abridged version of the *Kronprinzenwerk*. It too proceeded from physical to cultural geography. However, its German-nationalist authors abandoned Crown Prince Rudolf's multicultural ideal and instead insisted on the superiority of German culture at every turn. Fittingly, the accompanying photograph (figure 43) suggests a German scientist-explorer gazing down on an unpeopled "wilderness" of "eastern space."

Hanslik concluded that Europe possessed an east-west "cultural gradient." In *Inventing Eastern Europe*, Larry Wolff famously traced the roots of this conceit to the travel narratives of eighteenth-century western European writers. Yet the physical metaphor of a cultural "gradient" was first articulated circa World War One. Sieger used it in 1915 and appreciatively attributed it to Hanslik.[19] This stubborn concept, central to so much academic theorizing about eastern Europe in the twentieth century, was the direct result of a bastardization of Habsburg climatology. If earlier Habsburg thinkers had likened cross-cultural interactions to the physical equilibration of contrasting air masses, Hanslik was the first to assume that cultural contrasts could be graded on a linear scale like barometric readings of air pressure.

Hanslik's scholarship exemplifies the perversion of the Habsburg tradition of dynamic climatology to suit German-nationalist aims in the era of the First World War. Like Hassinger, Sieger, and Coudenhove-Kalergi, he tended toward environmental determinism. Although he borrowed empirical evidence of climatic diversity from Hann, Kerner, Suess, Supan, and their colleagues, he used it to support a naive map of civilization and backwardness. Missing entirely from his thinking was the new meaning that climate had acquired in the dynamic framework of the late nineteenth century. There climate had come to be understood as a multiscalar, dynamic system sensitive to small perturbations—and as a circulation that connects rather than divides, creating relations of mutual dependence. Climate as used by Hanslik was closer to its eighteenth-century meaning: static, regional in scale, and deterministically associated with discrete and hierarchically ordered human cultures. Missing too from the writings of these early twentieth-century Austrians was the implied audience of nineteenth-century climatography: the citizens ready to exercise their rights of mobility and free enterprise. This was physical geography written for leaders prepared to impose order by force on the human and natural complexity of central and eastern Europe.

In this way, the nineteenth century's naturalistic imagery of unity in diversity was co-opted to support German imperialism of the twentieth century. This transformation reflected the radicalizing experience of the Great War. Following Max Bergholz, one might tentatively describe this as a phenomenon of "sudden nationhood": a "counterintuitive dynamic in which violence creates antagonistic identities rather than antagonistic identities leading to violence."[20] We can glimpse something of this process in the private wartime correspondence of certain Habsburg geographers, particularly Brückner, Sieger, and Krebs. Early in the war, these men were buoyed by the potential of their research to inform not only military strategy but, ultimately, the future map of Europe. By 1915 Krebs was already speaking of framing his work "no longer for the field marshals," but already "for the diplomats." In 1919 he expressed the aim of serving what he called *Deutschtum*.[21] The Central powers' reversal of fate hit these men particularly hard. In the immediate wake of the war, Brückner and Sieger complained bitterly of the "unnatural" terms of the peace treaties. By this they meant that Austria had been shorn of its "productive regions" and was economically unfit for survival. The diplomats in Versailles had "cut Austria off by an unnatural border from its necessary sources of assistance"; they had "torn apart" a "unified economic region."[22] Violence crept into the very language of geography.

SCIENTIFIC PLURALISM

Yet there is no reason to believe that, before 1914, naturalists' celebrations of the environmental diversity of the Habsburg lands had hidden a German-nationalist agenda. On the contrary, the practice of imperial-royal science was highly pluralistic, as judged by the standards of the twenty-first century.[23] Consider that the practice of science today is restricted to the exceedingly narrow segment of the population who are accredited as experts; it is largely monolingual, conducted in scientific English and mathematical symbols; and its results are published in a limited number of elite journals, incomprehensible to general readers. By contrast, the Habsburg field sciences engaged a wide range of participants. Though weighted toward the Alpine and Bohemian lands, volunteer observers across the Monarchy produced meteorological, balneological, phenological, and seismological measurements, often in languages other than German. Indeed, even after the empire's collapse, stations across the former crown lands continued to send climatological observations to Vienna on a monthly basis.[24] Prominent among these observers were local teachers, physicians, pharmacists, civil servants, and proprietors of spas and resorts, among them a handful of women. To be sure, the directors of the ZAMG were invariably Catholic, male, native German speakers. But not all the scientists employed by the ZAMG were ethnically German. At war's end, one had taken up Czechoslovak citizenship and found employment at the observatory in Prague, while two others with Czech names were awaiting judgment on their petitions to remain in Austria. Meanwhile, scientists who had worked for Habsburg observatories outside Austria's new borders were in the process of nationalizing the infrastructure of imperial science. The eagerness with which the new nation-states seized meteorological stations and instruments suggests that this was never an exclusively German-Austrian enterprise.[25] True, the tributes we have seen to the climatic diversity of the Habsburg lands were published in German. But this does not imply that the authors identified as Germans. Rather, German was the principal language of international science in central Europe before World War One. A small number of east-central European scientists chose, under the banner of Pan-Slavism, to build their international reputations in Russian; otherwise, Habsburg scientists

FIGURE 44. Portrait of Julie Moscheles/Moschelesová (1892–1956), by Zdenka Landová.

tended to stake claims to original research in German, regardless of national affiliation.[26] This was all the more true in atmospheric science, for which the Vienna-based *Meteorologische Zeitschrift* was the leading international journal. These varied participants brought to climatology many different goals, and the ZAMG supported a wider range of them than national meteorological offices of its day.

In the nation-states of post-Habsburg central Europe, on the other hand, science was expected to serve, above all, the *national* interest. Under these circumstances, it was an effort to maintain a commitment to thinking at scales other than that of the nation. Few fought harder to do so than the geographer Julie Moscheles (see figure 44). Against all odds, she had begun a doctorate in geomorphology under Alfred Grund at the German university in Prague. During the war, her interests came to include climatology, which she pursued in the descriptive and statistical manner of climatography. Climate was one among several fields of research for this extraordinarily prolific scholar, who expanded from physical to human geography in the mid-1920s. Indeed, far more was expected of her than of her male colleagues. While one book-length study would normally suffice for the habilitation, it took her twenty years, five monographs, and roughly sixty scientific journal articles.[27]

But her gender was only one of the obstacles to a scientific career. Born in Prague to a wealthy Jewish family, she had lived much of her childhood with an English uncle, a peace activist with whom she traveled widely. In the new Czechoslovak state, she was a German among Czechs and a Brit among Germans. As she explained in 1919 (in English) to William Morris Davis, "You do not know how cruel some people can be on account of national hatred and how they can make suffer their victims!" Still, she remained optimistic and committed to the new Czechoslovak state, publishing increasingly in Czech rather than German. "So by blood and friendships I belong to two nations and live in the state of a third one," she wrote. "It is perhaps on account of that, that I have no sense for national feuds." She drew a revealing analogy between her identity and her scientific research: "It is just as in the landscape: we distinguish forms of the air, the glacial cycle, but everywhere we also find the forms worked out by the running water which I would compare to life and soul of all human beings only a little modified by circumstances but always to be recognized and loved."[28] Here was the lens through which Habsburg scientists had been trained to see the natural world: to track flows and exchanges across space and time even as they marked out the peculiarities of local landscapes. Ironically, after the Monarchy's collapse, the naturalist who most successfully embodied the Habsburg ideal of supranational science was a woman.

SCALING DOWN

And yet Julie Moscheles was not quite alone in her struggle to maintain this internationalist spirit after Austria-Hungary's defeat. As director of the ZAMG in the early 1930s, Wilhelm Schmidt argued to the Austrian ministry of education that "of all Austria's scientific establishments, the Central Institute embraces not merely the widest circle of beneficiaries, from the simplest farmer and tourists to researchers studying the conditions of life; it also certainly has the broadest and strongest *international* connections: our weather and climate are of course part of the general circulation of the whole earth."[29] This was far more than the boast of a scientist desperate for funding. As we have seen, Schmidt's research on atmospheric "exchange" aimed to represent those "connections" with mathematical precision.

In Czechoslovakia, meanwhile, a proposal was taking shape for the first "world center" for climatology. Its author, Leo Wenzel Pollak, was a Jew by birth who had trained in cosmic physics at the German university in Prague, studying with the climatologist Rudolf Spitaler. In 1927, inspired by the use of punch cards in the American business world, Pollak patented a calculating machine adapted to the needs of climatologists, geophysicists, and astronomers. Unlike the US scientists who began to digitize atmospheric physics in the immediate aftermath of World War Two, Pollak was not trying to solve the problem of weather forecasting. He wanted his machine to analyze periodicities in climate data, the kind of work that Julius Hann had so painstakingly carried out by hand.[30] In 1942, Pollak was credited with "the first practical introduction of the punched card system into meteorology."[31] By that time, however, he had been driven out of Nazi-occupied Czechoslovakia. He was fortunate to find refuge in Dublin, where his last publication was a handbook of statistical methods in climatology—coauthored with another Jewish refugee from the former Habsburg lands, Victor Conrad. The historian of science Paul Edwards goes so far as to suggest that Czechoslovakia might have become "the world's leading center for climatology," had economic and political crisis not intervened.[32]

However, internationalist science like this was not a priority for the interwar Austrian and Czechoslovak governments. Rather, the most successful domains of atmospheric research in the successor states of Austria and Czechoslovakia were the newly christened subfields of "local climatology" and "bioclimatology." These enterprises were, in part, a pragmatic response to postwar conditions. With the Monarchy's collapse, central European climatologists had lost their advantages of scale. As Sieger put it after the signing of the peace treaties, "It is all the more necessary to order observations of the immediate surroundings, since the trains must suspend service for weeks at a time and they are overcrowded and exorbitantly priced."[33] Moreover, atmospheric scientists needed to prove their value to the new governments. Up against the postwar economic crisis, the ZAMG's director, Felix Exner, addressed himself simultaneously to the ministries of transportation, defense, justice, trade, agriculture, and health. He heralded the value of his institute for air travel, agriculture and forestry, hydrography, mining, wind power, legal decisions in tort cases, hygiene, transportation, and tourism. Still, proving this potential would require fundamental shifts in methods.

Of necessity, climatology in interwar central Europe would supplement *Großraum* (large-scale) climatology with research at the micro- and mesoscales. Still, these local sciences did not represent a break with the Habsburg legacy. Rather, they adapted the scaling practices of imperial-royal science to postimperial states. They defined a gradation of scales of analysis, from the intimate to the global, along with the methods appropriate to each.

One focus of research in both interwar Austria and Czechoslovakia was the climatology of mountains. As Exner reminded the authorities at every opportunity, research on mountain climates was a good investment. At Davos, Switzerland, the site of Thomas Mann's 1924 novel, *The Magic Mountain*, the physician Carl Dorno offered sunlight cures for tuberculosis patients. Since postwar inflation made Swiss resorts prohibitively expensive for Austrian citizens, Exner and Ficker began to hunt for an "Austrian Davos." In the *Meteorologische Zeitschrift* in 1921, Ficker insisted that the eastern Alps contained enough sunny peaks to treat "the suffering people from all over Europe."[34] Meanwhile, Czechoslovak climatologists and physicians focused their attention on the High Tatras. Although spas and sanatoriums had spread through this region in the late nineteenth century, detailed studies of the climatic conditions were first carried out in the 1920s. The director of climatology at the Prague Observatory, Alois Gregor (1892–1972), took a particular interest in the study of mountain climates. He measured the local atmosphere's "thermal comfort" by means of a wet-bulb thermometer; the intensity of ultraviolet radiation using a dosimeter designed by his Prague colleague Leo Pollak; and the "purity" of the atmosphere visually, by inferring its "turbidity factor" using a standard scale of shades of blue developed by Frankfurt climatologist Franz Linke (1878–1944) in 1922. Writing in Czech for a popular audience, Gregor explained that the atmosphere is like a multistory building, and the first story is of a "higher biological quality" than the ground floor.[35]

As bioclimatological measurements became standardized, so did the process of certifying a climatically therapeutic site (*Luftkurort*). In the Republic of Austria, local communities could appeal to the ZAMG to be considered for this designation. A scientist would then travel to inspect the area. If his initial assessment were favorable, he would order the installation of a "climatic observation station," consisting of eight instruments to be operated by local volunteers. An analogous system existed in Czechoslovakia, administered jointly by the ministry of health and the national meteorological institute. Not every community got the results they hoped for, however. For instance, the Lower Austrian town of Mönichkirchen proved to have moderate summer and winter temperatures, average cloud cover, but heavy snowfall; it learned in 1929 that it had been approved as a destination for convalescents, but not for the seriously ill. In the words of the unfortunately named director of marketing for Austrian tourism, Dr. Erwin Naswetter ("Wetweather"), these stations supported "the modern view of advertising that sees the objectivity and truth of the commendation as the chief requirement of a good piece of publicity." As others in the tourist industry stressed, advertisements for climate therapy must exclude "speculation."[36]

Another focus of climatological research in interwar Austria and Czechoslovakia was the urban environment. "Red Vienna," under Social Democratic leadership in the 1920s, was an incubator for modernist projects to design a more egalitarian city. Thus the intended audience for urban climatology included city planners, hygienists, architects, and engineers. Wilhelm Schmidt, who pioneered this research in Vienna, was convinced that the city's climate was impairing the health of children and thus gradually weakening the urban population. In *The Artificial Climate of Human Surroundings* (1937), he and his longtime collaborator, hygienist Ernst Brezina, wrote that the task of urban climatology "is to correctly influence municipal housing politics and to create more bearable living conditions for the urban population, [conditions] which correspond better to the physical and spiritual nature of man than is usually the case for the urban dweller today."[37] Urban climatology thus developed as an empirical physical science of "artificial climates." In Prague, Alois Gregor warned forcefully that urban growth revealed the damaging "influence of man on climate" as starkly as did the Dust Bowl disaster in the United States.[38] As the Vienna climatologist Friedrich Steinhauser reflected in 1934, programs of "hygienic housing design and the functional design of communities" drove the need for "investigations of urban temperature patterns." "What was needed then," he added, "was to invent new methods of investigation."[39] Wilhelm Schmidt explained: "What remains of air movement within enclosed spaces—mostly in the form of drafts of light breezes—is of an essentially different magnitude than what we observe outdoors. The usual methods of meteorology fail here; those that replace them are indirect or highly sensitive."[40] For the study of urban environments, climatology would have to be rescaled.

Wilhelm Schmidt's "mobile climatological observatory" was one solution. It was no more than a large Opel fitted out with meteorological instruments, purchased with aid from the Notgemeinschaft der Deutschen Wissenschaften (Emergency Association for German Science) and the ministries of agriculture and education. But it allowed him and his colleagues to take measurements closely spaced in time at locations throughout the city and its outskirts. Thus Schmidt measured the heat-island effect and analyzed air samples for impurities. In the urban environment, less familiar meteorological variables also became significant, such as levels of carbon dioxide and dust. In Prague, Gregor emphasized the mesoscale of urban climate—that is, the influence of the built environment up to a height of roughly fifty meters. In Vienna, Schmidt and Brezina were more intent on the microscale, from a neighborhood down to a single building or room. They even considered the "climate" of the space between a person's clothing and his bare skin. They also introduced the notion of a "personal climate," which an individual could track in the course of her daily migrations from home to streetcar to laboratory and back. As practical measures, Gregor recommended expanding city parks, widening streets, and improving indoor ventilation systems.[41] Schmidt and Brezina, on the other hand, perhaps influenced by the increasingly *völkisch* tone of Austrian politics, endorsed the "green city" movement, which would build worker settlements just beyond the city limits. They looked forward to a time when urban dwellers would shed "their extreme urban character," would "no longer think and feel exclusively urbanly," and would be "tied to nature like their forefathers who worked the land."[42]

In 1934, scientists in Frankfurt and Vienna founded the journal *Bioklimatische Beiblätter* as an offshoot of the originally Austrian journal *Meteorologische Zeitschrift*. The goal of the editors, Linke and Schmidt, was to unite the various branches and methods of the sciences—physics, medicine, botany, and geography—that dealt with the biological significance of climate. In doing so, they expanded the meaning of "climate" beyond atmospheric phenomena to include those that had been known to an earlier age as "telluric": processes originating in the soil and in the layer of air closest to the ground.[43] Taken together, this area of study comprised the global domain that Suess had termed the biosphere. Appropriately, interwar Austria and Czechoslovakia were the recognized leaders of this new subfield. When Linke complained of the Weimar government's indifference to health-related climatological research, he pointed to the Austrians and Czechoslovaks as the pioneers of this valuable new field. In 1928, the Balneological Congress held in Baden officially thanked the Austrian and Czechoslovak governments for their support of medical climatology—or "bioclimatology," as it was increasingly known.[44]

As sciences of climate on the small scale, mountain, urban, and bioclimatology all built on methods developed by botanists to track minute spatial variations of atmospheric conditions. We have seen how Emanuel Purkyně pioneered such research in Bohemia in the 1860s. More immediately, in *Klima und Boden auf kleinstem Raum* (1911), the Würzburg botanist Gregor Kraus had published observations of a startling variety of conditions in air less than two meters from the ground. Yet it was primarily in the interwar period that central Europeans gave "local climatology" its theoretical and methodological grounding. The very first issue of *Bioklimatische Beiblätter* began by discussing the range of scales that were relevant. It provided what were likely the first climatological definitions of the terms "micro," "local," and "large scale."[45] In 1934, Schmidt and Rudolf Geiger, the German author of the widely read textbook *The Climate Near the Ground*, argued that the older methods of *large-*

scale climatology had grown out of man's experience of traveling the world. Its goal was the "study of the climate of a landscape." A "*local* climate," by contrast, covered only *part* of a landscape, such as a single valley or cliff. Even smaller, a "microclimate" could not even be recognized as part of a landscape; it was, to a first approximation, merely flat. Different methods needed to be used in each case.[46] Echoing one interpretation of the new quantum mechanics, Schmidt pointed out that one could not apply the instruments of large-scale climatology to the study of a microclimate, if only because they would generate "a noticeable disturbance in the air under investigation." The microscale was more sensitive to human activities, more "artificial," in Schmidt's parlance. Where large-scale climatology relied on long-term means to cancel out fluctuations of the meteorological elements, small-scale climatology was interested in short-term variability, in the atmospheric extremes that shaped the lives of plants, animals, and humans alike. As Gregor put it, "it would be impossible to judge the quality of such places for the stated purposes according to common macro-climatological values, such as monthly temperature and precipitation." After all, one early frost was meaningless in the long term but potentially life threatening in the short. Thus, as Gregor summed up, "the field of local climatology ('exposure') is the main asset of all of twentieth century climatology and deserves to be classed as an independent science."[47]

The rise of local climatology was a direct consequence of the geopolitical upheaval that followed the Great War, according to Wilhelm Schmidt. As he saw it, the shrunken territory of Austria had both invited and demanded a new approach to climate study. Within Austria's new borders, thanks to its predominantly mountainous terrain, weather "plays out spatially [*räumlich abspielt*]" as much as temporally. Scaling down climatology made a virtue of necessity. "The image of the climate," Schmidt wrote, "no longer fits into the familiar large scale; it demands methods of observation, analysis, and representation different from those we are used to."[48]

CLIMATE IN WORD AND IMAGE

This brings us to a final element of continuity in central European climatology across 1918: the pursuit of "the image of climate." In 1937, Bohuslav Hrudička (1904–42) in Brno argued programmatically for the value of the dynamic approach to climatology as a mode of representation and explanation in geography. In doing so, he echoed many of the values expressed by Habsburg climatologists of previous generations. Statistical values were inadequate for "tracing the effect of the climate on the organic and inorganic world . . . because meteorological elements do not function in living and nonliving nature in isolation, but rather as a complex whole." Underlining the need for a visualizable conception of climate, Hrudička insisted that weather, as the "foundation" of climate, "must not fully disappear in the climatic picture, as it does in statistical tables of values of meteorological elements." In order to explain climate on the small scale, in its "regional peculiarities," science could not dispense with "verbal description" and "abundant illustrations."[49]

As this suggests, climatology in interwar central Europe held fast to the "word and image" ideal. Indeed, Austrian literary writers of the 1920s and 1930s incorporated the vivid language of atmospheric dynamics to express a sense of connection to the space of the former empire. For instance, Hugo von Hofmannsthal, known for his personal sensitivity to weather, repeatedly used images of "waves" to evoke the effects of cultural and ideological clashes. In his 1917 essay "The Austrian Idea," he wrote of an "energy" radiating out from Austria "in renewed waves"; of Austria's "inner polarity," "genuine elasticity," and "flowing border"; and of the "cultural waves propagating toward the East, but also receiving and ready to receive the counter-wave striving westwards." Appealing to the "spiritual space of the nation," Hofmannsthal used atmospheric imagery to mediate between materialist and idealist conceptions of the political community.[50]

Climatology also shaped the imagination of central European space in novels of the war's aftermath. While writing *Strudlhofstiege* (1951), set in Vienna in 1925, Heimito von Doderer was in close contact with the ZAMG, "since he had to know whether and how it had rained at a certain time on the day in question in the 1920s, before he described it."[51] Such historical accuracy mattered to von Doderer because the atmosphere was a psychological force in his novel. It mediated the characters' shifting sense of near and far, past and present. In an early passage, for instance, certain qualities of light and air, perceived in combination with a hint of an accent in a woman's voice, elicit a flash of consciousness in the protagonist. He feels an air current from "a distant horizon," a sense of being "sucked in all four directions of the compass . . . into the green, the open," into spaces from which the city dweller had been "alienated." Longing for the imagined lands of an empire that exists only in memory, he thinks of those "remarkable southern lands, which had once belonged [to the city]. It had been five years now since those nerve tracts had been severed."[52] This motif recalls another exemplary novel of the First Republic, *Karl and the Twentieth Century*, by Rudolf Brunngraber. There we read that "the longing to be far away [*der Trieb ins Weite*] seemed to have become an epidemic in the '20s." The protagonist experiences a mounting wanderlust as he becomes acutely aware of the scale of the world, an experience he likens to the sensation of wind: "Then he suddenly felt the size of the world, as if he stood outside in the night wind." As the economic depression of the 1930s takes its toll, however, Vienna becomes motionless, lifeless, and Karl remarks on the absence of any "wind of existence."[53] Again, images of the atmosphere in motion convey the shifting and uncertain geography of interwar central Europe.

But surely the most famous literary memorial to Habsburg climatology is the opening passage of *The Man without Qualities*, Robert Musil's unfinished philosophical novel about the Monarchy's last years. It begins, somewhat perplexingly, with a meticulous description of the state of the atmosphere over central Europe:

> A barometric low hung over the Atlantic. It moved eastward toward a high-pressure area over Russia without as yet showing any inclination to bypass this high in a northerly direction. The isotherms and isotheres were functioning as they should. The air temperature was appropriate relative to the annual mean temperature and to the aperiodic monthly fluctuations of the temperature. The rising and setting of the sun, the moon, the phases of the moon, of Venus, of the rings of Saturn, and many other significant phenomena were all in accordance with the forecasts in the astronomical yearbooks. . . . In a word that characterizes the facts fairly accurately, even if it is a bit old-fashioned: It was a fine day in August 1913.[54]

Even as it parodies the statistical language of normality, the novel's central theme emerges immediately in this first paragraph: the tension between human experience and the scientific description of reality. Does this synoptic weather report—a fair imitation of those that had appeared in newspapers across the Habsburg Monarchy since 1877—actually refer to the same phenomenon as "a fine August day"? Why bother with precision if meaning can be expressed the "old-fashioned" way? Of what value are statistical averages if one is interested in individuals and their unique stories? In this case, the scientific description appears to float far above the realm of human experience, preoccupied with events at a scale utterly removed from that of human cares. To stop there, however, would miss the profound role that climatology plays in Musil's writings as a metaphor for the relationship between the individual and society. In "The Nation as Ideal and Reality" (1921), Musil recalled the advent of war in 1914 in atmospheric terms: "One suddenly became a tiny particle humbly dissolved in a suprapersonal event and, enclosed by the nation, sensed the nation in an absolutely physical way." He described the nation as an "enormous heterogeneous mass . . . that oscillates between solid and fluid," stressing the "atmospherically undefined nature" of the experience of patriotism.[55] Two years later, in "The German as Symptom," Musil envisioned the European as a particle of air, his path hinging on the infinite contingencies of pressure, temperature, and topography. The individual may be swept up on large-scale historical currents, but his life course depends on local contingencies and "competing influences," just as the chance of rain might depend on how a particular mountain diverts a particular current of air. In this sense, the European's identity is not a function of race or *Zeitgeist*; it can be specified only by a trajectory, one too complex to calculate in advance. The European is not "destined" but "situated."

In this geographic imagery, we can recognize the legacy of the interplay of Habsburg science and politics. The genre of climatography, in particular, had invited readers to imagine their relationship to the territory of the empire in dynamic terms—not as a "dwelling in place," as Martin Heidegger was then in the process of theorizing, but as a contingent, and potentially turbulent, circulation.

SCALING UP

A final legacy of the imperial-royal field sciences lies in the projects of scientific internationalism that they inspired. A number of scientists took the Habsburg state's supranational scientific institutions as a model for the internationalization of science beyond its borders. These men believed that the methods of collaboration and synthesis that worked within Austria could be scaled up to further the cause of worldwide scientific cooperation. Beginning with the First International Meteorological Congress hosted in Vienna in 1873, Austrian scientists led a series of endeavors to coordinate scientific research and natural resource management the world over. Among these, for instance, was the first International Polar Year of 1883, which aimed to round out the scientific picture of the earth system in three dimensions. It grew from a proposal by Karl Weyprecht, a Habsburg naval officer whose internationalism stemmed from frustration with late nineteenth-century nationalism and impatience with voyages of exploration that served merely "the vanity of national flags."[56] Driven by a vision of supranational science, Weyprecht set in motion a series of initiatives that culminated in the International Geophysical Year of 1957–58. Among other examples of Habsburg-led internationalist science, one might cite Eduard Suess's monumental geological survey *The Face of the Earth*; the International Map of the World, which was first proposed by the Vienna geographer Albrecht Penck in 1891; and the *Overviews of the World Economy* by Franz Neumann-Spallart (which, as we've seen, drew its methods in part from climatology), along with the International Statistical Institute, founded in 1885 and based on a proposal by Neumann-Spallart. Habsburg scientists also promoted the international management of natural resources. Josef Roman Lorenz von Liburnau, for instance, supported the coordination of forest-conservation policy across Europe. Similarly, Eduard Suess attempted to set international economic policy on more secure foundations by assessing the earth's total store of precious metals. Emanuel Herrmann likewise promoted international planning around the management of coal, petroleum, and other resources. Looking to the European peace movement for leadership, Herrmann even proposed transforming this "fragmented" continent into a common economic zone.[57] Little did he know that it would take two catastrophic world wars to bring such a plan to fruition. In all these cases, researchers extrapolated from their experiences with transnational science within Austria-Hungary in order to prescribe mechanisms for coordinating research at multiple scales around the world.

While these schemes often proved to be both Eurocentric and utopian, certain elements remain instructive today, in the face of the present climate crisis. As we have seen, Habsburg scientists insisted that environmental research could not be left entirely in the hands of nation-states, nor could the use of natural resources be left to the whims of the market. At the same time, many of these scientists foresaw the dangers of globalization. Some warned of the dire toll that empire-building could take on humans and nature alike, of the violence of a colonial *Raubwirtschaft* or pillage economy. From a different angle, some of these thinkers questioned conservation measures that were no more than a cover for an empire's appropriation of land and resources at the expense of indigenous populations. Likewise, they pointed out that imperial policies designed to protect inhabitants from natural disasters must not become substitutes for addressing the inequalities that left some populations more vulnerable than others.[58] These scientists worried too about the homogenizing effects of a global economy on the world's cultural diversity. In his influential 1906 study *The Territorial Development of European Colonies*, Alexander Supan pointed out how much Europeans had learned from the cultures of those they had colonized. Looking ahead to a world transformed by European imperialism, he predicted that "a uniform world culture will not emerge, and that is a good thing, for only in diversity lies movement and life."[59] Here we find the Habsburg principle of the motive force of local differences scaled up to a hopeful projection for the future of the world.

Although this pluralist ideal did not always guide Habsburg governance in practice, this book has argued that it did structure the practice of the field sciences. The institutions of imperial-royal science coordinated research across central Europe without imposing a single set of values nor demanding uniform methods. Today, in the face of the urgent problems associated with climate change, the main locus of international and interdisciplinary collaboration has become the Intergovernmental Panel on Climate Change. Researchers from eighty-five countries have contributed to the five assessment reports to date, although those from North America and Europe made up 75 percent of the authors of the first four reports (1990, 1995, 2001, 2007) and 62 percent of the fifth (2013).[60] What's more, the IPCC has been criticized for pressing this diversity of voices into a rigid mold of scientific consensus. Geographically, its research has been said to skew toward the impacts of climate change on industrialized nations; methodologically, it has emphasized economic analysis over the perspectives of the other social sciences; and epistemologically, it has avoided the kind of deliberative reflexivity that is the mark of effective pluralism.[61] In this context, it is worth recalling that imperial-royal science aimed not to "unify" its many moving parts but to "orchestrate" them—to borrow a term from the philosopher of science Otto Neurath, born in Vienna in 1882. Neurath foresaw that future generations would need to address "problems of world organization." Yet he was confident that this could be achieved while allowing "various ways of life" to "coexist." What was needed was not unification but "orchestration," not a uniform world government but "a gigantic 'muddling through'"—precisely the term (*fortwursteln*) commonly used to describe the style of rule of the late Habsburg state.[62] Neurath saw no reason why the international science of the future would need to sacrifice the principle of pluralism.

CLIMATE AND THE POLITICS OF SCALING

In the course of the nineteenth-century debates over the climatic impact of human activities, the authority of the scientist came to be defined in a new way. Scientists began to justify their interventions on the basis of their capacity to judge the significance of things according to a more enduring scale that that of the narrow spatial and temporal frame of a human life. To be sure, earlier men of science had prided themselves on their cosmic vision or their appreciation for nature's minutiae. What was different in the nineteenth century was that scientists were claiming the right to intervene on this basis in matters of public policy. One of the first assertions of this competence came from Julius Hann in 1869. Discussing the relationship between deforestation and desiccation,

Hann insisted on the need for expert knowledge, for only the man of science knew how to weigh "small causes" in nature: "The natural scientist is used to taking the effects of small causes into account; the uneducated mind disregards them, concerning itself only with forces that have compelled it to astonishment or fear."[63] As we have seen, Hann presented himself as a naturalist for whom no detail was too "paltry and minute," and yet as a leader immune to the narrow interests of "church tower politics."

As public debate over the forest-climate question widened over the following two decades, scientists elsewhere took up this argument—most famously, Eugène-Emmanuel Viollet-le-Duc. Though known for his restorations of medieval architecture, Viollet-le-Duc was also fascinated by the structure of the earth and its transformations by natural and human agency. A geologist could look at a mountainside and see its progressive modification through erosion, just as an architect could look at ruins and imagine the structure they had once been.[64] In this way, Viollet-le-Duc was drawn into the nineteenth-century debate over the reality of anthropogenic climate change. Could puny man really alter such a vast and ancient planet? "In the presence of great geological phenomena, what is man?" In answer to this question, Viollet-le-Duc argued that every large-scale change was the result of uncountable small causes, and thus by small means humanity too could have a planetary impact, for better or worse—the choice was theirs. The fate of the planet hinged on rethinking the meaning of "small": "There are no small measures in nature, or rather the actions of nature only result from the accumulation of small measures."[65] This pronouncement was repeated by French and Austrian foresters in the ensuing debates over the forest-climate question.[66] Here we find experts claiming the right to guide environmental policy due to their ability to weigh facts that might escape ordinary perception entirely. Implicitly, they claimed the capacity to judge the world according to the scale of the earth itself.

If anyone wields moral authority of this sort in the United States circa 2017, it is certainly not scientists. On the contrary, today's climate scientists tend to be wary of making any public claims that take them beyond the facts of their research. Few would suggest that their studies give them any special wisdom, nor even a more capacious way of viewing the impacts of human actions. Who would believe them if they did? To their skeptics, scientists' attempts to steer attention toward long-term impacts are a mystification of the more urgent problems of the here and now. Even many of their advocates would cringe at the thought of scientists transgressing the line between facts and values. Even the best human minds, they would say, are ill equipped to reflect on personal choices in terms of long-term, planetary consequences. Indeed, it's often said today that climate change is a phenomenon unfolding on such a superhuman scale that it exceeds our powers of cognition. It's not easy for us to recognize ourselves as the agents of the sixth mass extinction in our planet's history. We appear to be stumped by the problem of weighing the welfare of far-future generations in our calculations of costs and benefits. Social scientists have expressed doubt that humans possess ethical intuitions about the consequences of our actions across continents and generations. Psychologists are even gathering evidence that humans have a cognitive impediment to thinking with large magnitudes. To quote the philosopher Dale Jamieson, "The scale of a problem like climate change can be crippling."[67]

These two ways of thinking about climate science—as the product of particularly farsighted minds, or as a form of knowledge in tension with the nature of human cognition—share the same problem. It is a problem we are now in a position to understand. Both views ignore the *noncognitive* work behind the science of climate. As the historical analysis undertaken here suggests, scientific judgments about the significance of human actions in relation to planetary processes have not derived from a unique perceptual faculty, nor from personal wisdom. They come not as instantaneous flashes of insight, but as the result of a *process*, the process of scaling. Scaling is a learning process, but it is not exclusively cognitive. It can certainly be a process of calculation, of recalibrating size or duration according to new proportionalities. But this recalibration often depends on new modes of representation and new ways of looking at the world. It is, in this sense, an aesthetic process. It is also embodied: in order to revise our sense of the relationship between the near and the far, we often rely on kinesthesis, on the awareness of our own limbs moving through space. Scaling is therefore a somatic learning experience—but it is not performed by individuals in isolation. In order to orient ourselves with respect to distant places or past times, we rely on the knowledge of others. This makes scaling a social process as well, one often marked by conflict and negotiation. Least obviously, scaling is an affective process. Revising judgments of the relative significance of things in the world means letting go of some attachments while forming new ones. So scaling is often accompanied by feelings of longing and loss, by the seductions of the exotic, as well as by the pangs of homesickness.

From a historical perspective, then, modern climate science is a product of scaling, a process that is not only intellectual, but also sensuous, passionate, and politically charged. Scaling has been constitutive of the history of climate science, and it will be equally vital to its future, since global warming threatens the lives of communities that are drastically underrepresented in the highest echelons of international science. Climatology thrived on the institutionalized pluralism of Habsburg science, which generated skepticism from multiple standpoints toward every attempt at a synthetic overview. Future research might consider how other scientific and political institutions, coupled to different infrastructures and different aesthetic cultures, have supported the work of scaling, and with what consequences for other domains of environmental knowledge. Today, the history of imperial-royal science stands as a reminder that the Austrian Problem has no unique solution. Every means of representing unity in diversity inevitably masks some forms of difference at the expense of others. In the words of the first ruler to call himself emperor of all Austria, whether a matter is large or small is a question that cannot be judged a priori and for all time. And so the work of commensuration continues.

ACKNOWLEDGMENTS

The debts I have incurred in writing this book are beyond all measure. Early on, I was lucky to receive advice and encouragement from three pioneers of the history of atmospheric science: Jim Fleming, Vlad Janković, and Katey Anderson. In Vienna in the fall of 2001, Christa Hammerl generously shared her knowledge of the ZAMG. From 2004 to 2006, I was fortunate to be able to explore the history of climatology as a junior fellow of the Harvard Society of Fellows. It was there, too, that I retrained as a historian of the Habsburg world, and I thank the east-central Europeanists among the junior fellows for guidance, especially Tara Zahra. At Barnard and Columbia, where I taught from 2006 to 2017, I benefited from conversations with many wonderful colleagues who shared my interest in the history of empires, as well as from the chance to organize related events as acting director of the Center for International History in 2011–13. In 2014–15 I was a fellow at the Cullman Center of the New York Public Library, and I am especially grateful to Jean Strouse and the curators of the Maps Department for making it such a productive year. More recently, I have enjoyed sharing ideas with members of the environmental sciences and humanities research cluster at Columbia's Center for Science and Society. I am grateful, too, for having had the opportunity to present my research to helpful audiences at the University of Pennsylvania, Harvard, Cambridge, the University of Chicago, the New York Botanical Garden, and the Consortium for the History of Science, Technology, and Medicine's physical sciences seminar.

Several colleagues have gone to great lengths to answer questions or read portions of the manuscript and offer suggestions. Heartfelt thanks to Mitch

Ash, Mark Cane, Nuala Caomhanach, Holly Case, Dipesh Chakrabarty, Hasok Chang, Paula Sutter Fichtner, Isabel Gabel, Emily Greble, Mott Greene, Chris Harwood, Anna Henchman, Eva Horn, Fredrik Jonsson, Pieter Judson, Dan Kevles, Matthieu Kohl, Melissa Lane, Ben Orlove, Jerry Passannante, Dorothy Peteet, Steve Pincus, Adam Sobel, Jan Surman, Julia Adeney Thomas, Conevery Bolton Valencius, Andrea Westermann, Natasha Wheatley, and Nasser Zakariya. All of them have shown great patience with my clumsiness with subjects ranging from Renaissance kingship to paleobotany. I feel enormously fortunate to have had the chance to learn from these remarkable scholars and friends.

For helpful feedback I am also indebted to participants in the workshops "Late Imperial Epistemologies" (Columbia University, 2013), "Creative Commensuration" (University of Zurich, 2016), "Experiencing the Global Environment" (Max Planck Institute for History of Science, Berlin, 2016), and "Biodiversity and Its Histories" (Cambridge University, Columbia University, and the New York Botanical Garden, 2017).

For assistance with translations, I thank Chris Harwood, Bogdan Horbal, Daniel Mahla, and Dániel Margócsy. Manuela Krebser and Gerlinde Fichtinger contributed transcriptions of archival documents. It is also a pleasure to thank four excellent research assistants: Cathrin Hermann, Katya Motyl, John Raimo, and Sara Heiny. For shepherding this book into print with their usual attention to detail, I am grateful to my editor, Karen Darling, and manuscript editor, Mark Reschke.

Funding for the research, writing, and production of this book has been provided by the National Science Foundation (award #0848583), the Cullman Center of the New York Public Library, the American Council of Learned Societies, Barnard College, and Columbia's Center for Science and Society and Harriman Institute. Parts of the research for this book previously appeared in *Science in Context*, the *Journal of Modern History, Osiris,* the *Avery Review*, and *Intimate Universality: Local and Global Themes in the History of Weather and Climate*, ed. James R. Fleming, Vladimir Janković, and Deborah R. Coen (Sagamore Beach, MA: Science History Publications, 2006).

Size may be relative, but my largest debts are truly enormous. My husband, Paul Tuchmann, and my children, Amalia and Adam, have shown more patience with me than any mom has a right to ask. Whenever remotely possible, Paul has given me the gift of time, and done so with love and encouragement. We are blessed to have the support of our parents, Ruth and Stanley Coen and Naomi and Rob Tuchmann, and of our siblings and siblings-in-law. I would especially like this book to honor the memory of my brilliant, courageous, and loving sister Gwen Basinger, who lost her fight with cancer in August 2017.

Amalia and Adam, now eleven and eight, have learned to share me with a project that must at times have seemed as all-consuming as a third child. They have commiserated when I hit obstacles and cheered my progress. They have grown into delightful travel companions for my conference trips, and they have gamely joined in my long hunt for a title for this book. Most of all, I am grateful that they have refrained from hurting each other when I was too busy writing to break up their fights. This book is for them. I hope with all my heart that their generation will be able to look back on the history of human knowledge of climate change and see lessons learned in the nick of time and catastrophes safely averted.

NOTES

INTRODUCTION

1. Germany's most famous scientist, Hermann von Helmholtz, ventured the same explanation in a popular lecture that year, but it was Hann's account, published in the new journal of the Austrian Society for Meteorology, that carried the day. Wilhelm von Bezold, "Noch ein Wort zur Entwicklungsgeschichte der Ansichten über den Ursprung des Föhn," *Meteorologische Zeitschrift* 3 (1886): 85–87, on 86.

2. The conclusion to chapter 8 addresses disputes over the definition and genealogy of dynamic climatology.

3. After 1867, the legally correct term for joint Austro-Hungarian institutions was "k. und k.," emphasizing Franz Josef's identity as emperor of Austria and king of Hungary; "k.k." referred to institutions restricted to Cisleithania. I use the English translation "imperial-royal" without distinguishing between these meanings for the sake of simplicity and ease of pronunciation.

4. Hann, Diary C, 85, JH.

5. Yi-Fu Tuan, *Cosmos & Hearth: A Cosmopolite's Viewpoint* (Minneapolis: University of Minnesota Press, 1996). Compare the nineteenth-century geographer Carl Ritter's interpretation of homesickness (*Heimweh*), discussed in chapter 11.

6. For an introduction to scale interactions, see Günter Blöschl, Hans Thybo, and Hubert Savenije, *A Voyage through Scales: The Earth System in Space and Time* (Baden bei Wien: Lammerhuber, 2015).

7. Pitman et al., "Regionalizing Global Climate Models," *International Journal of Climatology* 32 (2012): 321–37, showed that many of the factors that determine the impact of climate change "at spatial scales relevant to policy makers, to impacts and adaptation" were excluded from global climate models.

8. Peter Cebon et al., eds., *Views from the Alps: Regional Perspectives on Climate Change* (Cambridge, MA: MIT Press, 1998).

9. Cleveland Abbe, review of Hann's *Handbuch der Klimatologie*, 3rd ed., *Science* 34 (1911): 155–56, on 155.

10. Hew C. Davies, "Vienna and the Founding of Dynamical Meteorology," in *Die Zentralanstalt für Meteorologie und Geodynamik, 1851–2001*, ed. Christa Hammerl et al., 301–12 (Graz: Leykam, 2001), 310.

11. Hans Schreiber, "Die Wichtigkeit des Sammelns volksthümlicher Pflanzennamen," *Zeitschrift für österreichische Volkskunde* 1 (1895): 36–43, on 43. All translations are my own unless otherwise noted.

12. Matthew Mulcahy, *Hurricanes and Society in the British Greater Caribbean, 1624–1783* (Baltimore: Johns Hopkins University Press, 2008); David Blackbourn, *The Conquest of Nature: Water, Landscape, and the Making of Modern Germany* (New York: Norton, 2007); Charles Walker, *Shaky Colonialism: The 1746 Earthquake-Tsunami in Lima, Peru, and Its Long Aftermath* (Durham, NC: Duke University Press, 2008).

13. Richard Grove, *Green Imperialism: Colonial Expansion, Tropical Island Edens and the Origins of Environmentalism* (Cambridge: Cambridge University Press, 1995); Tom Griffiths and Libby Robin, eds., *Ecology and Empire: Environmental History of Settler Societies* (Seattle: University of Washington Press, 1997); Peder Anker, *Imperial Ecology: Environmental Order in the British Empire, 1895–1945* (Cambridge, MA: Harvard University Press, 2001); Michael Osborne, "Acclimatizing the World: A History of the Paradigmatic Colonial Science," *Osiris* 15 (2000): 135–51.

14. Basalla, "The Spread of Western Science," *Science* 156 (1967): 611–22.

15. Kapil Raj, *Relocating Modern Science: Circulation and the Construction of Knowledge in South Asia and Europe, 1650–1900* (Basingstoke and New York: Palgrave Macmillan, 2007); Simon Schaffer et al., eds., *The Brokered World: Go-Betweens and Global Intelligence, 1770–1820* (Sagamore Beach, MA: Science History Publications, 2009); Londa Schiebinger and Claudia Swan, eds., *Colonial Botany: Science, Commerce, and Politics in the Early Modern World* (Philadelphia: University of Pennsylvania Press, 2005), chapters 5–9.

16. Robert E. Kohler, *All Creatures: Naturalists, Collectors, and Biodiversity, 1850–1950* (Princeton, NJ: Princeton University Press, 2006), chapter 1. On climate science and settler colonialism in Oceania, see James Beattie et al., eds., *Climate, Science, and Colonization: Histories from Australia and New Zealand* (New York: Palgrave, 2014).

17. Grove, *Green Imperialism*; Helen Tilley, *Africa as a Living Laboratory: Empire, Development, and the Problem of Scientific Knowledge* (Chicago: University of Chicago Press, 2011). See too Libby Robin, "Ecology, a Science of Empire," in Griffiths and Robin, *Ecology and Empire*, 63–75; Paul S. Sutter, "Nature's Agents or Agents of Empire? Entomological Workers and Environmental Change during the Construction of the Panama Canal," *Isis* 98 (2007): 724–54.

18. Griffiths and Robin, *Ecology and Empire*; Anker, *Imperial Ecology*; Denis E. Cosgrove, *Apollo's Eye: A Cartographic Genealogy of the Earth in the Western Imagination* (Baltimore: Johns Hopkins University Press, 2001). "Planetary consciousness" had a different meaning in Mary Louise Pratt's *Imperial Eyes: Studies in Travel Writing and Transculturation* (London: Routledge, 1992).

19. Cf. Rohan Deb Roy, ed., "Nonhuman Empires," special section of *Comparative Studies of South Asia, Africa and the Middle East* 35 (2015): 66–172.

20. Dr. Witte, "Über die Möglichkeit, das Klima zu beeinflussen," *Medicinische Blätter, Wochenschrift für die gesamte Heilkunde* 31 (1908): 1–2, on 1.

21. On the development of paleoclimatology, see John Imbrie and Katherine Palmer Imbrie, *Ice Ages: Solving the Mystery* (Cambridge, MA: Harvard University Press, 1979).

22. Alexander von Humboldt, *Cosmos*, trans. E. C. Otte (New York, 1858), 1:317.

23. Robert Marc Friedman, *Appropriating the Weather: Vilhelm Bjerknes and the Construction of a Modern Meteorology* (Ithaca, NY: Cornell University Press, 1989); Katharine Anderson, *Predicting the Weather: Victorians and the Science of Meteorology* (Chicago: University of Chicago Press, 2005); Lorraine Daston, "The Empire of Observation, 1600–1800," in *Histories of Scientific Observation*, ed. Daston and Elizabeth Lunbeck (Chicago: University of Chicago Press, 2011), Michael Reidy, *Ocean Science and Her Majesty's Navy* (Chicago: University of Chicago Press, 2008). Climate was conceived in still other ways beyond Europe and North America, where anthropologists have documented ideas of climate that draw "no sharp distinction" between "biophysical" and "social worlds." Julie Cruikshank, *Do Glaciers Listen? Local Knowledge, Colonial Encounters, and Social Imagination* (Vancouver: UBC Press, 2005), 258.

24. Anton Kerner, *Das Pflanzenleben der Donauländer* (Innsbruck: Wagner, 1863), 3; Albrecht Penck, "Das Klima Europas während der Eiszeit," *Naturwissenschaftliche Wochenschrift* 20 (1905): 593–97, on 594.

25. On climatological knowledge for agriculture, see Benjamin Cohen, *Notes from the Ground: Science, Soil, and Society in the American Countryside* (New Haven, CT: Yale, 2009); Denise Phillips and Sharon Kingsland, eds., *New Perspectives on the History of Life Sciences and Agriculture* (New York: Springer, 2015); Fredrik Jonsson, *Enlightenment's Frontier: The Scottish Highlands and the Origins of Environmentalism* (New Haven, CT: Yale University Press, 2013); David Moon, *The Plough That Broke the Steppes: Agriculture and Environment on Russia's Grasslands, 1700–1914* (Oxford: Oxford University Press, 2013).

26. This is what Gisela Kutzbach refers to as the discovery of meteorology's third dimension in *The Thermal Theory of Cyclones: A History of Meteorological Thought in the Nineteenth Century* (Boston: American Meteorological Society, 1979). On climatology's status between the natural and human sciences, see Deborah Coen, *Climate Change and the Quest for Understanding* (New York: Social Science Research Council, January 2018).

27. Frank Trentmann, *Free Trade Nation: Commerce, Consumption, and Civil Society in Modern Britain* (Oxford: Oxford University Press, 2008), 155.

28. James R. Fleming, *Historical Perspectives on Climate Change* (Oxford: Oxford University Press, 1998), chapter 1. However, eighteenth-century settler colonialists believed in their capacity to "improve" climate; see Anya Zilberstein, *A Temperate Empire: Making Climate Change in Early America* (Oxford: Oxford University Press, 2016).

29. Lisbet Koerner, *Linnaeus: Nature and Nation* (Cambridge, MA: Harvard University Press, 1999); Suman Seth, *Difference and Disease: Medicine, Locality, and Race in the Eighteenth Century* (Cambridge: Cambridge University Press, forthcoming).

30. Eric Jennings, *Curing the Colonizers: Hydrotherapy, Climatology, and French Colonial Spas* (Durham, NC: Duke University Press, 2006).

31. Spencer Weart, *Discovery of Global Warming* (Cambridge, MA: Harvard University Press, 2009), 10.

32. Mark Carey, "Inventing Caribbean Climates: How Science, Medicine, and Tourism Changed Tropical Weather from Deadly to Healthy," *Osiris* 26, no. 1, *Klima* (2011): 129–41.

33. Alexander Supan, *Statistik der unteren Luftströmungen* (Leipzig: Duncker & Humblot, 1881), 1.

34. Napier Shaw, "Address of the President to the Mathematical and Physical Section of the

BAAS," *Science* 28 (1908): 457–71, on 463, 464. For John Herschel's earlier critique of blind empiricism in British meteorology, see Vladimir Janković, "Ideological Crests versus Empirical Troughs: John Herschel's and William Radcliffe Birt's Research on Atmospheric Waves, 1843–50," *BJHS* 31, no. 1 (March 1998): 21–40.

35. On the failures of meteorological standardization in the British Empire before 1914, see Martin Mahony, "For an Empire of 'All Types of Climate': Meteorology as an Imperial Science," *Journal of Historical Geography* 51 (2016): 29–39. On the problem of centralizing British meteorology, see Simon Naylor, "Nationalizing Provincial Weather: Meteorology in Nineteenth-Century Cornwall," *BJHS* 39 (2006): 407–33.

36. Cited by Jim Endersby, *Imperial Nature: Joseph Hooker and the Practices of Victorian Science* (Chicago: University of Chicago Press, 2008), 155. See too Christophe Bonneuil, "The Manufacture of Species: Kew Gardens, the Empire and the Standardisation of Taxonomic Practices in Late 19th century Botany," in *Instruments, Travel and Science: Itineraries of Precision from the 17th to the 20th Century*, ed. M.-N. Bourguet, C. Licoppe, and O. Sibum, 189–215 (London: Routledge, 2002); Richard Drayton, *Nature's Government: Science, Imperial Britain and the "Improvement" of the World* (New Haven, CT: Yale University Press, 2000); Bruno Latour, *Science in Action: How to Follow Scientists and Engineers through Society* (Cambridge, MA: Harvard University Press, 1987), chapter 6.

37. This was the topic of a conference I organized at Columbia in 2013, and I am grateful to the participants for sharing their research and insights; the phrase in quotation marks is from Marina Mogilner's presentation.

38. James Scott, *Seeing Like a State: How Certain Schemes to Improve the Human Condition Have Failed* (New Haven, CT: Yale University Press, 1998); Karen Barkey, *Empire of Difference: The Ottomans in Comparative Perspective* (Cambridge: Cambridge University Press, 2008); Tilley, *Africa as a Living Laboratory*, 21, 130.

39. J. B. Harley, *The New Nature of Maps: Essays in the History of Cartography* (Baltimore: Johns Hopkins University Press, 2001); David Harmon, *In Light of Our Differences: How Diversity in Nature and Culture Makes Us Human* (Washington, DC: Smithsonian, 2002).

40. Pieter Judson, *The Habsburg Empire: A New History* (Cambridge, MA: Harvard University Press, 2016).

41. Quoted in Werner Telesko, *Kulturraum Österreich: Die Identität der Regionen in der bildenden Kunst des 19. Jahrhunderts* (Vienna: Böhlau, 2008), 15.

42. On Habsburg science and nationalism: Tatjana Buklijas and Emese Lafferton, introduction to the special section on "Science, Medicine and Nationalism in the Habsburg Empire from the 1840s to 1918," *SHPBBS* 38 (2007): 679–86; Mitchell Ash and Jan Surman, eds., *The Nationalization of Scientific Knowledge in the Habsburg Empire, 1848–1918* (New York: Palgrave, 2012), Jan Surman, *Biography of Habsburg Universities, 1848–1918* (West Lafayette, IN: Purdue University Press, forthcoming).

43. Schreiber, "Wichtigkeit des Sammelns," 41.

44. Julius Hann, "Die Temperatur-Abnahme mit der Höhe als eine Function der Windesrichtung," *Wiener Berichte II* 57 (1868) 740–65, on 749.

45. Ursula K. Heise, *Imagining Extinction: The Cultural Meanings of Endangered Species* (Chicago: University of Chicago Press, 2016), 50.

46. Friedrich Kenner, "Karl Kreil, eine biographische Skizze," *Österreichische Wochenschrift* 1 (1863): 289–366, on 360–61.

47. Illuminating counterexamples are provided by James Bergman, "Climates on the Move: Climatology and the Problem of Economic and Environmental Stability in the Career of C. W. Thornthwaite, 1933–1963" (PhD diss., Harvard University, 2014); Jamie Pietruska, "US Weather Bureau Chief Willis Moore and the Reimagination of Uncertainty in Long-Range Forecasting," *Environment and History* 17 (2011): 79–105.

48. Nailya Tagirova, "Mapping the Empire's Economic Regions from the Nineteenth to the Early Twentieth Century," in *Russian Empire: Space, People, Power, 1700–1930*, ed. Jane Burbank et al., 125–38 (Bloomington: Indiana University Press, 2007). See too Marina Loskutova, "Mapping Regions, Understanding Diversity: Russian Economists Confront Natural Scientists, ca. 1880s–1910s," Encounters of Sea and Land (6th ESEH conference), Turku, 1 June 2011.

49. Henry Francis Blanford, *A Practical Guide to the Climates and Weather of India, Ceylon and Burmah* (London: Macmillan, 1889), 95.

50. Anderson, *Predicting the Weather*, chapter 6. Mahony, "Empire of All Types of Climate," suggests that the British did not support the regionalization of climatology until the peak of colonial nationalism after World War One.

51. Wladimir Köppen, "Die gegenwärtige Lage und die neueren Fortschritte der Klimatologie," *Geographische Zeitschrift* 1 (1895): 613–28. Cf. A. Kh. Khrgian, *Meteorology: A Historical Survey*, ed. Kh. P. Pogosyan (Jerusalem: Israel Program for Scientific Translations, 1970), vol. 1. On the imperial logic of Russian science, see Gordin, *A Well-Ordered Thing: Dmitrii Mendeleev and the Shadow of the Periodic Table* (New York: Basic, 2004).

52. Quoted in Ellsworth Huntington, review of Voeikov's *Le Turkestan Russe*, *Bulletin of the American Geographical Society* 47 (1915): 708. Cf. Voeikov, "De l'influence de l'homme sur la terre," pt. 2, *Annales de Géographie* 10 (1901): 193–215, esp. 193–95.

53. Moon, *The Plough That Broke the Steppes*.

54. Catherine Evtuhov, *Portrait of a Russian Province: Economy, Society and Civilization in Nineteenth-Century Nizhnii Novgorod* (Pittsburgh: University of Pittsburgh Press, 2011), 160; Khrgian, *Meteorology*, chapter 16; Olga Elina, "Between Local Practices and Global Knowledge: Public Initiatives in the Development of Agricultural Science in Russia in the 19th Century and Early 20th Century," *Centaurus* 56 (2014): 305–29.

55. Lorin Blodget, *Climatology of the United States* (Philadelphia: J. B. Lippincott and Co., 1857), 25.

56. Ibid., 208–9.

57. On the neglect of climatology and seismology at the federal level, see Deborah R. Coen, *The Earthquake Observers: Disaster Science from Lisbon to Richter* (Chicago: University of Chicago Press, 2013), chapter 9.

58. Rajmund Przybylak et al., eds., *The Polish Climate in the European Context: An Historical Overview* (Dordrecht: Springer, 2010); Simron Jit Singh et al., eds., *Long Term Socio-Ecological Research: Studies in Society-Nature Interactions across Spatial and Temporal Scales* (Dordrecht: Springer, 2013); Lajos Rácz, *The Steppe to Europe: An Environmental History of Hungary in the Traditional Age* (Cambridge: White Horse Press, 2013).

59. Quoted in Eva Wiedemann, *Adalbert Stifters Kosmos: Physische und experimentelle*

Weltbeschreibung in Adalbert Stifters Roman Der Nachsommer (Frankfurt am Main: Lang, 2009), 685.

60. Komlosy, *Grenze und ungleiche regionale Entwicklung: Binnenmarkt und Migration in der Habsburgermonarchie* (Vienna: Promedia, 2003); David F. Good, *The Economic Rise of the Habsburg Empire, 1750–1914* (Berkeley: University of California Press, 1984).

61. E.g., on Bosnia, Voeikov, "De l'influence de l'homme," 202.

62. Julius Hann, *Die Vertheilung des Luftdruckes über Mittel- und Süd-Europa* (Vienna: Hölzel, 1887), 5.

63. *Der Kaiserstaat Oesterreich unter der Regierung Kaiser Franz I*, vol. 2 (Stuttgart: Hallberger, 1841), 263.

64. I gratefully acknowledge the help of Andrea Westermann and Nils Güttler in composing this definition. The following discussions of scale have been particularly helpful: Jacques Revel, ed., *Jeux d'échelles: La micro-analyse à l'expérience* (Paris: Gallimard, 1996); Francesca Trivellato, "Is There a Future for Italian Microhistory in the Age of Global History?," *California Italian Studies* 2 (2011): 1–26; Wendy Espeland and Mitchell L. Stevens. "Commensuration as a Social Process," *Annual Review of Sociology* 24 (1998): 313–43; Nicholas B. King, "Scale Politics of Emerging Diseases," *Osiris*, 2nd ser., 19 (2004): 62–76; Dipesh Chakrabarty, "The Climate of History: Four Theses," *Critical Inquiry* 35 (2009): 197–222; Julia Adeney Thomas, "History and Biology in the Anthropocene: Problems of Scale, Problems of Value," *AHR* 119 (December 2014): 1587–607.

65. John Tresch, "Cosmologies Materialized: History of Science and History of Ideas," in *Rethinking Modern European Intellectual History*, ed. Darrin M. McMahon and Samuel Moyn, 153–72, (Oxford: Oxford University Press, 2014), 162.

66. Benedict Anderson, *Imagined Communities* (London: Verso, 1991), chapter 2.

67. Richard White, *Railroaded: The Transcontinentals and the Making of Modern America* (New York: London, 2011), chapter 4.

68. Jürgen Osterhammel, *The Transformation of the World: A Global History of the Nineteenth Century*, trans. Patrick Camiller (Princeton, NJ: Princeton University Press, 2014), 573.

69. Jennifer Raab, *Frederic Church: The Art and Science of Detail* (New Haven, CT: Yale University Press, 2015).

70. Anna Henchman, *The Starry Sky Within: Astronomy and the Reach of the Mind in Victorian Literature* (Oxford: Oxford University Press, 2014), 3. See too Adelene Buckland, *Novel Science: Fiction and the Invention of Nineteenth-Century Geology* (Chicago: University of Chicago Press, 2013).

71. Jesse Oak Taylor, *The Sky of Our Manufacture: The London Fog in British Fiction from Dickens to Woolf* (Charlottesville: University of Virginia Press, 2016), 11.

72. Allen MacDuffie, *Victorian Literature, Energy, and the Ecological Imagination* (Cambridge: Cambridge University Press, 2014), esp. 79–80.

73. Preface to *Živa* 1 (1853), iv.

74. Eduard Suess, *Das Antlitz der Erde*, vol. 1, 2nd ed. (Vienna: Tempsky, 1892), 25. Quoted and translated in A. M. Celâl Şengör, "Eduard Suess and Global Tectonics: An Illustrated 'Short Guide,'" *Austrian Journal of Earth Sciences* 107 (2014): 6–82, on 30.

75. Karl Kreil, *Die Klimatologie von Böhmen* (Vienna: Gerold's Sohn, 1865), 2–3.

76. E.g., Jan Patočka, *Body, Community, Language, World*, trans. Erazim Kohák (Chicago:

Open Court, 1998), 54–56. Michael Gubser emphasizes such metaphors of distance and proximity in his analysis of the political legacy of central European phenomenology in *The Far Reaches: Phenomenology, Ethics, and Social Renewal in Central Europe* (Stanford, CA: Stanford University Press, 2014).

77. Ludwig Landgrebe, *The Phenomenology of Edmund Husserl*, ed. Donn Welton (Ithaca, NY: Cornell University Press, 1981), 191.

78. David Woodruff Smith, *Husserl*, 2nd ed. (New York: Routledge, 2013), 329.

79. Simon Schaffer, "Late Victorian Metrology and Its Instrumentation: A Manufactory of Ohms," in *Invisible Connections: Instruments, Institutions, and Science*, ed. R. Bud and S. E. Cozzans, 23–56 (Bellingham: SPIE Press, 1991); Ken Alder, *The Measure of All Things: The Seven-Year Odyssey and Hidden Error That Transformed the World* (New York: Free Press, 2002).

CHAPTER ONE

1. Anton Kerner, "Die Geschichte der Aurikel," *Z. d. ö. AV* 6 (1875): 39–65, on 58.

2. On Maximilian, see Paula Sutter Fichtner, *Emperor Maximilian II* (New Haven, CT: Yale University Press, 2001).

3. Anton Kerner von Marilaun, "Die Geschichte der Aurikel," 4.

4. The consensus today is that the garden auricula is *Primula pubescens*, which in turn is a cross between *Primula auricula* and *Primula hirsuta*, apparently hybridized in the days of Clusius.

5. Kerner, "Die Geschichte der Aurikel," 46.

6. Marjorie Hope Nicolson, *Mountain Gloom and Mountain Glory: The Development of the Aesthetics of the Infinite* (Ithaca, NY: Cornell University Press, 1959).

7. Quoted in Kerner, "Die Geschichte der Aurikel," 55.

8. Kerner, *Die Botanischen Gärten, ihre Aufgabe in der Vergangenheit, Gegenwart und Zukunft* (Innsbruck: Verlag der Wagnerschen Universitätsbuchhandlung, 1874), 3–4.

9. Werner Telesko, *Geschichtsraum Österreich: Die Habsburger und ihre Geschichte in der bildenden Kunst des 19. Jahrhunderts* (Vienna: Böhlau, 2006); Christine Ottner, "Historical Research and Cultural History in Nineteenth-Century Austria: The Archivist Joseph Chmel (1798–1858)," *Austrian History Yearbook* 45 (2014): 115–33; Natasha Wheatley, "Law, Time, and Sovereignty in Central Europe: Imperial Constitutions, Historical Rights, and the Afterlives of Empire" (PhD diss., Columbia University, 2015).

10. Chmel, "Über die Pflege der Geschichtswissenschaft in Oesterreich," *Wiener Berichte* Phil-Hist. Kl. 1 (1850): 29–42, on 29.

11. Chmel, *Die Aufgabe einer Geschichte des österreichischen Kaiserstaates* (Vienna: Hof- und Staatsdrückerei, 1857), 13.

12. Joseph Chmel, "Ueber die Pflege der Geschichtswissenschaft in Oesterreich (Fortsetzung)," *Wiener Berichte* Phil-Hist. Kl. 1 (1850): 122–43, on 127–28.

13. Ottner, "Historical Research," 119, 126, 129.

14. Kerner, *Die Botanischen Gärten*; Alix Cooper, *Inventing the Indigenous: Local Knowledge and Natural History in Early Modern Europe* (Cambridge: Cambridge University Press, 2007).

15. Robert Kann, *The Habsburg Empire: A Study in Integration and Disintegration* (New York: Praeger, 1957), 4.

16. Fichtner, *Maximilian II;* Howard Louthan, *The Quest for Compromise: Peacemakers in Counter-Reformation Vienna* (Cambridge: Cambridge University Press, 1997).

17. Selma Krasa-Florian, *Die Allegorie der Austria: Die Entstehung des Gesamtstaatsgedankens in der österreichisch-ungarische Monarchie und die bildende Kunst* (Vienna: Böhlau, 2007).

18. Pamela H. Smith, *The Body of the Artisan: Art and Experience in the Scientific Revolution* (Chicago: University of Chicago Press, 2004).

19. Thomas DaCosta Kaufmann, *The Mastery of Nature: Aspects of Art, Science, and Humanism in the Renaissance* (Princeton, NJ: Princeton University Press, 1993), 181; Paula Findlen, *Possessing Nature: Museums, Collecting, and Scientific Culture in Early Modern Italy* (Berkeley: University of California Press, 1994).

20. Lorraine Daston and Katharine Park, *Wonders and the Order of Nature, 1150–1750* (Cambridge, MA: Zone Books, 1998).

21. Bruce Moran, "Patronage and Institutions: Courts, Universities, and Academies in Germany; An Overview, 1550–1750," in *Patronage and Institutions: Science, Technology and Medicine at the European Court, 1500–1750*, ed. Bruce Moran, 169–83 (Rochester, NY: Boydell Press, 1991), 174.

22. Marlies Raffler, *Museum—Spiegel der Nation? Zugänge zur Historischen Museologie am Beispiel der Genese von Landes- und Nationalmuseen in der Habsburgermonarchie* (Vienna: Böhlau, 2008), 165; Findlen, "Courting Nature," in *Cultures of Natural History*, ed. N. Jardine, J. A. Secord, and E. C. Spary, 57–74 (Cambridge: Cambridge University Press, 1996).

23. Fichtner, *Maximilian II*, 96.

24. Eliška Fučiková, "Cabinet of Curiosities or Scientific Museum?," in *The Origins of Museums: The Cabinet of Curiosities in Sixteenth- and Seventeenth-Century Europe*, ed. O. Impey and A. MacGregor (Oxford: Clarendon Press, 1985).

25. Thomas DaCosta Kaufmann, "Remarks on the Collections of Rudolf II: The Kunstkammer as a Form of *Representatio*," *Art Journal* 38 (1978): 22–28; Thomas DaCosta Kaufmann, *Court, Cloister, and City: The Art and Culture of Central Europe, 1450–1800* (Chicago: University of Chicago Press, 1995), 179.

26. Erik A. De Jong, "A Garden Book Made for Emperor Rudolf II in 1593: Hans Puechfeldner's 'Nützliches Khünstbüech der Gartnereij,'" *Studies in the History of Art* 69 (2008): 186–203, on 200.

27. Rita Krueger, *Czech, German, and Noble: Status and National Identity in Habsburg Bohemia* (Oxford: Oxford University Press, 2009), chapter 4.

28. See Kaufmann, "Remarks on the Collections," 25–26, and Smith, "Body of the Artisan," 77.

29. Thomas DaCosta Kaufmann, *Arcimboldo: Visual Jokes, Natural History, and Still-Life Painting* (Chicago: University of Chicago Press, 2009), 163.

30. Ibid., 115, 66.

31. Peter Marshall, *The Magic Circle of Rudolf II: Alchemy and Astrology in Renaissance Prague* (New York: Walker, 2006), 156.

32. Peter Barker, "Stoic Alternatives to Aristotelian Cosmology: Pena, Rothmann and Brahe," *Revue d'histoire des sciences* 61 (2008): 265–86.

33. Liba Taub, *Ancient Meteorology* (London: Routledge, 2003); Craig Martin, *Renaissance Meteorology: Pomponazzi to Descartes* (Baltimore: Johns Hopkins Press, 2011).

34. Patrick J. Boner, *Kepler's Cosmological Synthesis: Astrology, Mechanism and the Soul* (Boston: Brill, 2013).

35. Katharine Park, "Observation in the Margins, 500–1500," in Daston and Lunbeck, *Histories of Scientific Observation*, 15–44.

36. Christian Pfister et al., "Daily Weather Observations in Sixteenth-Century Europe," in *Climatic Variability in Sixteenth-Century Europe and Its Social Dimension*, ed. Pfister et al., 111–50 (Dordrecht: Springer, 1999).

37. Fritz Klemm, "Die Entwicklung der meteorologischen Beobachtungen in Österreich einschließlich Böhmen und Mähren bis zum Jahr 1700," *Annalen der Meteorologie* 21 (Offenbach am Main: Deutscher Wetterdienst, 1983), 14–16.

38. Ibid., 21.

39. Geoffrey Parker, *Global Crisis: War, Climate Change and Catastrophe in the Seventeenth Century* (New Haven, CT: Yale University Press, 2013).

40. Brahe's observations are now considered valuable because he "recorded items such as wind directions, along with noting several observations a day and not using a restricted terminology" (Pfister et al., "Weather Observations," 130). In the nineteenth century, his records were used to investigate the stability of the climate in the region of the Øresund Sound where Brahe made his home on the island of Hven. In 1876 Poul La Cour noted a higher frequency of snow days in Brahe's times than in his own, as well as differences in wind direction. Yet he stressed that cloud and rain patterns had not changed noticeably. However, when the Swedish meteorologist Nils Ekholm carried out his own calculations, based on observations at Uraniborg in the 1880s, he found a warming trend in the intervening three centuries, namely, a rise in the average February temperature of 1.4 degrees. Nils Ekholm, "On the Variations of the Climate of the Geological and Historical Past and Their Causes," *Quarterly Journal of the Meteorological Society* 27 (1901): 1–61, on 52–55.

41. Sigmund Fellöcker, *Geschichte der Sternwarte der Benediktiner-Abtei Kremsmünster* (Linz: Verlag des Stiftes, 1864), 95.

42. Andreas von Baumgartner, "Der Zufall in den Naturwissenschaften," *Almanach der kaiserlichen Akademie der Wissenschaften* 5 (1855): 55–76, on 64; Josef Durdík, "Kopernik a Kepler," *Osvěta* 3 (1873): 123–34.

43. Norbert Herz, *Keplers Astrologie* (Vienna: Gerold's Sohn, 1895), 61.

44. Ibid., 80.

45. Romuald Lang, "Das unbewußte im Menschen," Programm des k.k. Gymnasiums zu Kremsmünster für das Schuljahr 1859: 3–22, on 17. In the early twentieth century, the German physician and politician Willy Hellpach attracted popular interest to his inquiry into what he called the *Geopsyche*.

46. Anderson, *Predicting the Weather*, chapter 2; Jamie Pietruska, "Propheteering: A Cultural History of Prediction in the Gilded Age" (PhD diss., MIT, 2009), chapter 4.

47. Bohuslav Hrudička, "Meteorologie v české populární literature prvé polovice XIX. století," *Říše hvězd* 14 (1931): 109–14.

48. Coen, *Earthquake Observers*, 53–55.

49. Anderson, *Predicting the Weather*, 267; Mike Davis, *Late Victorian Holocausts: El Nino Famines and the Making of the Third World* (London: Verso, 2001).

50. Fleming, "James Croll in Context: The Encounter between Climate Dynamics and

Geology in the Second Half of the Nineteenth Century," *History of Meteorology* 3 (2006): 43–54, on 43.

51. Aleksandar Petrovic and Slobodan B. Markovic, "*Annus mirabilis* and the End of the Geocentric Causality: Why Celebrate the 130th Anniversary of Milutin Milanković?," *Quaternary International* 214 (2010): 114–18.

52. Vanessa Ogle, *The Global Transformation of Time, 1870–1950* (Cambridge, MA: Harvard University Press, 2015).

53. R. J. W. Evans, *Rudolf II and His World: A Study in Intellectual History, 1576–1612* (Oxford: Clarendon Press, 1973), 243.

54. H. W. Reichardt, "Ueber das Haus, in welchem Carl Clusius während seines Aufenthaltes in Wien (1573–1588) wohnte," *Blätter des Vereines für Landeskunde von Niederösterreich* 2 (1868): 72–73, on 72.

55. Evans, *Rudolf II*, 244, 172–73.

56. Fichtner, *Maximilian II*, 104.

57. Pamela Smith, *Body of the Artisan*, 64.

58. Evans, *Rudolf II*, 217–18.

59. Franz von Hauer, "Die Geologie und ihre Pflege in Österreich," *Almanach der Kaiserlichen Akademie der Wissenschaften* 11 (1861): 199–230, on 209.

60. Carina L. Johnson, *Cultural Hierarchy in Sixteenth-Century Europe: The Ottomans and Mexicans* (New York: Cambridge University Press, 2011), chapter 6.

61. Kaufmann, *Arcimboldo*, 120.

62. Dóra Bobory, *The Sword and the Crucible: Count Boldizsár Batthyány and Natural Philosophy* (Newcastle upon Tyne: Cambridge Scholars, 2009), 90.

63. Unlike nineteenth-century naturalists, however, those of the sixteenth century did not dwell on the provenance of their natural specimens. Princely collections tended to be organized, if they were organized at all, without regard to geography.

64. Krueger, *Czech, German, and Noble*, 164–65; Monika Sommer, "Zwischen flüssig und fest: Metamorphosen eines steirischen Gedächtnisortes," in *Das Gewebe der Kultur: Kulturwissenschaftliche Analysen zur Geschichte und Identität Österreichs in der Moderne* (Innsbruck: Studien-Verlag, 2001), 105–26, on 111.

65. Christa Riedl-Dorn, *Das Haus der Wunder: Zur Geschichte des Naturhistorischen Museums in Wien* (Vienna: Holzhausen, 1998); Michael Hochedlinger, *Österreichische Archivgeschichte vom Spätmittelalter bis zum Ende des Papierzeitalters* (Vienna: Böhlau, 2013), 109.

66. Hochedlinger, *Österreichische Archivgeschichte*, 88–90; Raffler, *Museum*, 181–89; Telesko, *Kulturraum*, chapter 14.

67. Telesko, *Kulturraum*, 380.

68. Kaspar von Sternberg, *Umrisse einer Geschichte der böhmischen Bergwerke* (Prague: Gottlieb Haase Söhne, 1836), xiii, i, v–vi. On Sternberg and the Bohemian National Museum, see Rita Krueger, *Czech, German, and Noble*, chapter 5.

69. Verein für Landeskunde von Niederösterreich, *Topographie von Niederösterreich*, vol. 1 (Vienna: Verein für Landeskunde von Niederösterreich, 1877), 559.

70. Eduard Suess, *Die erdbeben Nieder-Österreich's* (Vienna: k.k. Hof- und Staatsdruckerei, 1873); M. Porkorný, "Astronomie a meteorologie," *Památník druhého sjezdu českých lékařův a přírodozpytcův* (Prague: Komitét sjezdu českých lékařův a přírodozpytcův, 1882), 38–41, on 38.

71. Josef Schwerdfeger, *Die historischen Vereine Wiens, 1848–1908* (Vienna: Braumüller, 1908), 75; F. A. Slavík, ed., *Vlastivěda Moravská*, vol. 1 (Brno: Moravské akciové knihtiskárny, 1897), 8; Jindřich Metelka, "J. A. Komenského mapa Moravy," *Časopis Matice Moravské*, vol. 16 (1892), 144–51.

72. Ad. Horčička, "Dr. Wenzel Katzerowsky," *Mitteilungen des Vereins für Geschichte der Deutschen in den Sudetenländern* 40 (1901): 303–4.

73. Klemm, "Entwicklung," 11–13.

74. Reichardt, "Ueber das Haus," 72; H. W. Reichardt, *Carl Clusius' Naturgeschichte der Schwämme Pannoniens* (Vienna: k.k. Zoologisch-Botanische Gesellschaft, 1876), 3, my emphasis.

75. Hauer, "Die Geologie," 209.

76. Ibid., 230.

CHAPTER TWO

1. A. J. P. Taylor, *The Habsburg Monarchy, 1809–1918* (Chicago: University of Chicago Press, 1976), 175. See too Claudio Magris, *Der habsburgische Mythos in der modernen österreichischen Literatur*, trans. Madeleine von Pásztory (Vienna: Zsolnay, 2000); Mark Cornwall, *The Undermining of Austria-Hungary: The Battle for Hearts and Minds* (Basingstoke: Macmillan, 2000); Daniel Unowsky, *The Pomp and Politics of Patriotism: Imperial Celebrations in Habsburg Austria, 1848–1916* (West Lafayette, IN: Purdue University Press, 2005).

2. Paul de Lagarde, "Über die gegenwärtigen Aufgaben der deutschen Politik," in *Deutsche Schriften*, 22–46 (Göttingen: Dieterich, 1886), 45.

3. Julius Andrássy, *Ungarns Ausgleich mit Österreich vom Jahre 1867* (Leipzig: Duncker & Humblot, 1897), 41; cf. Alfons Danzer, *Unter den Fahnen: Die Völker Österreich-Ungarns in Waffen* (Vienna: Tempsky, 1889), 4.

4. Tamara Scheer, "Habsburg Languages at War," in *Languages and the First World War: Communicating in a Transnational War*, ed. Julian Walker and Christophe Declercq, 62–78 (London: Macmillan, 2016), 62; Christa Hämmerle, "Allgemeine Wehrpflicht in der multinationalen Habsburgmonarchie," in *Der Burger als Soldat: Die Militarisierung europäischer Gesellschaften im langen 19. Jahrhundert: Ein internationaler Vergleich*, ed. Christian Jansen, 175–213 (Essen: Klartext, 2004). István Deák, *Beyond Nationalism: A Social and Political History of the Habsburg Officer Corps, 1848–1918* (New York: Oxford, 1990).

5. Croat, Czech, German, Hungarian, Italian, Polish, Romanian, Ruthenian/Ukrainian, Slovak, Slovene, and Serbian; Bosnian was an unofficial twelfth (Scheer, "Languages at War," 65).

6. Franz/František Palacký, "Eine Stimme über Österreichs Anschluß an Deutschland," in *Oesterreichs Staatsidee*, 79–86 (Prague: J. L. Kober, 1866), 83.

7. David Luft, ed., *Hugo Von Hofmannsthal and the Austrian Idea: Selected Essays and Addresses, 1906–1927* (West Lafayette, IN: Purdue University Press, 2007).

8. David F. Lindenfeld, *The Practical Imagination: The German Sciences of State in the Nineteenth Century* (Chicago: University of Chicago Press, 1997), Lisbet Koerner, *Linnaeus: Nature and Nation* (Cambridge, MA: Harvard University Press, 1999).

9. Quoted and translated in Isaac Nachimovsky, *The Closed Commercial State: Perpetual*

Peace and Commercial Society from Rousseau to Fichte (Princeton, NJ: Princeton University Press, 2011), 83.

10. Werner Drobesch, "Die ökonomischen Aspekte der Bruck-Schwarzenbergschen 'Mitteleuropa,'" in *Mitteleuropa—Idee, Wissenschaft und Kultur im 19. und 20. Jahrhundert*, 19–42 (Vienna: Austrian Academy of Sciences, 1997), 24.

11. An echo of the question posed by Benedict Anderson in *Imagined Communities*, except that he excluded multinational communities from consideration.

12. Stifter, *Nachsommer*, 118.

13. Karl Winternitz, *Länderspiel vom Kaiserstaate Oesterreich. In 21 Stücken sammt der Karte* (Vienna: Rudolf Lechner, 1861); Johannes Dörflinger, *Descriptio Austriae: Österreich und seine Nachbarn im Kartenbild von der Spätantike bis ins 19. Jahrhundert* (Vienna: Edition Tusch, 1977), 146.

14. Quoted in Drobesch, "Die ökonomischen Aspekte," 25.

15. Wolfgang Göderle, *Zensus und Ethnizität: Zur Herstellung von Wissen über soziale Wirklichkeiten im Habsburgrreich, 1848–1910* (Göttingen: Wallstein, 2016). Göderle provides an important history of the Habsburg state's "production of diversity" through the technology of the census, but does not consider statistical surveys of the nonhuman world.

16. Sander Gliboff, "Gregor Mendel and the Laws of Evolution," *History of Science* 6 (1999): 217–35.

17. Albrecht Penck, foreword to *Geographischer Jahresbericht aus Österreich* 4 (1906): 1–8, on 4.

18. Norbert Krebs, *Länderkunde der österreichischen Alpen* (Stuttgart: Engelhorn, 1913), 3.

19. Adler cited in Michael Steinberg, *Austria as Theater and Ideology: The Meaning of the Salzburg Festival* (Ithaca, NY: Cornell University Press, 2000), 120; Kraus cited in Edward Timms, *Karl Kraus: Apocalyptic Satirist*, vol. 1, *Culture and Catastrophe in Habsburg Vienna* (New Haven, CT: Yale University Press 1986), 10; Jászi and Masaryk cited in Mark Mazower, *Dark Continent: Europe's Twentieth Century* (New York: Knopf, 1998), ix, 45.

20. Tatjana Buklijas, "Surgery and National Identity in Late Nineteenth-Century Vienna," *Studies in History and Philosophy of Biological and Biomedical Sciences* 38 (2007), 756–74; Lafferton, "The Magyar Moustache: The Faces of Hungarian State Formation, 1867–1918," *Studies in History and Philosophy of Biological and Biomedical Sciences* 38 (2007): 706–32; Bojan Baskar, "Small National Ethnologies and Supranational Empires: The Case of the Habsburg Monarchy," in *Everyday Culture in Europe*, ed. Ullrich Kockel (Aldershot: Ashgate, 2008). The Czech-speaking physician Emanuel Rádl argued that small nations contribute to world science not in spite of but by virtue of their unique histories and languages—a position that Jan Surman terms a "philosophy of science of small nations." Jan Surman, "Imperial Knowledge? Die Wissenschaften in der späten Habsburg-Monarchie zwischen Kolonialismus, Nationalismus und Imperialismus," *Wiener Zeitschrift zur Geschichte der Neuzeit* 9 (2009): 119–33.

21. Diana Reynolds Cordileone, *Alois Riegl in Vienna, 1875–1905: An Institutional Biography* (Burlington, VT: Ashgate, 2014), shows the influence of natural science on the Vienna School of Art History. Here, I draw on her analysis but also show a more fundamental common ground in the conditions for spatial research that derived from the structure of the supranational state.

22. Matthew Rampley, *The Vienna School of Art History: Empire and the Politics of Scholarship, 1847–1918* (University Park: Penn State Press, 2013), 84.

23. T. G. Masaryk, *Otázka Sociální* (Prague: Leichter, 1898), 647. See chapter 3, below.

24. Peter Stachel, "Die Harmonisierung national-politischer Gegensätze und die Anfänge der Ethnographie in Österreich," in *Geschichte der österreichischen Humanwissenschaften*, vol. 4, *Geschichte und fremde Kulturen*, ed. Karl Acham (Vienna: Passagen Verlag, 2002), 323–67; Brigitte Fuchs, *"Rasse," "Volk," Geschlecht: Anthropologische Diskurse in Österreich, 1850–1960* (Frankfurt: Campus, 2003), chapter 10.

25. Matthew Rampley, "Peasants in Vienna: Ethnographic Display and the 1873 World's Fair," *Austrian History Yearbook* 42 (2011): 110–32.

26. Rudolf von Eitelberger, *Gesammelte kunsthistorische Schriften*, vol. 2 (Vienna: Braumüller, 1879), 333.

27. Cited in Cordileone, *Alois Riegl*, 99.

28. Rampley, "World's Fair," 132.

29. Riegl quoted in Bernd Euler-Rolle, "Der 'Stimmungswert' im spätmodernen Denkmalkultus: Alois Riegl und die Folgen," *Österreichische Zeitschrift für Kunst und Denkmalpflege* 59 (2005): 27–34, on 30.

30. Max Dvorak, "Einleitung," in *Die Denkmale des Politischen Bezirkes Krems*, ed. Hans Tietze (Vienna: Anton Schroll, 1907), xvii.

31. E.g., Thomas M. Lekan, *Imagining the Nation in Nature: Landscape Preservation and German Identity, 1885–1945* (Cambridge, MA: Harvard University Press, 2004).

32. Johannes Straubinger, *Sehnsucht Natur: Geburt einer Landschaft* (Norderstedt: Books on Demand, 2009), 239, 264–67.

33. Quoted and translated in Rampley, *Vienna School*, 203.

34. Quoted and translated in Cordileone, *Alois Riegl*, 276.

35. Ibid., xviii, my emphasis.

36. Rampley, *Vienna School of Art History*, chapter 9.

37. Richard Charmatz, *Minister Freiherr von Bruck, der Vorkämpfer Mitteleuropas: Sein Lebensgang und seine Denkschriften* (Leipzig: S. Hirzel, 1916), 24.

38. Ibid., 53.

39. Quoted in ibid., 188.

40. Ibid., 189. Lynn Nyhart, "The Political Organism: Carl Vogt on Animals and States in the 1840s and '50s," *Historical Studies in the Natural Sciences* 47, no. 5 (Fall 2018): forthcoming, analyzes the consequences of organism-state analogies for biological research on lower organisms; I am pointing to the role of such analogies in motivating the study of the state's territory as an organic unit with its own metabolism.

41. *Die Denkschriften des österreichischen Handelsministers über die österreichisch-deutsche Zoll- und Handelseinigung* (Vienna: Carl Gerold, 1850), 94.

42. *Denkschriften des österreichischen Handelsministers*, 257.

43. Charmatz, *Minister von Bruck*, 227.

44. "Ueber die Weltstellung Oesterreichs," *Innsbrucker Zeitung*, 15 January 1850, 52.

45. Ferdinand Stamm, *Verhältnisse der Volks, Land und Forstwirthchaft des Königreiches Böhmen* (Prague: Rohlíček, 1856).

46. Ferdinand Stamm, "Landwirtschaftliche Briefe," *Die Presse*, 14 December 1855. Even the Prussians acknowledged this climatic advantage: see, e.g., Ernst Von Seydlitz, *Handbuch der Geographie* (Breslau: F. Hirt, 1914), 79.

47. Maureen Healy, *Vienna and the Fall of the Habsburg Empire: Total War and Everyday Life in World War I* (Cambridge: Cambridge University Press, 2004).

48. Quoted in John Deak, *Forging a Multinational State: State Making in Imperial Austria from the Enlightenment to the First World War* (Stanford, CA: Stanford University Press, 2015), 103.

49. David Good, *Economic Rise*. Max-Stephan Schulze and Nikolaus Wolf counter that, from the late 1880s, nationalism had the economic effect of tying regions "of similar ethno-linguistic composition" more closely to each other than to the rest of the Monarchy; "Economic Nationalism and Economic Integration: The Austro-Hungarian Empire in the Late Nineteenth Century," *Economic History Review* 65 (2011): 652–73.

50. Andrea Komlosy, "State, Regions, and Borders: Single Market Formation and Labor Migration in the Habsburg Monarchy, 1750–1918," *Review* (Fernand Braudel Center) 27 (2004): 135–77.

51. A. Zeehe, F. Heiderich, and J. Grunzel, *Österreichische Vaterlandskunde für die obserte Klasse der Mittelschulen*, 3rd ed. (Ljubljana: Kleinmayr & Bamberg, 1910), 8.

52. Good, *Economic Rise*, 246.

53. "Volkswirtschaft," *Oesterreichische Neuigkeiten und Verhandlungen* 53 (1850): 417–19, on 418.

54. Alexander von Bally, *Das neue Österreich, seine Handels- und Geldlage* (Vienna: Beck, 1850), 8.

55. Dominique K. Reill, *Nationalists Who Feared the Nation: Adriatic Multi-Nationalism in Habsburg Dalmatia, Trieste, and Venice* (Stanford, CA: Stanford University Press, 2012), 177.

56. Margaret Schabas, *The Natural Origins of Economics* (Chicago: University of Chicago Press, 2005), 150.

57. Carl Menger, *Principles of Economics*, trans. J. Dingwall and B. F. Hoselitz (Auburn, AL: Institute for Humane Studies, 1976), 167.

58. *Die österreichisch-ungarische Monarchie in Wort und Bild*, vol. 15, *Böhmen*, vol. 2, (Vienna: k.k. Hof- und Staatsdruckerei 1896), 464.

59. As Fredrik Jonsson has pointed out to me, Menger's British counterpart William Stanley Jevons was likewise of two minds, considering population growth as a factor in his forecast of coal exhaustion but not in his theory of political economy.

60. Quinn Slobodian, "How to See the World Economy: Statistics, Maps, and Schumpeter's Camera in the First Age of Globalization," *Journal of Global History* 10 (2015): 307–32, on 316.

61. Eugen von Philippovich, *Grundriss der politischen Oekonomie*, vol. 1 (Freiburg i. B.: J. C. B. Mohr, 1893), 86. Compare Karl Polanyi's critique of free trade in *The Great Transformation*, in which he attacked neoclassical economics for having obscured the many ways in which prosperity rises and falls with natural conditions that in no way respond to the market.

62. See, for instance, Karl von Rokitansky's remarks on the environmental origins of racial difference at the opening of the Vienna Anthropological Society, February 1870.

63. Franz Heiderich, "Die Wirtschaftsgeographie und ihre Grundlagen," in *Karl Andrees Geographie des Welthandels*, vol. 1, ed. Franz Heiderich and Robert Sieger (Frankfurt am Main: H. Keller, 1910), 39.

64. *Beiträge zur Wirtschaftskunde Österreichs: Vorträge des 4. International Wirtschaftskurses* (Vienna: A. Hölder, 1911), 1–39.

65. See, e.g., Jennings, *Curing the Colonizers*; Michael A. Osborne and Richard S. Fogarty, "Medical Climatology in France: The Persistence of Neo-Hippocratic Ideas in the First Half of the Twentieth Century," *Bulletin of the History of Medicine* 86 (2012): 543–63.

66. "Sterblichkeit," *Militär-Zeitung*, 10 July 1863, 17–18, on 17; Hämmerle, "Allgemeine Wehrpflicht," 202; Teodora Daniela Sechel, "Contagion Theories in the Habsburg Monarchy," in *Medicine Within and Between the Habsburg and Ottoman Empires, 18th–19th Centuries,* ed. Sechel, 55–77 (Bochum: D. Winkler, 2011), esp. 73.

67. "Einfluss des Klimas, der Orts- und Landesverhältnisse so wie der Lebensweise der Soldaten auf den Gesundheitszustand," *Allgemeine Militärärztliche Zeitung*, 25 August 1867, 276–80.

68. E.g., Alois Fessler, *Klimatographie von Salzburg* (Vienna: Gerold & Co., 1912), 17.

69. August von Härdtl et al., *Die Heilquellen und Kurorte des oestreichischen Kaiserstaates und Ober-Italien's* (Vienna: Braumüller, 1862), iv–v.

70. Enoch Kisch, *Klimatotherapie* (Berlin: Urban and Schwarzenberg, 1898), 641.

71. Alison Frank, "The Air Cure Town: Commodifying Mountain Air in Alpine Central Europe," *Central European History* 44, no. 2 (June 2012), 185–207; Jill Steward, "Travel to the Spas: The Growth of Health Tourism in Central Europe, 1850–1914," in *Journeys into Madness: Mapping Mental Illness in the Austro-Hungarian Empire* (New York: Berghahn, 2012), 72–89.

72. Adalbert Stifter, "Zwei Schwestern," in *Studien*, vol. 2, 6th ed. (Pest: Heckenast, 1864), 388.

CHAPTER THREE

1. Karl Kreil, "Über die k.k. Zentralanstalt für Meteorologie und Erdmagnetismus" (Vienna: k.k. Hof- und Staatsdruckerei, 1852), 85.

2. See Kreil's letters to Humboldt from Milan, published in *Annalen der Physik und Chemie* 13 (1838): 292–303; 16 (1839): 443–58.

3. Adalbert Stifter, *Der Nachsommer: Eine Erzählung*, vol. 1 (Pest: Heckenast, 1865), 177.

4. Ibid., 337.

5. Eduard Fenzl, "Eröffnungsrede," *Verh. Zool.-Bot. Ver.* 2 (1852): 1–5, on 4, original emphasis.

6. Ulrich L. Lehner, *Enlightened Monks: The German Benedictines, 1740–1803* (New York: Oxford University Press, 2011), 5.

7. William Clark, "The Death of Metaphysics in Enlightened Prussia," in *The Sciences in Enlightened Europe*, ed. William Clark, Jan Golinski, and Simon Schaffer (Chicago: University of Chicago Press, 1999), 423–73, on 434; Katharine Park, "Observation in the Margins, 500–1500," in Daston and Lunbeck, *Histories of Scientific Observation*, 15–44, on 23.

8. Fellöcker, *Geschichte der Sternwarte*, 241.

9. P. Augustin Reslhuber, "Die Sternwarte zu Kremsmünster," *Unterhaltungen im Gebiete der Astronomie, Meteorologie und Geographie* 10 (1856): 382–88, 392–96.

10. Cf. Cooper, *Inventing the Indigenous*.

11. Marian Koller, *Ueber den Gang der Wärme in Oesterreich ob der Enns* (Linz: F. Eurich, 1841), 7.

12. Karl Fritsch, autobiographical sketch, *Zs. Ö. G. Meteo.* 15 (1880): 105–19, on 106.

13. In Moravia, similar efforts grew with the support of the imperial state. Rudolf Brazdíl, Hubert Valášek, et al., *History of Weather and Climate in the Czech Lands: Instrumental Measurements in Moravia up to the End of the Eighteenth Century* (Brno: Masaryk University, 2002), 2–23.

14. Monika Baar, *Historians and Nationalism: East-Central Europe in the Nineteenth Century* (Oxford: Oxford University Press, 2010), 264.

15. Jan Janko and Soňa Štrbáňová, *Věda Purkyňovy doby* (Prague: Academia, 1988), 193; Karel Krška and Ferdinand Šamaj, *Dějiny meteorologie v českých zemích a na Slovensku* (Prague: Karolinium 2001), 87.

16. Strbanova, *Věda Purkyňovy doby*, 118–19.

17. Pseudonym of Heinrich Landesmann, *Die Muse des Glücks und moderne Einsamkeit* (Dresden: H. Linden, 1893), 14.

18. Jan Evangelista Purkyně, "Čtenářům ku konci roku," *Živa* 1 (1853): iii–iv, on iv.

19. Surman, *Biography of Habsburg Universities.*

20. Quoted in E. M. Kronfeld, *Anton Kerner von Marilaun* (Leipzig: Tauchnitz, 1908), 306.

21. Baar, *Historians and Nationalism*, 11.

22. Fasz. 683/Sig. 4A/Nr. 7757/1868: 27 June 1868, MCU.

23. Michael von Kast et al., *Geschichte der Österreichischen Land- und Forstwirtschaft und ihrer Industrien*, vol. 1 (Vienna: Moritz Perles, 1899), 558.

24. Josef Wessely to Emanuel Purkyně, undated, ca. 1878, EP.

25. *Die österreichisch-ungarische Monarchie in Wort und Bild*, 1:135.

26. E.g., *Osiris* 11 (1996): "Science in the Field," ed. Henrika Kuklick and Robert E. Kohler.

27. Larry Wolff, *The Idea of Galicia: History and Fantasy in Habsburg Political Culture* (Stanford, CA: Stanford University Press, 2010).

28. When a researcher at the ZAMG fell ill with tuberculosis, the director attempted to send him to South Tyrol for a climatic cure. The tubercular scientist refused the offer, insisting that the only mountain therapy he would undergo would be research at the 3,105-meter Sonnblick observatory in Carinthia. Fasz. 684/Sig. 4A/Nr. 45052: 30 November 1905, MCU.

29. Hammerl, *Zentralanstalt*, 37.

30. Mary Louise Pratt, *Imperial Eyes*; Edney, *Mapping an Empire: The Geographical Construction of British India, 1765–1843* (Chicago: University of Chicago Press, 1997).

31. Adalbert Stifter, *Bunte Steine*, 4th ed. (Pest: G. Hackenast, 1870), 56, 61. The surreal effect of Kafka's *The Castle* comes in part from the very ordinariness of the protagonist's occupation: he is a land surveyor.

32. Denise Phillips, *Acolytes of Nature: Defining Natural Science in Germany, 1770–1850* (Chicago: University of Chicago Press, 2012), 80–82; and Krueger, *Czech, German, and Noble*, 37.

33. Quoted in Inge Franz, "Eduard Suess im ideengeschichtlichen Kontext seiner Zeit," *Jahrbuch der Geologischen Bundesanstalt* 144 (2004): 53–65, on 64.

34. F. K. Branky, "Die Exkursionen des geographischen Seminars der k.k. Wiener Universität," *Zeitschrift für Schul-Geographie* 26 (1904): 65–72, on 62.

35. Andreas Helmedach, *Das Verkehrssystem als Modernisierungsfaktor: Strassen, Post, Fuhrwesen und Reisen nach Triest und Fiume vom Beginn des 18. Jahrhunderts bis zum Eisenbahnzeitalter* (Munich: Oldenbourg, 2002), 479.

36. Margarete Girardi, "Bericht über die Feier des 90 jährigen Jubiläums der ehemaligen k.k. Geologischen Reichsanstalt," *Verhandlungen der Zweigstelle Wien der Reichsstelle für Bodenforschung* (1939): 243–54, on 247.

37. Eduard Suess, *Erinnerungen* (Leipzig: Hirzel, 1916), 100.

38. Karl Kreil and Karl Fritsch, *Magnetische und geographische Ortsbestimmungen im österreichischen Kaiserreich*, vol. 1 (Prague: G. Haase, 1848), 3.

39. Hann, Diary B, 65a, 65b, JH.

40. Hann, Diary B, 102–3, JH.

41. Suess, *Erinnerungen*, 161.

42. Kenner, "Karl Kreil," 334.

43. Helmedach, *Verkerhrssystem*, 267–73.

44. Quoted in Christina Bachl-Hofmann, ed., *Die Geologische Bundesanstalt in Wien: 150 Jahre Geologie im Dienste Österreichs* (Vienna: Böhlau, 1999), 77.

45. Vejas Gabriel Liulevicius, *The German Myth of the East: 1800 to the Present* (Oxford: Oxford University Press, 2009), 7.

46. Marie Petz-Grabenhuber, "Anton Kerner von Marilaun," in *Anton Kerner von Marilaun (1831–1898)*, ed. Grabenbauer and Michael Kiehn, 7–23 (Vienna: Academy of Sciences, 2004), 10.

47. Anton Kerner, *Das Pflanzenleben der Donauländer* (Innsbruck: Wagner, 1863), 23.

48. Bachl-Hofmann, ed., *Die Geologische Bundesanstalt*, 76.

49. Surman, *Biography of Habsburg Universities*, 14, 237.

50. E.g., Kapil Raj, *Relocating Modern Science: Circulation and the Construction of Knowledge in South Asia and Europe, 1650–1900* (Basingstoke and New York: Palgrave Macmillan, 2007).

51. Eduard Brückner, "Dr. Josef Roman Lorenz von Liburnau, Sein Leben und Wirken," *Mitt. Geog. Ges.* 56 (1912): 523–51, on 541.

52. Karl Fritsch, "Nachruf an Anton Kerner von Marilaun," *Verh. Zool.-Bot. Ver.* 48 (1898): 694–700, on 696.

53. Vittoria Di Palma, *Wasteland: A History* (New Haven, CT: Yale University Press, 2014).

54. See esp. Michael S. Reidy, "Mountaineering, Masculinity, and the Male Body in Mid-Victorian Britain," *Osiris* 30 (2015): 158–81.

55. See drawings in Ficker, "Untersuchungen über die meteorologischen Verhältnisse der Pamirgebiete (Ergebnisse einer Reise in Ostbuchara)," *Wiener Berichte* IIa 97 (1921; submitted June 1919): 151–255.

56. Ficker, "Föhnuntersuchungen im Ballon," *Wiener Berichte* IIa 121 (1912): 829–73, on 830.

57. Quoted in Jennifer Tucker, *Nature Exposed: Photography as Eyewitness in Victorian Science* (Baltimore: Johns Hopkins University Press, 2005), 154.

58. Ficker, "Wirbelbildung im Lee des Windes," *MZ* 28 (1911): 539.

59. Hann to Wladimir Köppen, 28 October 1886, WK.

60. Voeikov, *Le Turkestan Russe* (Paris: Colin, 1914), vi.

61. Robert DeCourcy Ward, "The Value of Non-Instrumental Weather Observations," *Popular Science Monthly* 80 (1912): 129–37, on 131.

62. Alfred Hettner, "Methodische Zeit- und Streitfragen: Die Wege der Klimaforschung," *Geographische Zeitschrift* 30 (1924): 117–20, on 117.

63. Hann, *Klimatographie von Niederösterreich* (Vienna: Braumüller, 1904), 4.

64. Ficker, "Pamirgebiete," 153.

65. Hasok Chang, *Inventing Temperature: Measurement and Scientific Progress* (Oxford: Oxford University Press, 2004).

66. Wilhelm Schmidt, "Zur Frage der Verdunstung," *Ann. Hyd.* 44 (1916): 136–45, on 142.

67. Edmund Husserl, *Ideas Pertaining to a Pure Phenomenology and to a Phenomenological Philosophy*, vol. 2, trans. R. Rojcewicz and A. Schuwer (Dordrecht: Springer, 1990), 61, 166; Patočka, *Body, Community*.

68. Hann, "Über die monatlichen und jährlichen Temperaturschwankungen in Österreich-Ungarn," *Wiener Berichte* IIa 84 (1881): 965–1037; Hann, "Untersuchungen über die Veränderlichkeit der Tagestemperatur," *Wiener Berichte* II 71 (1875): 571–657.

69. Quoted from Rudolf's Nachlass in Christiane Zintzen, "Vorwort," in *Die österreichisch-ungarische Monarchie in Wort und Bild. Aus dem "Kronprinzenwerk" des Erzherzog Rudolf* (Vienna: Böhlau, 1999), 9–20, 10.

70. Suess, *Erinnerungen*, 101.

71. Ibid., 130.

72. "Farewell Lecture by Professor Eduard Suess on Resigning His Professorship," *Journal of Geology* 12 (1904): 264–75, on 267.

73. Mott Greene, *Geology in the Nineteenth Century*, chapter 7; Şengör, "Eduard Suess."

74. Norman Henniges, "Human Recording Machines? The German Geological Survey and the Moral Economy of Scale," paper for the workshop "Creative Commensuration," Zurich, 2016.

75. "Erinnerungen von Albrecht Penck," AP.

76. Penck, "Das Klima Europas während der Eiszeit," *Naturwissenschaftliche Wochenschrift* 20 (1905): 593–97; Penck, foreword to *Geographischer Jahresbericht aus Österreich* (Vienna, 1906), 4:4.

77. Norman Henniges, "'Sehen lernen': Geographische (Feld-)Beobachtung in der Ära Albrecht Penck," *Mitteilungen der Österreichischen Geographischen Gesellschaft* 156 (2014): 141–70, esp. 163.

78. Cvijić, *La Geographie des Terrains Calcaires* (1960), reprinted in *Cvijić and Karst*, ed. Zoran Stevanović and Borivoje Mijatović (Belgrade: Serbian Academy of Science and Arts, 2005), 147–304, on 173.

79. Cvijić, "Forschungsreisen auf der Balkan-Halbinsel," *Zeitschrift der Gesellschaft für Erdkunde zu Berlin* (1902): 196–214, on 197.

80. Cvijić, *La Péninsule balkanique* (Paris: Colin, 1918), 13–14, 18; Karl Kaser, "Peoples of the Mountains, Peoples of the Plains: Space and Ethnographic Representation," in *Creating the Other: Ethnic Conflict and Nationalism in Habsburg Central Europe*, ed. Nancy M. Wingfield, 216–30 (New York: Berghahn, 2003).

81. Suess, *Erinnerungen*, 125.

82. The *Beamte*, in the words of the fin de siècle writer Hermann Bahr, was the individual who took responsibility for preserving the Austrian Idea, "the trustee of the old *Staatsidee*." Bahr, *Austriaca* (Berlin: S. Fischer, 1911), 33. Bahr had little sympathy for these "trustees," whom he viewed as anachronistic and naive. But he did not consider the ways in which natural scientists were reinventing the *Beamte* as a modern persona.

83. Kreil, *Klimatologie von Böhmen*, 2.

84. Josef Durdík, *Rozpravy filosofické* (Prague: Kober, 1876), 49.

85. Tomáš Garrigue Masaryk, *Česká otázka* (Prague: Čas, 1895), 240.

86. Cf. Barry Smith, "Von T. G. Masaryk bis Jan Patočka: Eine philosophische Skizze," in *T. G. Masaryk und die Brentano-Schule*, ed. J. Zumr and T. Binder, 94–110 (Prague: Czech Academy of Sciences, 1993).

87. Eduard Suess, *Das Bau und Bild Österreichs* (Vienna: Tempsky, 1903), xiv.

88. Henniges, "Human Recording Machines?"

89. Marianne Klemun, "National 'Consensus' as Culture and Practice: The Geological Survey in Vienna and the Habsburg Empire (1849–1867)," in Ash and Surman, *Nationalization of Scientific Knowledge*, 83–101.

90. Hann, *Die Vertheilung des Luftdruckes über Mittel- und Süd-Europa* (Vienna: Hölzel, 1887), 5.

91. Hann, "Der Pulsschlag der Atmosphäre," *MZ* 23 (1906): 82–86, on 83; Hann, *Lehrbuch der Meteorologie*, 3rd ed. (Leipzig: Tauchnitz, 1915), 637.

CHAPTER FOUR

1. Kreil, "Einleitung," *Jb. ZAMG* 1 (1848–49): 1–32, on 2–3.

2. For the legal language of "doubleness" in this sense, see, e.g., Georg Jellinek, *Ueber Staatsfragmente* (Heidelberg: Gustav Koester, 1896), 28–29, cited in Wheatley, "Law, Time, and Sovereignty," 61.

3. Fabien Locher, "The Observatory, the Land-Based Ship and the Crusades: Earth Sciences in European Context, 1830–50," *BJHS* 40 (2007): 491–504.

4. Quoted in Kenner, "Karl Kreil," 332.

5. Fritsch, autobiographical sketch, 112.

6. Hedwig Kopetz, *Die Österreichische Akademie der Wissenschaften: Aufgaben, Rechtsstellung, Organisation* (Vienna: Böhlau, 2006), 34.

7. Christine Ottner, "Zwischen Wiener Localanstalt und Centralpunct der Monarchie," *Anzeiger der Akademie der Wissenschaften, phil.-hist. Kl.* 143 (2008): 171–96, on 174, 178.

8. Reprinted in Hammerl, *Zentralanstalt*, 21, 23.

9. Fasz. 677/Sig. 4A/Nr. 6015/694: 20 July 1850, MCU; Fasz. 677/Sig. 4A/Nr. 2372/167: 10 March 1852, MCU.

10. Kreil, "Einleitung," *Jb. ZAMG* 1 (1848–49): 1–32, on 1, 2.

11. Fasz. 677/Sig. 4A/Nr. 9369/609: 6 September 1852, MCU.

12. Kenner, "Karl Kreil," 360–61.

13. Egon Ihne, "Geschichte der pflanzenphänologischen Beobachtungen in Europe," *Beiträge zur Phänologie* 1 (1884): 1–176, on 36.

14. Quoted and translated in Gliboff, "Mendel and the Laws of Evolution," 225.

15. Franz Unger, *Versuch einer Geschichte der Pflanzenwelt* (Vienna: Braumüller, 1852), 5.

16. Kreil, "Einleitung," *Jb. ZAMG* 1 (1848–49): 1–32, on 2–3, my emphasis.

17. *Die Markgrafschaft Mähren und das Herzogthum Schlesien in ihren geographischen Verhältnissen* (Vienna: Hölzel, 1861), iii.

18. František Augustin, *O potřebě zorganisovati meteorologická pozorování v Čechách* (Prague: Otty, 1885), 6, 13–16.

19. Jindřich Metelka, review of *Zeměpisný Sborník*, *Hlídka Literarní* 4 (1887): 44–48.

20. Fasz. 680/Sig. 4A/Nr. 1605: 7 January 1914, MCU.

21. Fasz. 680/Sig. 4A/Nr. 8888: 25 February 1914, MCU.

22. David Aubin, Charlotte Bigg, and H. Otto Sibum, "Introduction," *The Heavens on Earth: Observatory and Astronomy in Nineteenth-Century Science and Culture*, ed. Aubin, Bigg, and Sibum (Durham, NC: Duke University Press, 2010), 7.

23. Simony, "Das meteorologische Element in der Landschaft," *Zs. Ö. G. Meteo.* 5 (1870): 49–60.

24. Kreil, "Einleitung," 9.

25. Jelinek, *Anleitung zur Anstellung meteorologischer Beobachtungen* (Vienna: k.k. Hof- und Staatsdruckerei, 1869), 1.

26. Ibid., 64.

27. Kenner, "Karl Kreil," 362.

28. Ibid.

29. Fritsch, autobiographical sketch, 115.

30. Wilhelm von Haidinger, *Das Kaiserlich-Königliche Montanistische Museum und die Freunde der Naturwissenschaften in Wien in den Jahren 1840–1850* (Vienna: Braumüller, 1869), 72, 115.

31. Haidinger, "Gesellschaft der Freunde der Naturwissenschaften," *Berichte über die Mittheilungen von Freunden der Naturwissenschaften in Wien* 5 (1848): 274–78, on 275.

32. Karl Fritsch, "Nekrologie [W. v. Haidinger]," *Zs. Ö. G. Meteo.* 6 (1871): 205–8, on 207.

33. Haidinger, "Historische Entwicklung und Plan der Gesellschaft," *Berichte über die Mittheilungen von Freunden der Naturwissenschaften in Wien* 5 (1848): 280–87; cf. Karl Kadletz, "Krisenjahre zwischen 1849 und 1861," in Christina Bachl-Hofmann, *Geologische Bundesanstalt*, 78–92.

34. *Verhandlungen des österreichischen verstärkten Reichsrathes* 1 (1860): 305. Cf. Böhm, "Erinnerungen an Franz von Hauer," *Abhandlungen der k.k. Geographischen Gesellschaft in Wien* 1 (1899): 100.

35. Advertisement, *Zeitschrift der k.k. Gesellschaft der Aerzte zu Wien* 17 (1861): 392.

36. "Dr. Carl Jelinek," *Zs. Ö. G. Meteo.* 12 (1877): 69–80, on 71.

37. Hammerl, *Zentralanstalt*, 58.

38. Anderson, *Predicting the Weather*, 143–44.

39. Hann, "Arthur Schuster über Methoden der Forschung in der Meteorologie," *MZ* 20 (1903): 19–30, on 28.

40. Josef Chavanne, *Die Temperatur-Verhältnisse von Österreich-Ungarn dargestellt durch Isothermen* (Vienna: Gerold's Sohn, 1871),

41. Ibid., 21.

42. Fasz. 677/Sig. 4A/Nr. 3128/478: 19 April 1849, MCU.

43. János/Johann Hunfalvy, "Die klimatischen Verhältnisse des ungarischen Länderkomplexes," *Zs. Ö. G. Meteo.* 2 (1867): 273–79, 289–98.

44. Jelinek, "Meteorologische Stationen in Ungarn," *Zs. Ö. G. Meteo.* 1 (1866): 171–72.

45. Josef Chavanne, *Physikalisch-Statistisches Hand-Atlas* (Vienna: Hölzel, 1887). The yearbook of the Hungarian Central Station was published in Hungarian and German.

46. Fasz. 677/Sig. 4A/Nr. 8208: 19 August 1864, MCU.

47. Josef Roman Lorenz, *Physikalische Verhältnisse und Vertheilung der Organismen im Quarnerischen Golfe* (Vienna: Karl Gerold's Sohn, 1869), 2–3.

48. "Korespondencya Komisyi," *Sprawozdanie Komisyi Fizyograficnej* 28 (1893): vii–xxv. Thanks to Jan Surman for this reference and to Daniel Mahla for the translation.

49. Moriz Rohrer, *Beitrag zur Meteorologie und Klimatologie Galiziens* (Vienna: Carl Gerold's Sohn, 1866), 1.

50. Janina Bożena Trepińska, "The Development of the Idea of Weather Observations in Galicia," in *Acta Agrophysica* 184 (2010): 9–23, on 13; see too Przybylak et al., *The Polish Climate*.

51. Jan Hanik, *Dzieje meteorologii i obserwacji meteorologicznych w Galicji od XVIII do XX wieku* (Wrocław: Zakład Narodowy im. Ossolińskich, 1972), 87, 89–94. I am grateful to Bogdan Horbal for translation of this material.

52. Ibid., 157–59. For comparison: Carniola, another of the poorer crown lands, was better integrated into the ZAMG's network than Galicia. By 1891, Carniola boasted twenty-six weather stations, enough for "The Climate of Carniola," which appeared serially from 1891 to 1893, in German, in the journal of the Museum Society of Ljubljana. Its author was Ferdinand Seidl, a teacher at the Realschule in Görz/Gorica/Gorizia. Seidl collected meteorological observations made by locals primarily in the Slovenian language, and he also published scientific work in Slovenian. After 1918, he went on to direct the meteorological observatory in Ljubljana, in the Republic of Yugoslavia. (Tanja Cegnar, "Beginnings of Instrumental Meteorological Observations in Slovenia," http://cagm.arso.gov.si/posters/Beginnings_instrumental_meteorological_observations_in%20_slovenia.pdf.)

53. J. Valentin, "Der tägliche Gang der Lufttemperatur in Österreich," *Denk. Akad. Wiss. math-nat.* 73 (1901): 133–229, on 201.

54. Ibid., 210.

55. See Conrad, *Methods in Climatology* (Cambridge, MA: Harvard University Press, 1944), 2.

56. Ibid., 129.

57. Fasz. 679/Sig. 4A/Nr. 22093: 18 May 1910; Fasz. 679/Sig. 4A/Nr. 30079: 23 June 1913, MCU. The Zentralanstalt offered to supply instruments for new stations in eastern Galicia but refused to pay construction costs. Fasz. 679/Sig. 4A/Nr. 23858: 22 May 1911, MCU.

58. Fasz. 679/Sig. 4A/Nr. 52972/1913: 19 November and 31 December 1913, MCU.

59. Victor Conrad, *Klimatographie der Bukowina* (Vienna, 1917), 20.

60. Ibid., 25.

61. Conrad, "Beiträge zu einer Klimatographie von Serbien," *Wiener Berichte* IIa 125 (1916): 1377–417, on 1411.

62. Ibid., 1377.

63. Ibid., 1380, 1400.

64. Ibid., 1410–11.

65. Ludwig Dimitz, *Die forstlichen Verhältnisse und Einrichtungen Bosniens und der Hercegovina* (Vienna: W. Frick, 1905), 11; see too Alfred Grund, *Die Karsthydrographie: Studien aus Westbosnien* (Leipzig: Teubner, 1903).

66. Hann, "Über die klimatischen Verhältnisse von Bosnien und der Herzegowina," *Wiener Berichte* II 88 (1884): 96–116, on 96; "Das meteorologische Beobachtungsnetz von Bosnien und der Hercegovina und dessen Gipfelstation auf der Bjelašnica," *MZ* 13 (1896): 41–49, on 41.

67. Philipp Ballif, *Wasserbauten in Bosnien und der Hercegovina* (Vienna: Adolf Holzhausen, 1896).

68. Srećko M. Džaja, *Bosnien-Herzegowina in der österreichisch-ungarischen Epoche, 1878–1918* (Munich: Oldenbourg, 1994), 82.

69. J. Moscheles to W. Morris Davis, 15 August 1919, folder 336, WMD.

70. On the complicity of Austrian ethnographers in Habsburg colonialism in Bosnia, see Christian Marchetti, "Scientists with Guns: On the Ethnographic Exploration of the Balkans by Austrian-Hungarian Scientists before and during World War I," *Ab Imperio* (2007).

71. J. Moscheles, *Das Klima von Bosnien und der Hercegovina*, vol. 20 of *Kunde der Balkanhalbinsel* (Sarajevo: J. Studnička & Co., 1918), 3.

72. Conrad, *Methods*, 140–49.

73. Robert Klein, *Klimatographie von Steiermark* (Vienna: ZAMG, 1909), 4–5.

74. "Bedeutung des Sonnwendstein als Wetterwarte für den praktischen Wetterdienst," *MZ* 20 (1903): 268–70, on 268, 269.

75. Leopold von Sacher-Masoch, "Auf der Höhe," *Auf der Höhe* 1 (1881): iii–v, on iii.

76. On mountain science, see Charlotte Bigg, David Aubin, and Philipp Felsch, eds., "The Laboratory of Nature—Science in the Mountains," special issue of *Science in Context* 22, no. 3 (2009).

77. For references, see Coen, "The Storm Lab: Meteorology in the Austrian Alps," *Science in Context* 22 (2009): 463–86, on 473–75.

78. Patrice Dabrowski, "Constructing a Polish Landscape: The Example of the Carpathian Frontier," *AHY* 39 (2008): 45–65.

79. Fasz. 678/Sig. 4A/Nr. 15530: 5 May 1902, MCU; Fasz. 680/Sig. 4A/Nr. 47480: 18 October 1913, MCU.

80. See references in Coen, "Storm Lab," 470–71.

81. See references in ibid., 475–77.

82. Fasz. 680/Sig. 4A/Nr. 52972: 19 November und 31 December 1913, MCU. For a map showing a proposed subdivision of forecasting districts from 1897, see Coen, "Climate and Circulation in Imperial Austria," *Journal of Modern History* 82 (2010): 839–75, on 872.

83. Based on documents in Fasz. 677, 678, and 679/Sig. 4A, MCU. See too A. E. Forster, "Die Fortschritte der klimatologischen Forschung in Österreich in den Jahren 1897–1905," *Geographischer Jahresbericht aus Österreich* 5 (1905): 156–91.

84. Johann Gottfried Sommer, quoted in Josef Emanuel Hibsch, "Der Donnersberg," *Erzgebirgs-Zeitung* 50 (1929): 26–28.

85. Maximilian Dormitzer and Edmund Schebek, *Die Erwerbsverhältnisse im böhmischen Erzgebirge* (Prague: H. Merch, 1862), 1.

86. Fasz. 678/Sig. 4A/Nr. 21535: 29 June 1902, MCU.

87. Eduard Brückner, "Bericht über die Fortschritte der geographischen Meteorologie," *Geographisches Jahrbuch* 21 (1898): 255–416, on 257.

88. Hann, "Die meteorologische Verhältnisse auf der Bjelašnica," *MZ* 20 (1903): 1–19, on 1.

CHAPTER FIVE

1. Gerhard Mandl, *Die frühen Jahre des Dachsteinpioniers Friedrich Simony, 1813–1896* (Vienna: Geologische Bundesanstalt, 2013), 124.

2. Franz Grims, "Das wissenschaftliche Wirken Friedrich Simonys im Salzkammergut," in *Ein Leben für das Dachstein: Friedrich Simony zum 100. Todestag*, ed. Franz Speta (Linz: Francisco-Carolinum, 1996). Simony would return to climatological questions throughout his career, including the dating of the Ice Ages, the variability of the climate of Vienna, and the climatic impact on deforestation. On geography, see Petra Svatek, "'Natur und Geschichte': Die Wissenschaftsdisziplin 'Geographie' und ihre Methoden an den Universitäten Wien, Graz und Innsbruck bis 1900," in *Wissenschaftliche Forschung in Österreich, 1800–1900: Spezialisierung, Organisation, Praxis*, ed. Christine Ottner, Gerhard Holzer, and Petra Svatek, 45–71 (Göttingen: V & R, 2015).

3. *Mémoires Metternich*, vol. 6 (Paris: E. Plon, 1883), 659; Hedwig Kadletz-Schöffel and Karl Kadletz, "Metternich und die Geowissenschaften," *Berichte der Geologischen Bundesanstalt* 51 (2000): 49–52. It was also at Metternich's residence that Simony met Stifter.

4. Kadletz-Schöffel and Kadletz, "Metternich," 51.

5. Albrecht Penck, *Friedrich Simony: Leben und Wirken eines Alpenforschers* (Vienna: Hölzel, 1898), 8.

6. Franz Wawrik and Elisabeth Zeilinger, eds., *Austria Picta: Österreich auf alten Karten und Ansichten* (Graz: Akademische Druck- und Verlagsanstalt, 1989), 70.

7. Madalina Valeria Veres, "Putting Transylvania on the Map: Cartography and Enlightened Absolutism in the Habsburg Monarchy," *AHY* 43 (2012): 141–64.

8. Wawrik and Zeilinger, *Austria picta*, 86.

9. Veres, "Putting Transylvania."

10. Komlosy, "State, Regions, and Borders," 148–49, my emphasis; Cooper, *Inventing*, 97.

11. Komlosy, *Grenze und ungleiche regionale Entwicklung*, 65, 67, 76.

12. Wawrik and Zeilinger, *Austria picta*, 97.

13. On this expansion, see Helmedach, *Verkehrssystem als Modernisierungsfaktor*.

14. Johannes Dörflinger, *Descriptio Austriae: Osterreich und seine Nachbarn im Kartenbild v.d. Spatantike bis ins 19. Jahrhundert* (Vienna: Edition Tusch, 1977), 190, plate 63.

15. Ingrid Kretschmer, Johannes Dörflinger, and Franz Wawrik, *Österreichische Kartographie: Von den Anfängen im 15. Jahrhundert bis zum 21. Jahrhundert,* Wiener Schriften zur Geographie und Kartographie 15 (Vienna: Institut für Geographie und Regionalforschung, 2004), 91, 137.

16. Ibid., 139–41.

17. Veres, "Putting Transylvania," 147.

18. Paula Sutter Fichtner, *The Habsburgs: Dynasty, Culture, and Politics* (Chicago: Reaktion, 2014), 158. His "new" crown had actually belonged to Rudolph II.

19. Telesko, *Geschichtsraum Österreich*, 47–48, 203.

20. Franz Sartori, *Länder- und Völker-Merkwürdigkeiten des österreichischen Kaiserthumes*, 4 vols. (Vienna: A. Doll, 1809); Sartori, *Naturwunder des österreichischen Kaiserthumes*, 4 vols. (Vienna: A. Doll, 1807).

21. Sartori, *Historisch-ethnographische Übersicht der wissenschaftlichen Cultur, Geistes-*

thätigkeit, und Literatur des österreichischen Kaiserthums, vol. 1 (Vienna: C. Gerold, 1830), ix., xiv; cf. Telesko, *Geschichtsraum*, 52–54.

22. Andrian-Werburg, *Österreich und dessen Zukunft*, 2nd ed. (Hamburg, 1843), 201.

23. Penck, *Simony*, 12.

24. Cf. Charlotte Bigg, "The Panorama, or La Nature à Coup d'Œil," in *Observing Nature—Representing Experience: The Osmotic Dynamics of Romanticism, 1800–1850*, 73–95 (Berlin: Reimer, 2007).

25. Penck, *Simony*, 10–12.

26. Ibid., 29.

27. Thomas Hellmuth, "Die Erzählung des Salzkammerguts: Entschlüsselung einer Landschaft," in *Die Erzählung der Landschaft*, ed. Dieter Binder et al., 43–68 (Vienna: Böhlau, 2011).

28. Charlotte Klonk, *Science and the Perception of Nature: British Landscape Art in the Late Eighteenth and Early Nineteenth Centuries* (New Haven, CT: Yale University Press, 1996).

29. Simony, "Das wissenschaftliche Element in der Landschaft II. Luft und Wolken," *Schr. d. Ver. z. Verbr. naturw. Kenntn.* 17 (1877): 511–47, on 522.

30. Ibid., 511. He also delivered an address titled "Das meteorologische Element in der Landschaft" to the Austrian Society for Meteorology, making the point that art and science should learn from each other.

31. Stifter, *Der Nachsommer*, 2:48.

32. Wilhelm Haidinger, *Bericht über die geognostische Übersichts-Karte der Österreichischen Monarchie* (Vienna: Hof- und Staatsdruckerei, 1847), 22.

33. Ibid., 24.

34. Ibid., 42–43, 32.

35. Haidinger, "Die K.K. Geologische Reichsanstalt in Wien und ihre bisherigen Leistungen," *Mittheilungen aus Justus Perthes' Geographischer Anstalt* (1863): 428–44, on 432, 443.

36. Haidinger, "Die Aufgabe des Sommers 1850 für die k.k. geologische Reichsanstalt in der geologischen Durchforschung des Landes," *Jahrbuch der Geologischen Bundesanstalt* 1 (1850): 6–16, on 7.

37. A. H. Robinson and H. M. Wallis, "Humboldt's Map of Isothermal Lines: A Milestone in Thematic Cartography," *Cartographic Journal* 4 (1967): 119–23.

38. Mott Greene, "Climate Map," in *History of Cartography*, ed. Mark Monmonier, vol. 6, *Cartography in the Twentieth Century* (Chicago: University of Chicago Press, 2015).

39. Hettner, *Die Gewässer des Festlandes: Die Klimate der Erde* (Leipzig: Teubner, 1934), 158.

40. Alexander Supan, *Grundzüge der physischen Erdkunde* (Leipzig: Veit, 1911), 231.

41. On early weather (not climate) mapping, see Mark Monmonier, *Air Apparent: How Meteorologists Learned to Map, Predict, and Dramatize Weather* (Chicago: University of Chicago Press, 1999), chapter 2; Eckert, *Kartenwissenschaft*, vol. 2, esp. 336.

42. Friedrich Umlauft, ed., "Länderkunde von Österreich-Ungarn," in *Die Pflege der Erdkunde in Österreich, 1848–1898*, 132–60 (Vienna: Lechner, 1898), 132.

43. *Physikalisch-statistischer Handatlas von Österreich-Ungarn* (Vienna: E. Hölzel, 1882–87). A physical-statistical atlas of the German Empire had been published in 1876–78.

44. C. H. Haskins and R. H. Lord, *Some Problems of the Peace Conference* (Cambridge, MA: Harvard University Press, 1922), 228.

45. *Physikalisch-statistischer Handatlas*, ix.

46. Eckert, *Kartenwissenschaft*, vol. 1, pt. 4.

47. Ingrid Kretschmer, "The First and Second Austrian School of Layered Relief Maps in the Nineteenth and Early Twentieth Centuries," *Imago Mundi* 40 (1988): 2, 9–14, on 11; Kretschmer, Dörflinger, and Wawrik, *Österreichische Kartographie*, 261–63.

48. William Rankin, *After the Map: Cartography, Navigation, and the Transformation of Territory in the Twentieth Century* (Chicago: University of Chicago Press, 2016), 35–38.

49. *Physikalisch-statistischer Handatlas*, xv, my emphasis.

50. Ibid., viii, my emphasis.

51. Chavanne, *Die Temperatur-Verhältnisse*, 19.

52. Klein, *Klimatographie von Steiermark*, 7.

53. Alexander Supan, "Die Vertheilung der jährlichen Wärmeschwankung auf der Erdoberfläche," *Zeitschrift für wissenschaftliche Geographie* 1 (1880): 141–56, on 146.

54. Eckert, *Kartenwissenschaft*, 2:339.

55. "Versuch einer Übersicht der geographischen Verbreitung der Gewitter," in *Physikalischer Atlas*, 2nd ed. (Gotha: Berghaus, 1852), xxxv.

56. Julius Hann, *Atlas der Meteorologie* (Gotha: Justus Perthes, 1887), 5.

57. Ibid., 3.

58. Valentin, "Der tägliche Gang der Lufttemperatur in Österreich," *Denk. Akad. Wiss. math-nat.* 73 (1901): 133–229, on 133.

59. Ibid., 201.

60. Rudolf Spitaler, *Klima des Eiszeitalters* (Prague: self-published, 1921); cf. John E. Kutzbach, "Steps in the Evolution of Climatology: From Descriptive to Analytic," in *Historical Essays on Meteorology, 1919–1995*, 353–77 (Boston: American Meteorological Society, 1996), 358.

CHAPTER SIX

1. Heinrich von Ficker, *Die Zentralanstalt für Meteorologie und Geodynamik in Wien, 1851–1951* (Vienna: Springer, 1951), 6.

2. *Almanach der Akademie der Wissenschaften* (Vienna, 1902), 371–74.

3. Fasz. 680/Sig. 4A/Nr. 12192: 17 March 1911, MCU; *Österreichische Statistik* 65 (1904): xli.

4. Fasz. 681/Sig. 4A/Nr. 29356: 1 August 1918, SAU.

5. On Carniola, Fasz. 680/Sig. 4A/Nr. 42581: 17 September 1914, MCU.

6. J. M. Pernter, foreword to Hann, *Klimatographie von Niederösterreich*, i.

7. For comparison, the US Weather Bureau issued a one-volume *Climatology of the United States in 1906*, but its methods were suspect: it was not even clear whether the data had been reduced to a uniform time period. See Robert DeCourcy Ward, *BAGS* 38 (1906): 709–11.

8. John Frow, *Genre: The New Critical Idiom* (New York: Routledge, 2006), 16. Geoffrey C. Bowker and Susan Leigh Star, *Sorting Things Out: Classification and Its Consequences* (Cambridge, MA: MIT Press, 1999).

9. Paul N. Edwards, *A Vast Machine: Computer Models, Climate Data, and the Politics of Global Warming* (Cambridge, MA: MIT Press, 2010), 32–33; David Cassidy, "Meteorology in Mannheim: The Palatine Meteorological Society, 1780–1795," *Sudhoffs Archiv* 69 (1985): 8–25.

10. Mitchell Thomashow, *Bringing the Biosphere Home* (Cambridge, MA: MIT Press, 2002), 98.

11. William Morris Davis, "The Relations of the Earth Sciences in View of their Progress in the Nineteenth Century," *Journal of Geology* 12 (1904): 669–87. A related term was "topography," used in German to mean a description of a land and its inhabitants.

12. Rob Nixon, *Slow Violence and the Environmentalism of the Poor* (Cambridge, MA: Harvard University Press, 2011), 10.

13. Michael Gamper attends to the role of "weather knowledge" in Stifter's fiction, concluding that it is neither strictly local folk knowledge nor universal science. Gamper, "Literarische Meteorologie: Am Beispiel von Stifters 'Das Haidedorf,'" in *Wind und Wetter: Kultur—Wissen—Ästhetik*, ed. Georg Braungart and Urs Büttner, 247–63 (forthcoming), 262; "Wetterrätsel: Zu Adalbert Stifters 'Kazensilber,'" in *Literatur und Nicht-Wissen: Historische Konstellationen, 1730–1930*, ed. Michael Bies and Michael Gamper, 325–38 (Zurich: Diaphanes, 2012).

14. María M. Portuondo, *Secret Science: Spanish Cosmography and the New World* (Chicago: University of Chicago Press, 2009), 9; Ayesha Ramachandran, *The Worldmakers: Global Imagining in Early Modern Europe* (Chicago: University of Chicago Press, 2015).

15. Humboldt, *Cosmos*, 1:3. Useful analyses of Humboldtian cosmography include Joan Steigerwald, "The Cultural Enframing of Nature: Environmental Histories during the Early German Romantic Period," *Environment and History* 6 (2000): 451–96, and Laura Dassow Walls, *The Passage to Cosmos: Alexander von Humboldt and the Shaping of America* (Chicago: University of Chicago Press, 2009).

16. Humboldt, *Kosmos: Entwurf einer physischen Weltbeschreibung*, vol. 1 (Philadelphia: F. W. Thomas, 1869), iv, my translation.

17. It's worth noting, however, that *Cosmos* furnished no concrete model for a physical description of the atmosphere, since Humboldt's projected final volume dealing with "air and oceans" was incomplete at his death.

18. Hann, Diary A, 73; Diary B, 56, JH.

19. Humboldt, *Kosmos*, 37.

20. Ibid., 24, my translation.

21. Adalbert Stifter, *Wien und die Wiener in Bildern aus dem Leben*, ed. Elisabeth Buxbaum (Vienna: LIT, 2005), 1.

22. Stifter, "Aussicht und Betrachtungen von der Spitze des St. Stephansthurms," in *Wien und die Wiener*, 3–21, on 9, 13, my emphasis.

23. Ibid., 3, 11, 17.

24. Stifter, "Wiener=Wetter," in *Wien und die Wiener*, 263–80, on 263, 267; cf. Vladimir Janković, "A Historical Review of Urban Climatology and the Atmospheres of the Industrialized World," *WIREs Climate Change* 4 (2013): 539–53.

25. "Wiener=Wetter," 263, 265, 269.

26. Stifter, "Die Sonnenfinsternis am 8. Juli 1842," *Schweizer Monatshefte* 72 (1992): 603–10, on 604, 605, 606.

27. Stifter, *Bunte Steine: Eine Festgeschenk*, vol. 1 (Pest: Heckenast, 1853), 1.

28. Wiedemann, *Stifters Kosmos*, 85n272.

29. Kenner, "Karl Kreil," 360.

30. Kreil, *Klimatologie von Böhmen*, 4.

31. Ibid., 2.

32. Wladimir Köppen, *Klimakunde*, vol. 1 (Leipzig: G. J. Göschen, 1906), 8.

33. Blanford, *Practical Guide*, viii.

34. Komlosy, *Grenze*, 164.

35. Kreil, *Klimatologie von Böhmen*, 2.

36. Coen, *Vienna in the Age of Uncertainty* (Chicago: University of Chicago Press, 2007), chapter 8.

37. Kreil, *Klimatologie von Böhmen*, 3.

38. Stifter remarked on the similarity between Simony's images and descriptions and his own. Michael Kurz, "Maler—Dichter—Pädagoge—Konservator: Adalbert Stifter und das Salzkammergut," *Oberösterreichische Heimatblätter* 3 (2005): 115–59, on 120–21.

39. Stifter, *Nachsommer*, 1:175–82, 337.

40. Amitav Ghosh, *The Great Derangement: Climate Change and the Unthinkable* (Chicago: University of Chicago Press, 2016), pt. 1.

41. Stifter, "Der Hagestolz" (1844) in *Studien*, vol. 3, 5th ed., 1–110 (Pest: Hackenast, 1863), 4.

42. Stifter, "Zwei Schwester" (1850) in *Studien*, 3:169–204, on 193.

43. Elisabeth Strowick, "Poetological-Technical Operations: Representation of Motion in Adalbert Stifter," *Configurations* 18 (2011): 273–89.

44. Stifter, "Der Kuss von Sentze," http://gutenberg.spiegel.de/buch/der-kuss-von-sentze-200/1.

45. Quoted and translated in Strowick, "Poetological-Technical Operations," 274.

46. Rilke to Helmuth Westhoff, 12 November 1901, in *Letters of Rainer Maria Rilke*, 1892–1910, trans. Jane Bannard Greene and M. D. Herter Norton (New York: Norton, 1945), 59.

47. Rainer Maria Rilke, *Rilke's Book of Hours: Love Poems to God*, trans. Anita Barrows and Joanna Macy (New York: Penguin, 1996), 171.

48. Robert DeCourcy Ward, review of Hann, *Klimatographie von Niederösterreich*, *BAGS* 36 (1904): 569.

49. Hann, *Klimatographie von Niederösterreich*, 3.

50. Heinrich von Ficker, *Klimatographie von Tirol und Vorarlberg* (Vienna: Gerold, 1909), 2, 7, 135; see too Hann, *Klimatographie von Niederösterreich*, 18.

51. Dana Phillips, *The Truth of Ecology: Nature, Culture, and Literature in America* (Oxford: Oxford University Press, 2003).

52. Stifter, *Nachsommer*, 2:135.

53. *Die österreichisch-ungarische Monarchie in Wort und Bild*, 1:158.

54. Joseph Roth, "The Bust of the Emperor," in *The Collected Stories of Joseph Roth*, trans. Michael Hofmann (New York: Norton, 2002), 228.

55. *Die österreichisch-ungarische Monarchie in Wort und Bild*, 1:148, 149, 153.

56. Klein, *Klimatographie von Steiermark*, 1.

57. Umlauft, *Wanderungen durch die Oesterreichisch-Ungarische Monarchie* (Wien: Carl Graeser, 1879), v, vi, 34.

58. Ficker, *Klimatographie von Tirol*, 1, 39, 96, 107, 116.

59. Klein, *Klimatographie von Steiermark*, 4–5.

60. A. Hahlmann et al., "A Reanalysis System for the Generation of Mesoscale Climatographies," *Journal of Applied Meteorology and Climatology* 49 (2010): 954–72.

61. Intergovernmental Panel on Climate Change, *Managing the Risks of Extreme Events and Disasters to Advance Climate Change*, ed. C. B. Field et al. (Cambridge: Cambridge University Press, 2012), 39. On the reception of impact assessments, see Michael Bravo, "Voices from the Sea

Ice: The Reception of Climate Impact Narratives," *Journal of Historical Geography* 35 (2009): 256–78.

62. Yates McKee, "On Climate Refugees: Biopolitics, Aesthetics, and Critical Climate Change," *Qui Parle* 19 (2011): 309–25, on 313; https://www.amazon.com/Climate-Refugees-Press-Collectif-Argos/dp/0262514397, accessed 17 May 2017.

CHAPTER SEVEN

1. Supan, *Grundzüge*, 63. Supan was born in Tyrol and educated in Ljubljana; he taught at Czernowitz/Chernivtsi/Cernăuţi from 1877 to 1909.

2. Hans-Günther Körber, *Vom Wetteraberglaube zur Wetterforschung* (Innsbruck: Pinguin, 1987), 59.

3. Thomas Stevenson, "The Intensity of Storms Referred to a Numerical Value by the Calculation of Barometric Gradients," *Meteorological Magazine* 3 (1869): 184.

4. A. Achbari and F. van Lunteren, "Dutch Skies, Global Laws: The British Creation of 'Buys Ballot's Law,'" *HSNS* 46 (2016): 1–43.

5. Wladimir Köppen, "Untersuchungen von Prof. Erman und Dr. Dippe aus den Jahren 1853 und 1860 über das Verhältniss des Windes zur Vertheilung des Luftdruckes," *Zs. Ö. G. Meteo.* 13 (1878): 374–79, on 379.

6. Hann, *Vertheilung des Luftdruckes*, 24.

7. Wladimir Köppen, "Ueber die Abhängigkeit des klimatischen Charakters der Winde von ihrem Ursprunge," *Repertorium für Meteorologie* 4 (1874).

8. Hann, *Vertheilung des Luftdruckes*, 2.

9. Ibid., 5.

10. Ibid., 25–28.

11. Hann, *Klimatographie von Niederösterreich*, 4.

12. Ficker, *Zentralanstalt*, 21.

13. *Salzburger Volksblatt*, 13 November 1886, 2.

14. Josef Roman Lorenz and Carl Rothe, *Lehrbuch der Klimatologie mit besonderer Rücksicht auf Land- und Forstwirthschaft* (Vienna: Braumüller, 1874), 7.

15. Ibid., 198.

16. Dr. Samuely, "Die Meteorologische Stationen, deren Wesen und Bedeutung," *Teplitzer Anzeiger*, 31 July 1880, 2–4; 7 August 1880, 2–7.

17. F. Wařéka, "Ueber Wettertelegraphie," *Wiener Landwirtschaftliche Zeitung*, 16 May 1885, 314–15, on 315.

18. Mach and Odstrcil, *Grundrisse der Naturlehre* (1886), quoted in Ernst Kaller, "Das Teschner Wetter im Zusammenhange mit der allgemeinen Wetterlage," *Programm der k.k. Staatsoberrealschule in Teschen* 28 (1900): 3–23, on 7.

19. *Instructionen für den Unterricht an den Realschulen in Österreich* (1899), 15, quoted in Kaller, "Teschner Wetter," 8.

20. See, e.g., Otto Rühle, "Drei gestrenge Herren," *Linzer Tagespost*, 7 May 1899, 1–2.

21. For a recent evaluation of the evidence for "singularities" like this one, see Michaela Radová and Jan Kyselý, "Temporal Instability of Temperature Singularities in a Long-Term Series at Prague-Klementinum," *Theoretical Applied Climatolology* 95 (2009): 235–43.

22. Robert Billwiller, "Die Kälterückfälle im Mai," *Zs. Ö. G. Meteo.* 19 (1884): 245–46; August Petermann, "Die Kälterückfälle im Mai," *Die Presse*, 21 May 1885, 1–2.

23. "Die Eismänner," *Innsbrucker Nachrichten*, 12 May 1887, 7–8, on 7.

24. Dove, "Über die kalte Tage im diesjährigen Mai," *Monatsberichte der Königlich Preussischen Akademie der Wissenschaften zu Berlin* (1859): 426–31.

25. Sigmund Günther, *Lehrbuch der Geophysik und physikalischen Geographie*, vol. 2 (Stuttgart: F. Enke, 1884), 204, 207. A lengthy debate over competing explanations was recorded in the *Zs. Ö. G. Meteo.* in 1884.

26. "Die Kälterückfälle zu Beginn des Sommers," *Linzer Tagespost* 2 July 1884, 1–2 on 2.

27. "Die Eismänner," *Innsbrucker Nachrichten*, 12 May 1887, 7–8, on 7.

28. Ludwig Reissenberger, "Ueber die Kälte-Rückfälle im Mai mit Beziehung auf Hermannstadt und Siebenbürgen," *Verhandlungen und Mitteilungen des Siebenbürgischer Vereins für Naturwissenschaften* 37 (1887): 6–26, on 15.

29. W. Prausnitz, *Grundzüge der Hygiene* (Munich: Lehmann, 1892), "Vorwort."

30. Carl Odehnal, "Ein Besuch in der Centralanstalt für Meteorologie und Erdmagnetismus," *Drogisten-Zeitung* 15 (July 1901): 378–79, on 378.

31. Wilhelm Schmidt and Ernst Brezina, "Relations between Weather and Mental and Physical Condition of Man," *MWR* 49 (1917): 293–94; Schmidt and Brezina, "Witterung und Befinden des Menschen," *MZ* 32 (1915): 43–44.

32. Carl Sigmund, "Unsere Ziele. Einleitendes Wort an den Leser," *Vierteljahrschrift für Klimatologie* 1 (1876): 1.

33. Kisch, *Klimatotherapie*, 654; Prausnitz, *Grundzüge der Hygiene*, 111.

34. Marcel Chahrour, "'A civilizing mission'? Austrian Medicine and the Reform of Medical Structures in the Ottoman Empire, 1838–1850," *SHPBBS* 38 (2007): 687–705.

35. Kisch, *Klimatotherapie*, 660.

36. Ibid., 661.

37. Ibid., 661; Karl Weyprecht, "Bilder aus dem hohen Norden: Unser Matrose im Eise," *Mittheilungen aus Justus Perthes' Geographischer Anstalt* 22 (1876): 341–47, on 341.

38. Prausnitz, *Grundzüge der Hygiene*, 111.

39. Kisch, *Klimatotherapie*, 655.

40. Lorenz and Rothe, *Lehrbuch der Klimatologie*, 190, 413–20, 422.

41. Friedrich Umlauft, *Die osterreichisch-ungarische Monarchie: Geographisch-statistisches Handbuch* (Vienna: Hartleben, 1876), 1.

42. Ibid., 376, 374, 2.

43. Felix Exner, *Dynamische Meteorologie*, 2nd ed. (Vienna: Springer, 1925), 131.

44. Carl Ritter, *Einleitung zur allgemeinen vergleichenden Geographie* (Berlin: Reimer, 1852), 160–61.

45. Schmidt, "Ausfüllende, im Sinne des Druckgefälles verlaufende Luftströmungen unter verschiedenen Breiten," *Ann. Hyd.* 46 (1918): 130–32.

46. Supan, *Statistik der unteren Luftströmungen* (Leipzig: Duncker & Humblot, 1881). Hann's *Handbuch der Klimatologie* appeared in 1883; Voeikov's *Climates of the Earth* in 1887.

47. Hettner, *Gewässer des Festlandes*, 94. One might compare Guldberg and Mohn's *Les mouvements de l'atmosphère* (1876), which was a mathematical treatment of atmospheric dynamics with an application to storms, but with no attempt to explain long-term climate.

48. V. Lenin, *Imperialism: The Highest Stage of Capitalism* (Sydney: Resistance Books, 1999; orig. 1916), 82; David T. Murphy, *The Heroic Earth: Geopolitical Thought in Weimar Germany, 1918–1933* (Kent, OH: Kent State University Press, 1997), 141.

49. Supan, "Über die Aufgaben der Spezialgeographie und ihre gegenwärtige Stellung in der geographischen Litteratur," *Verhandlungen des 7. Deutschen Geographentages zu Karlsruhe* (Berlin: Dietrich Reimer, 1887), 76–85, on 83.

50. Alexander Supan, *Österreich-Ungarn* (Vienna, 1889), 324.

51. Supan, "Über die Aufgaben," 85.

52. Cited in Cordileone, *Alois Riegl*, 99.

53. Andrássy, *Ungarns Ausgleich*, 41, 124.

54. Emanuel Herrmann, *Miniaturbilder aus dem Gebiete der Wirthschaft* (Halle: L. Nebert, 1872), 60; Heinrich Wiskemann, *Die antike Landwirtschaft und das von Thünen'sche Gesetz* (Leipzig: Hirzel, 1859), 3; Wilhelm Roscher, *Ansichten der Volkswirthschaft*, vol 2., 3rd ed. (Leipzig: Winter, 1878), 27–30.

55. Slobodian, "How to See the World Economy: Statistics, Maps, and Schumpeter's Camera in the First Age of Globalization," *Journal of Global History* 10 (2015): 307–32.

56. Neumann-Spallart, *Übersichten über Produktion, Verkehr und Handel in der Weltwirthschaft* (Stuttgart: Julius Maier, 1878), 19.

57. Herrmann, *Miniaturbilder*, 59.

58. On Herrmann's invention of the postcard, see ibid., chapter 2; note the cultural significance he attached to the circulation of letters in Europe.

59. Ibid., 69.

60. Emil Sax, *Die Verkehrsmittel in Volks- und Staatswirtschaft*, vol. 1 (Vienna: Hölder, 1878), 48.

61. Quoted and translated in Alexander Gerschenkron, *An Economic Spurt That Failed* (Princeton, NJ: Princeton University Press, 1977), 30.

62. Rudolf Springer (pseud. Karl Renner), *Grundlagen und Entwicklungsziele der österreichisch-ungarischen Monarchie* (Vienna: Deuticke, 1906), 172. Renner contended that the railway had not displaced Vienna from the center of Habsburg trade (ibid., 171).

63. Ibid., 202–3.

64. Karl Rabe, "Zur Apologie der stehenden Heere," *Militär-Zeitung*, 2 June 1866, 351–53, on 352.

65. Heiderich, *Beiträge zur Wirtschaftskunde Österreichs*, 2–3.

66. Alexander von Peez, *Europa aus der Vogelschau* (Vienna, 1916 [1889]), 119. See too Norbert Krebs, *Länderkunde der österreichischen Alpen* (Stuttgart, 1913), 3.

67. Emanuel Herrmann, *Sein und Werden in Raum und Zeit: Wirthschaftliche Studien*, 2nd ed. (Berlin: Allgemeiner Verein für Deutsche Litteratur, 1889), 337.

68. Wilhelm Schmidt, "Ausfüllende, im Sinne des Druckgefälles verlaufende Luftströmungen."

69. He is also known among physical chemists for his theory of the mixing of liquid solutions, a problem similar to the mixing of air masses, described below. Jaime Wisniak, "Max Margules: A Cocktail of Meteorology and Thermodynamics," *Journal of Phase Equilibria* 24 (2003): 103–9.

70. John M. Wallace and Peter V. Hobbs, *Atmospheric Science: An Introductory Survey*, 2nd ed. (Amsterdam: Elsevier, 2006), 294.

71. "Bericht über die Leistungen der österreichischen Staats-Institute und Vereine im Gebiete der geographischen oder verwandten Wissenschaften für das Jahr 1885," *Mitteilungen der Geographischen Gesellschaft Wien* 29 (1886): 290–312, on 295; Max Margules, "Errichtung meteorologischer Beobachtungsstationen in Russisch-Polen," *Zs. Ö. G. Meteo.* 20 (1885): 534–35.

72. [Max Margules], "Ergebnisse aus den Regenaufzeichnungen der Forstlich-Meteorologischen Stationen," *Jb. ZAMG* 28 (1891): 62–70, on 62.

73. Fasz. 683/Sig. 4A/Nr. 14516: 7 July 1888; Fasz. 684/Sig. 4A/Nr. 32454: 28 October 1901, MCU.

74. [Max Margules], "Niederschlagsbeobachtungen in Crkvice," *MZ* 14 (1897): 156–57.

75. Max Margules, "Temperatur-Mittel aus den Jahren 1881–1885 and 30 jährige Temperatur-Mittel 1881–1880 für 120 Stationen in Schlesien, Galizien, Bukowina, Ober-Ungarn und Siebenbürgen," *Jb. ZAMG* 23 (1886): 109–26.

76. Gerhard Oberkofler and Peter Goller, "Von der Lehrkanzel für kosmische Physik zur Lehrkanzel für Meteorologie und Geophysik," in *100 Jahre Institut für Meteorologie und Geophysik*, Veröffentlichungen der Universität Innsbruck 178 (Innsbruck: Universität Innsbruck, 1990), 11–96, on 24.

77. Chavanne, *Temperatur-Verhältnisse*, 13.

78. Cf. Kutzbach, *Thermal Theory*, 195. Earlier analyses of squall lines on this scale by Köppen and Durand-Gréville relied on preexisting stations.

79. Max Margules, "Über die Beziehung zwischen Barometerschwankungen und Kontinuitätsgleichung," *Festschrift Ludwig Boltzmann* (Leipzig: J. A. Barth, 1904), 585–89. Peter Lynch, "Max Margules and His Tendency Equation," *Irish Meteorological Service Historical Notes* 5 (2001): 1–18. "Margules' Tendency Equation and Richardson's Forecast," *Weather* 58 (2003): 186–93.

80. Exner, "Über eine erste Annäherung zur Vorausberechnung synoptischer Wetterkarten," *MZ* 25 (1908): 57–67.

81. Quoted in Heinz Fortak, "Felix Maria Exner und die österreichische Schule der Meteorologie," in Hammerl, *Zentralanstalt*, 354–86.

82. Max Margules, "On the Energy of Storms," in *The Mechanics of the Earth's Atmosphere*, ed. and trans. Cleveland Abbe, 533–95 (Washington, DC: Smithsonian, 1910 [1903]).

83. Ibid., 538–39; Wisniak, "Margules." The analogy between baroclinic instability and convection remained popular into the mid-twentieth century but is no longer considered apt. See Isaac Held, "The Macroturbulence of the Troposphere," *Tellus* (1999): 51A-B, 59–70, on 64.

84. Wilhelm Trabert, "Der tägliche Luftdruckgang in unserer Atmosphäre," *MZ* 25 (1908): 39–40, on 40.

85. Margules nuanced that claim in his last meteorological publication, "Zur Sturmtheorie," *MZ* 23 (1906): 481–97.

86. Napier Shaw, *Manual of Meteorology*, vol. 4 (Cambridge: Cambridge University Press, 1919), 297, 347.

87. On the priority dispute over cyclogenesis, see Friedman, *Appropriating*, 199; Coen, *Vienna in the Age of Uncertainty*, 289–92.

88. Edward N. Lorenz, "Available Potential Energy and the Maintenance of the General Circulation," *Tellus* 7 (1955): 157–67.

89. Fasz. 683/Sig. 4A/Nr. 7971: 2 May 1885; Fasz. 683/Sig. 4A/Nr. 14516/1888: 7 July 1888; Fasz. 684/Sig. 4A/Nr. 29371: 4 October 1901; Fasz. 684/Sig. 4A/Nr. 32454: 28 October 1901, MCU.

90. Margules, "Zur Sturmtheorie," 483.

91. Oberkofler and Goller, "Von der Lahrkanzel," 18.

92. Fasz. 684/Sig. 4A/Nr. 25971: 20 June 1906; Fasz. 684/Sig. 4A/Nr. 43999: 14 November 1906, MCU.

93. Oberkofler and Goller, "Von der Lahrkanzel," 24.

94. Quoted and translated in Wisniak, "Margules," 104.

95. Wisniak, "Margules," 104.

CHAPTER EIGHT

1. Turbulence can be defined as fluid motion so complex that the velocity of the flow does not vary continuously from one point to another; it appears random to our eyes.

2. Körber, *Vom Wetteraberglaube*, 167.

3. William Ferrel, *The Motions of Fluids and Solids on the Earth's Surface* (Washington, DC: Office of the Chief Signal Officer, 1882), 38, cited in Kutzbach, *Thermal Theory*, 39.

4. Supan, *Statistik*, 12.

5. Hann, *Atlas der Meteorologie*, 5; see also Hann, *Vertheilung des Luftdruckes*, 1.

6. Hann, *Lehrbuch der Meteorologie*, 1st ed. (Leipzig: Tauchnitz, 1901), 578.

7. Quoted in and translated by Kutzbach, *Thermal Theory*, 138.

8. Davis, "Notes on Croll's Glacier Theory," *American Meteorological Journal* 11 (1895): 441–44, on 442.

9. Supan, *Statistik*, 12.

10. Trabert, "Die Luftdruckverhältnisse in der Niederung und ihr Zusammenhang mit der Verteilung der Temperatur," *MZ* 25 (1908): 103–8, on 104.

11. J. Hann et al., *Allgemeine Erdkunde* (Prague: Tempsky, 1872), 61.

12. Wilhelm Schmidt, *Der Massenaustausch in freier Luft und verwandte Erscheinungen* (Hamburg: Henri Grand, 1925), 5.

13. Hann, "Studien über die Luftdruck- und Temperaturverhältnisse auf dem Sonnblickgipfel," *Wiener Berichte* IIa 100 (1891): 367–452, on 444.

14. Hann, *Lehrbuch der Meteorologie*, 1st ed., 485–86, original emphasis.

15. In fact, twentieth-century scientists found sources of variation in the tropics that Hann had not suspected, such as intraseasonal oscillations.

16. "The energy of these atmospheric disturbances finds its equivalent in the loss of rotational velocity of the upper air circulation" (Hann, *Lehrbuch der Meteorologie*, 1st ed., 585). The actual contribution of eddies to the general circulation is still debated today.

17. Schmidt, *Massenaustausch*, 5.

18. Olivier Darrigol, *Worlds of Flow: A History of Hydrodynamics from the Bernoullis to Prandtl* (Oxford: Oxford University Press, 2005), 172–73.

19. Quoted in Olivier Darrigol, "Turbulence in 19th-Century Hydrodynamics," *Historical Studies in the Physical and Biological Sciences* 32 (2002): 207–62, on 247.

20. Ibid., 259–60.

21. Peter Galison, *Image and Logic: A Material Culture of Microphysics* (Chicago: University of Chicago Press, 1997), chapter 2.

22. Naomi Oreskes, "From Scaling to Simulation: Changing Meanings and Ambitions of Models in Geology," in *Science without Laws: Model Systems, Cases, Exemplary Narratives*, ed. A. Creager, E. Lunbeck, and M. N. Wise, 93–124 (Durham, NC: Duke University Press, 2007).

23. Schmidt, "Gewitter und Böen, rasche Druckanstiege," *Wiener Berichte* IIa 119 (1910): 1101–213, on 1135.

24. Vettin, "Experimentelle Darstellung von Luftbewegungen unter dem Einflusse von Temperatur-Unterschieden und Rotations-Impulsen," *MZ* 1 (1884): 227–30, 271–76. Schmidt read this only as his first study was going to press. In 1924 Friedrich Ahlborn published a review of simulations of the general circulation, which Schmidt mentioned to Prandtl two years later.

25. Coen, *Vienna in the Age of Uncertainty*, chapter 8.

26. Schmidt, "Gewitter und Böen," on 1135, my emphasis.

27. Schmidt, "Zur Mechanik der Böen," *MZ* 28 (1911): 355–62, on 355.

28. Schmidt, "Weitere Versuche über den Böenvorgang und das Wegschaffen der Bodeninversion," *MZ* 48 (1913): 441–47.

29. Ibid., 447.

30. Ficker, *Die Zentralanstalt für Meteorologie und Geodynamik in Wien, 1851–1951* (Vienna: Springer, 1951), 8.

31. *Jb. ZAMG* 50 (1913, printed 1917): 10.

32. Alon Rachamimov, *POWs and the Great War: Captivity on the Eastern Front* (Oxford: Berg, 2002), 37.

33. Michael Eckert, *The Dawn of Fluid Dynamics: A Discipline between Science and Technology* (Weinheim: Wiley, 2006), chapters 2–3.

34. Schmidt to Prandtl, 5 June 1926, LP.

35. Cf. Galison, *Image and Logic*, chapter 3.

36. Exner, "Über die Bildung von Windhosen und Zyklonen," *Wiener Berichte* IIa 132 (1923): 1–16, on 2–3.

37. Ibid., 4.

38. Ibid., 2, 6. On Bjerknes's theory at this time, see Friedman, *Appropriating*, chapter 11.

39. Mott Greene, *Alfred Wegener: Science, Exploration, and the Theory of Continental Drift* (Baltimore: Johns Hopkins University Press, 2015), 340–41, 516–17.

40. James Rodger Fleming, *Inventing Atmospheric Science: Bjerknes, Rossby, Wexler, and the Foundations of Modern Meteorology* (Cambridge, MA: MIT Press, 2016), chapter 3.

41. Felix M. Exner, "Dünen und Mäander, Wellenformen der festen Erdoberfläche, deren Wachstum und Bewegung," *Geografiska Annaler* 3 (1921): 327–35.

42. Wilhelm Schmidt, "Modellversuche zur Wirkung der Erddrehung auf Flußläufe," in *Festschrift der Zentralanstalt für Meteorologie und Geodynamik zur Feier ihres 75 jährigen Bestandes* (Vienna: ZAMG, 1926), 187–95, on 195.

43. Felix M. Exner, "Zur Wirkung der Erddrehung auf Flussläufe," *Geografiska Annaler* 9 (1927): 173–80.

44. Albert Einstein, "Die Ursache der Mäanderbildung der Flußläufe und des sogenannten Baerschen Gesetzes," *Die Naturwissenschaften* 11 (1926): 223–24.

45. Subhasish Dey, *Fluvial Hydrodynamics: Hydrodynamic and Sediment Transport Phenomena* (Berlin: Springer, 2014), 539–42.

46. Exner, "Zur Wirkung der Erddrehung," esp. 173, 178.

47. Johanna Vogel-Prandtl, *Ludwig Prandtl: Ein Lebensbild; Erinnerungen, Dokumente* (Göttingen: Universitätsverlag, 2005), 94–95.

48. Prandtl to Schmidt, 11 June 1926, LP.

49. Schmidt to Prandtl, 17 June 1926, LP. In response to the mention of Exner's studies, Prandtl scribbled "Wo?" in the margin.

50. For Prandtl's challenge to Bjerknes' theory of cyclogenesis, see Eckert, *Dawn of Fluid Dynamics*, 168.

51. Schmidt to Prandtl, 28 October 1926, LP.

52. Schmidt to Prandtl, 17 June 1926, LP.

53. Schmidt, "Der Massenaustausch bei der ungeordneten Strömung in freier Luft und seine Folgen," *Wiener Berichte* IIa 126 (1917): 757–804, on 757.

54. Prandtl, "Meteorologische Anwendungen der Strömungslehre," *Beiträge zur Physik der freien Atmosphäre* 19 (1932): 188–202, reprinted in Ludwig Prandtl, *Gesammelte Abhandlungen* 3, ed. Walther Tollmien et al. (Berlin: Springer, 1961), 1081–97, on 1106.

55. Richardson, *Weather Prediction by Numerical Process* (Cambridge: Cambridge University Press, 1922), 220.

56. Dave Fultz, Robert R. Long, et al., "Studies of Thermal Convection in a Rotating Cylinder with some Implications for Large-Scale Atmospheric Motions," *Meteorological Monographs* 4 (1959): 1–105; Fleming, *Inventing*, 81.

57. Fultz et al., "Rotating Cylinder," 2.

58. Ibid., 3. Edward Lorenz reflected on the significance of such experiments as a strategy for the "idealization of the atmosphere" in "Large-Scale Motions of the Atmosphere: Circulation" (Cambridge. MA: MIT Press, 1966), 95–109, on 99.

59. Fultz et al., "Rotating Cylinder," 4.

60. Isaac Held, "The Gap between Simulation and Understanding in Climate Modeling," *BAMS* 86 (2005): 1609–14, on 1610.

61. Schmidt, "Luftwogen im Gebirgstal," *Wiener Berichte* IIa 122 (1913): 835–911, on 839.

62. Wilhelm Schmidt, "Zur Frage der Verdunstung," *Ann. Hyd.* 44 (1916): 136–45, on 443.

63. Schmidt, "Der Massenaustausch bei der ungeordneten Strömung."

64. It was later found that A is not the same for transfers of heat and of momentum in the ocean; Bernhard Haurwitz, *Dynamic Meteorology* (New York: McGraw Hill, 1941), 220. For a derivation of *Austausch* in English, see ibid., chapter 11.

65. Schmidt, *Massenaustausch*, 113.

66. Ibid., 111. These values were obtained mainly by means of anemometers or observations of the dispersion of pollen or smoke.

67. John M. Lewis, "The Lettau-Schwerdtfeger Balloon Experiment: Measurement of Turbulence via Austausch Theory," *BAMS* 78 (1997): 2619–35.

68. Anders Ångström, review of Schmidt, *Massenaustausch*, *Geografiska Annaler* 8 (1926): 250–51.

69. Prandtl to Schmidt, 29 June 1926, LP.

70. Schmidt, *Massenaustausch*, 26.

71. Henri Grand, review of Schmidt, *Massenaustausch, Quarterly Journal of the Royal Meteorological Society* 53 (1927): 93–94, on 93.

72. Schmidt, *Massenaustausch*, 110.

73. Schmidt, "Messungen des Staubkerngehalts der Luft am Rande einer Großstadt," *Meteorologische Zeitschrift* 35 (1918): 281–85.

74. Schmidt, *Massenaustausch*, 109.

75. Schmidt, "Der Massenaustausch bei der ungeordneten Strömung," 804.

76. Albert Defant, "Die Zirkulation der Atmosphäre in den Gemässigten Breiten der Erde," *Geografiska Annaler* 3 (1921): 209–66.

77. Harold Jeffreys, "On the Dynamics of Geostrophic Winds," *Quarterly Journal of the Royal Meteorological Society* 52 (1926): 85–104.

78. Defant, "Zirkulation der Atmosphäre," 212.

79. Ibid., 213; Greene, *Wegener*, 316–17.

80. Defant, "Zirkulation der Atmosphäre," 213.

81. Ibid., 218–22, 214.

82. Eduard Brückner, *Klimaschwankungen seit 1700, nebst Bemerkungen über die Klimaschwankungen der Diluvialzeit* (Vienna: Hölzel, 1890); James Croll, *Climate and Time in Their Geological Relations; A Theory of Secular Changes of the Earth's Climate* (London: Daldy, Isbister, 1875).

83. Defant, "Zirkulation der Atmosphäre," 260.

84. Ibid., 264.

85. Ibid., 232.

86. Trabert, "Luftdruckverhältnisse in der Niederung," 107.

87. Tor Bergeron, "Richtlinien einer dynamischen Klimatologie," *MZ* 4 (1930): 246–62.

88. Kenneth Hare, "Dynamic and Synoptic Climatology," *Annals of the Association of American Geographers* 45 (1955): 152–62.

89. Sergei Chromow, "'Dynamische Klimatologie' und Dove," *Zeitschrift für angewandte Meteorologie, Das Wetter*, (1931): 312–14, on 313.

90. Arnold Court, "Climatology: Complex, Dynamic, and Synoptic," *Annals of the Association of American Geographers* 47 (1957): 125–36, on 134–35.

91. Hare, "Dynamic and Synoptic," 1955.

92. Köppen, "Die gegenwärtige Lage und die neueren Fortschritte der Klimatologie," 627.

CHAPTER NINE

1. James Strachey, ed., *The Standard Edition of the Complete Psychological Works of Sigmund Freud*, vol. 21 (London: Hogarth, 1961), 68.

2. Friedrich Simony, *Schutz dem Walde!* (Vienna: Verein zur Verbreitung naturwissenschaftlicher Kenntnisse, 1878), 19.

3. Brückner, *Klimaschwankungen seit 1700*, 290.

4. Max Endres, *Handbuch der Forstpolitik* (Berlin: Spring, 1905), 137.

5. Emanuel Purkyně, "Ueber die Wald und Wasserfrage," pt. 1, *Oesterreichische Monatsschrift für Forstwesen* 25 (1875): 479–525, on 488.

6. Review of Lorenz, *Wald, Klima, und Wasser*, *Neue Freie Presse* 19 March 1879, 4.

7. Endres, *Handbuch der Forstpolitik*, 160.

8. Ludwig Landgrebe, "The World as a Phenomenological Problem," *Philosophy and Phenomenological Research* 1 (1940): 38–58, on 47–49.

9. Grove, *Green Imperialism*, chapter 4; Jorge Cañizares-Esguerra, "How Derivative Was Humboldt?," in *Nature, Empire, and Nation: Explorations of the History of Science in the Iberian World* (Stanford, CA: Stanford University Press, 2006), 112–28.

10. Fabien Locher and Jean-Baptiste Fressoz, "Modernity's Frail Climate: A Climate History of Environmental Reflexivity," *Critical Inquiry* 38 (2012): 579–98; Aaron Sachs, *The Humboldt Current: Nineteenth-Century Exploration and the Roots of American Environmentalism* (New York: Viking, 2006); Diana Davis, *Resurrecting the Granary of Rome: Environmental History and French Colonial Expansion in North Africa* (Athens: Ohio University Press, 2007).

11. Ferdinand Wang, *Grundriss der Wildbachverbauung*, vol. 1 (Vienna: Hirzel, 1901), 78.

12. Moon, *Plough That Broke*; see too A. A. Fedotova and M. V. Loskutova, "Forests, Climate, and the Rise of Scientific Forestry in Russia: From Local Knowledge and Natural History to Modern Experiments (1840s–Early 1890s)," in Phillips and Kingsland, *Life Sciences and Agriculture*, 113–38; A. A. Fedetova, "Forestry Experimental Stations: Russian Proposals of the 1870s," *Centaurus* 56 (2014): 254–74.

13. *Die österreichisch-ungarische Monarchie in Wort und Bild*, vol. 15, *Böhmen*, vol. 2 (1896), 502–3.

14. Review of Lorenz, *Wald, Klima, und Wasser*, *Neue Freie Presse*, 19 March 1879, 4.

15. Brückner, *Klimaschwankungen seit 1700*, 29.

16. Holly Case, "The 'Social Question,' 1820–1920," *Modern Intellectual History* 13 (2016): 747–75, on 753.

17. Joachim Radkau, "Wood and Forestry in German History: In Quest of an Environmental Approach," *Environment and History* 2 (1996): 63–76, on 67.

18. Gerhard Weiss, "Mountain Forest Policy in Austria: A Historical Policy Analysis on Regulating a Natural Resource," *Environment and History* 7 (2001): 335–55.

19. Ibid., 343–44; Feichter, "Öffentliche und private Interessen an der Waldbewirtschaftung im Zusammenhang mit der Entstehung des österreichischen Reichsforstgesetzes von 1852," *Forstwissenschaftliche Beiträge* 16 (1996): 42–63.

20. A. C. Becquerel, *Mémoire sur les forêts et leur influence climatérique* (Paris: Academie des sciences, 1865).

21. Killian, *Der Kampf gegen Wildbäche und Lawinen im Spannungsfeld von Zentralismus und Föderalismus*, vol. 2, *Das Gesetz*, Mitteilungen der forstlichen Bundesversuchsanstalt 164 (Vienna: Bundesforschungszentrum für Wald, 1990).

22. An exception is Adolph Hohenstein, *Der Wald sammt dessen wichtigem Einfluss auf das Klima der Länder, Wohl der Staaten und Völker, sowie die Gesundheit der Menschen* (Vienna: Carl Gerold's Sohn, 1860).

23. Hann, "Ueber den Wolkenbruch, der am 25. Mai 1872 in Böhmen niederging," *Zs. Ö. G. Meteo.* 8 (1873): 234–35.

24. Micklitz, "Die Forstwirtschaft," in *Die Bodenkultur auf der Wiener Weltausstellung*, vols. 2–3, ed. Josef Roman Lorenz (Vienna: Faesy und Frick, 1874), 4. Micklitz was impressed that Purkyně made these measurements with no support other than the aid of his students.

25. Walter Schiff, *Geschichte der Österreichischen Land- und Forstwirtschaft und ihrer Industrien, 1848–1898* (Jena: Fischer, 1901), 557. Killian, *Kampf gegen Wildbäche*, vol. 2.

26. Killian, *Der Kampf gegen Wildbäche und Lawinen im Spannungsfeld von Zentralismus und Föderalismus*, vol. 1, *Die historischen Grundlagen* (Vienna: Bundesforschungszentrum für Wald, 1990), 95–96.

27. Cf. Kieko Matteson, *Forests in Revolutionary France: Conservation, Community, and Conflict, 1669–1848* (New York: Cambridge University Press, 2015), 11.

28. Killian, *Kampf gegen Wildbäche*, 2:76.

29. Stenographische Protokolle des Abgeordnetenhauses 1882, 9 March, 7347; see too Stenographische Protokolle des Abgeordnetenhauses 1876, 17 December, 7639.

30. Stenographische Protokolle des Abgeordnetenhauses 1907, 21 December, 3877.

31. Killian, *Kampf gegen Wildbäche*, 2:63.

32. Walter Schiff, *Österreichs Agrarpolitik seit der Grundentlastung*, vol. 1 (Tübingen: H. Laupp, 1898), 618.

33. Endres, *Handbuch der Forstpolitik*, 306.

34. Ekaterina Pravilova argues that the concept of "public property" coalesced in Russia in this period in part around arguments for the benefits of the forest as a public good. *A Public Empire: Property and the Quest for the Common Good in Imperial Russia* (Princeton, NJ: Princeton University Press, 2014), esp. 51.

35. "Zweite Sitzung," *Verhandlungen des Forstvereins der österreichischen Alpenländer* 1 (1852): 33–75, on 35.

36. David Ricardo, *On the Principles of Political Economy, and Taxation* (London: John Murray, 1821), 56.

37. Alexandre Moreau de Jonnès, *Quels sont les changements que peut occasioner le déboisement de forêts?* (Bruxelles: P. J. de Mat, 1825).

38. Gottlieb von Zötl, *Handbuch der Forstwirtschaft im Hochgebirge* (Vienna: C. Gerold, 1831), 54–61.

39. Josef Roman Lorenz, *Über Bedeutung und Vertretung der land- und forstwirthschaftlichen Meteorologie* (Vienna: Faesy & Frick, 1877), 4.

40. "Zur forstlichen Standortslehre," *Allgemeine Land- und Forstwirthschaftliche Zeitung*, 14 May 1853, 157.

41. Hann, "Thatsachen und Bemerkungen über einige schädliche Folgen der Zerstörung des natürlichen Pflanzkleides . . . ," *Zs. Ö. G. Meteo.* 4 (1869): 18–22, on 21.

42. Ernst Ebermayer, *Die physikalischen Einwirkungen des Waldes auf Luft und Boden und seine klimatologische und hygienische Bedeutung* (Aschaffenberg: C. Krebs, 1873).

43. Lorenz, *Über Bedeutung und Vertretung*, 18, 23.

44. Lorenz to Purkyně, 22 September 1876, EP.

45. Ibid.

46. "Propositions of the Fourth Section of the International Statistical Congress at Buda-Pesth, in 1876, Relative to Agricultural Meteorology," *Report of the Permanent Committee of the First International Meteorological Congress at Vienna* (London: J. D. Potter, 1879), 13.

47. "Zum dritten Programmspunkte der V. Versammlung deutscher Forstwirthe in Eisenach," *Centralblatt für das gesamte Forstwesen* 2 (1876): 480–82, on 480, 481.

48. Lorenz von Liburnau, *Resultate Forstlich-Meteorologischer Beobachtungen, Mittheilun-*

gen aus dem forstlichen Versuchswesen Oesterreichs XII, vol. 1 (Vienna: k.k. Hof- und Staatsdrückerei, 1890), 4

49. Cited in Lorenz, *Bedeutung und Vertretung*, 34.

50. Lorenz to Purkyně, undated, EP.

51. Wessely to Purkyně, undated, EP.

52. Ibid.

53. Bernhard Eduard Fernow, *Economics of Forestry: A Reference Book for Students of Political Economy* (New York: Thomas Crowell, 1902), 495.

54. Lorenz, *Wald, Klima, und Wasser*, 49.

55. Ibid., 272–74.

56. Ibid., 275–83.

57. Review of Lorenz, *Wald, Klima, und Wasser*, *Neue Freie Presse*, 19 March 1879, 4.

58. Anon., review of *Wald, Klima, und Wasser*, *Wiener Landwirtschaftliche Zeitung*, 22 March 1879, 5.

59. Jan Evangelista Purkyně, *Austria Polyglotta* (Prague: Ed. Grégr, 1867); simultaneously published in Czech and German.

60. Bernard Borggreve, "Dr. Emanuel Ritter von Purkyně, Nekrolog," *Forstliche Blätter* 19 (1882): 214–18, on 214.

61. Janko and Štrbáňová, *Věda Purkyňovy doby*, 200.

62. Krška and Šamaj, *Dějiny meteorologie*, 187.

63. Borggreve, "Nekrolog," 214.

64. V. Krečmer, "Přispěvek k Historii Užité Meteorologie," *Meteorologické zprávy* 16 (1963): 8–12, on 9.

65. In fact, Ebermayer turned to an Austrian colleague for advice: Ebermayer to Kerner, 1 March 1865, in Kronfeld, *Anton Kerner von Marilaun*, 292–94.

66. "Plenar-Versammlung des böhmischen Forstvereines in Böhmisch-Skalitz am 7. August 1878," *Vereinsschrift für Forst-, Jagd- und Naturkunde* 105 (1879): 5–27, on 12.

67. Krečmer, "Přispěvek k Historii Užité Meteorologie," 10.

68. Purkyně to Engelmann, 27 May 1878, EP-GE.

69. "Plenar-Versammlung am 7. August 1878," 15, 16. Cf. Matthew Maury, *Investigations of the Wind Currents of the Sea* (Washington, DC: C. Alexander, 1851), 8.

70. Purkyně, "Wald und Wasserfrage," pt. 1, 500–501, 520–21.

71. Ibid., 521.

72. Ibid., 495.

73. Purkyně to Engelmann, 2 February 1877, EP-GE.

74. Hann, Diary B, 40, 54, JH.

75. Purkyně to Engelmann, 20 August 1875, EP-GE.

76. *Písemná pozůtalost Emanuel Purkyně* (Prague: Literární Archiv PNP, 1988), 4.

77. F. J. Studnička, *Z pozemské přírody: Sebrané výklady a úvahy* (Prague: Dr. Frant. Bačkovský, 1893), 7.

78. Ibid., 30.

79. Emanuel Purkyně, "Vylet do Tater," *Živa* 1 (1853): 245–53, on 245.

80. Studnička, *Z pozemské přírody*, 27, 100.

81. Ibid., 31.

82. Ibid., 54.

83. Hann, "Ueber den Wolkenbruch, der am 25. Mai 1872 in Böhmen niederging," on 235.

84. Josef Roman Lorenz, ed., *Die Bodencultur auf der Wiener Weltaustellung 1873*, vol. 2, *Das Forstwesen* (Vienna: Faesy & Frick, 1874), 4.

85. Purkyně to Engelmann, 18 March 1876, EP-GE.

86. Purkyně to Engelmann, 27 May 1878, EP-GE.

87. Tomás Hermann, "Originalita vědy a problém plagiátu (Tři výstupy E. Rádla k jazykové otázce ve vědě z let 1902–1911)," in *Místo národních jazyků ve výchově, školství a vědě v Habsburské monarchii 1867–1918*, ed. Harald Binder et al. (Prague: Výzkumné centrum pro dějiny vědy, 2003), 441–68.

88. Hann to Purkyně, 28 August 1873, 15 December 1874, 7 December 1875, 11 July 1877, EP.

89. Jelinek to Purkyně, 21 January 1874, 5 May 1875, EP.

90. Wessely to Purkyně, 9 May 1875, EP.

91. Lorenz to Purkyně, all undated, EP.

92. Lorenz to Purkyně, undated, EP.

93. Wessely to Purkyně, 17 January 1874 and undated, EP.

94. Lorenz to Purkyně, undated, EP.

95. Wessely to Purkyně, 15 August 1875, EP.

96. Emanuel Purkyně, "Ueber die Wald und Wasserfrage," *Oesterreichische Monatsschrift für Forstwesen* 25 (1875): 479–525; 26 (1876): 136–51; 161–204; 209–51; 267–91; 327–49; 405–26; 473–98; 27 (1877): 102–43.

97. Wessely to Purkyně, 15 August 1875, EP.

98. E.g., Julius Micklitz, "Über die Einwirkungen des Waldes auf Luft und Boden," *Centralblatt für das gesammte Forstwesen* 3 (1877): 495–503.

99. Purkyně to Engelmann, 27 May 1878, EP-GE.

100. Lorenz to Purkyně, 24 September 1876, EP.

101. It has even been claimed that Studnička burned Purkyně's original climatic observations; Krška and Šamaj, *Dějiny meteorologie*, 89.

102. Steven Beller, "Hitler's Hero: Georg von Schönerer and the Origins of Nazism," in *In the Shadow of Hitler: Personalities of the Right in Central and Eastern Europe*, ed. Rebecca Haynes and Martyn Rady, 38–54 (New York: Palgrave Macmillan, 2011).

103. "Generalversammlung des Manhartsberger Forstvereines in Gmünd," *Landwirthschaftliches Vereinsblatt*, 1 August 1876, 61–62.

104. Stenographische Protokolle des Abgeordnetenhauses, vol. 7, 16 December 1876, 7623.

105. Killian, *Kampf gegen Wildbäche*, 2:99–112.

106. Ibid., 2:102.

107. Schiff, *Österreichs Agrarpolitik seit der Grundentlastung*; Otto Bauer, *Der Kampf um Wald und Weide: Studien zur österreichischen Agrargeschichte und Agrarpolitik* (Vienna: Volksbuchhandlung, 1925).

108. Killian, *Kampf gegen Wildbäche*, 2:69. On the devolution of power from Vienna to the crown land and municipal governments, and on municipal projects of modernization, all in the last quarter of the nineteenth century, see Judson, *Habsburg Empire*, 341–63.

109. "Allgemeiner Operations- und Organisationsplan für das forstliche Versuchswesen," in *Taschenausgabe der österreichischen Gesetze*, vol. 8, *Forstwesen*, 778–92 (Vienna: Manz, 1906), 786.

110. Jürgen Büschenfeld, *Flüsse und Kloaken: Umweltfragen im Zeitalter der Industrialisierung* (Stuttgart: Klett-Cotta, 1997), 415.

111. Lorenz von Liburnau, *Resultate forstlich-meteorologischer Beobachtungen*, 1: 3, 139.

112. Eckert, "Die Vegetationsdecke als Modificator des Klimas mit besonderer Rücksicht auf die Wald- und Wasserfrage," *Österreichische Vierteljahresschrift für Forstwesen* 11 (1893): 254–70, on 258, 269, 270.

113. Frank Uekötter, *The Age of Smoke: Environmental Policy in Germany and the United States* (Pittsburgh: University of Pittsburgh Press, 2009), 18.

114. Büschenfeld, *Flüsse und Kloaken.*

115. Ernst Brezina, "Die Donau vom Leopoldsberge bis Preßburg, die Abwässer der Stadt Wien und deren Schicksal nach ihrer Einmündung in den Strom," *Zeitschrift für Hygiene und Infektionskrankheiten* 53 (1906): 369–503, on 490.

116. Christiane W. Runyan et al., "Physical and Biological Feedbacks of Deforestation," *Reviews of Geophysics* 50 (2012): 1–32, on 5.

117. Roger G. Barry, "A Framework for Climatological Research with Particular Reference to Scale Concepts," *Transactions of the Institute of British Geographers* 49 (1970): 61–70, on 65.

118. Locher and Fressoz, "Modernity's Frail Climate."

119. Edwards, *Vast Machine.*

120. Ghosh, *Great Derangement*, 30.

CHAPTER TEN

1. Anton Kerner, "Beiträge zur Geschichte der Pflanzenwanderungen," *Deutsche Revue* 2 (1879): 104–13, on 107.

2. On the historical relationship between plant ecology and knowledge of climate change, see Christophe Masutti, "Frederic Clements, Climatology, and Conservation in the 1930s," *Historical Studies in the Physical and Biological Sciences* 37 (2006): 27–48.

3. Alexander von Humboldt, review of Thaddäus Haenke, *Beobachtungen auf Reisen nach dem Riesengebirge* (Dresden: Walther, 1791), *Annalen der Botanick* 1 (1791): 78–83, on 79.

4. Kerner, *Pflanzenleben der Donauländer*, 3. In the same vein, Friedrich Simony wrote that dendrochronology told "the whole story of the life and travails of old trees and shrubs" and disclosed "the climatic character of places and times where no meteorological observations have yet been made." *Zs. Ö. G. Meteo.* 1 (1866): 52.

5. Otto Sendtner, "Bemerkungen über die Methode, die periodischen Erscheinungen an den Pflanzen zu beobachten," reprinted in *Jb. ZAMG* 4 (1856): 30–48, on 30.

6. Klemun, "National 'Consensus,'" 96; she points out that this resulted in dating quaternary layers as tertiary.

7. Quoted in Maria Petz-Grabenbauer and Michael Kiehn, eds., *Anton Kerner von Marilaun* (Vienna: Österreichische Akademie der Wissenschaften, 2004), 21.

8. Kerner saw little profit from this publication, although it would be reprinted in 1951 with the momentous title "The Background to Plant Ecology." By contrast, Kerner's monumental popular treatise *Das Pflanzenleben*, originally published in two volumes in 1888 and 1890, was a commercial success (Kronfeld, *Anton Kerner von Marilaun*, 368).

9. Charles Darwin to Kerner, quoted in Kronfeld, *Kerner von Marilaun*, 156–57; and Darwin to William Ogle, 17 August 1878, Darwin Correspondence.

10. Fritz Kerner von Marilaun, *Die Paläoklimatologie* (Berlin, 1930). For an example of continuity between the work of father and son, see Fritz Kerner, "Untersuchungen über die Schneegrenze im Gebiete des Mittleren Innthales," *Denk. Akad. Wiss. math-nat.* 54 (1889), esp. 17.

11. Dvorak, *Denkmale Krems.*

12. Karl Fritsch, "Nachruf an Anton Kerner von Marilaun," 694.

13. Anton Kerner, "Ueber eine neue Weide, nebst botanische Bemerkungen," *Verh. Zool.-Bot. Ver.* 2 (1852): 61–64, on 62.

14. Cf. Georg Grabherr, "Vegetationsökologie und Landschaftsökologie," in *Geschichte der österreichischen Humanwissenschaften*, vol. 2, ed. Karl Acham, 149–85 (Vienna: Passagen, 2001), 150–60.

15. Marie Petz-Grabenhuber, "Anton Kerner von Marilaun," in Grabenbauer and Kiehn, *Kerner von Marilaun*, 7–23, on 9.

16. Kerner, "Ueber eine neue Weide," 63.

17. Petz-Grabenhuber, "Anton Kerner von Marilaun," 9.

18. Kronfeld, *Anton Kerner von Marilaun*, 249.

19. Rácz, *Steppe to Europe*, 182–226.

20. Anton Kerner, "Die Steppenvegetation des ungarischen Tieflandes," *Wiener Zeitung*, 27 January 1859, 6.

21. Anton Kerner, "Die Entsumpfungsbauten in der Nieder-Ungarischen Ebene und ihre Rückwirkung auf Klima und Pflanzenwelt," *Wiener Zeitung* 8 April 1859, 4–5, and 17 April 1859, 6.

22. Anton Kerner, "Studien über die oberen Grenzen der Holzpflanzen in den österreichischen Alpen," in *Der Wald und die Alpenwirtschaft in Österreich und Tirol*, ed. Karl Mahler (Berlin: Gerdes & Hödel, 1908), 20–121, on 22; originally published in the *Österreichische Revue*, 1865.

23. Anton Kerner, "Niederösterreichische Weiden," pt. 1, *Verh. Zool.-Bot. Ver.* 10 (1860): 3–56, on 40.

24. Kerner, *Das Pflanzenleben* (Leipzig and Vienna: Bibliographisches Institut, 1891), 2:815.

25. Kerner, "Österreichs waldlose Gebiete," in *Wald und Alpenwirtschaft*, 5–19, on 8; originally published in *Österreichische Revue*, 1863.

26. Ibid., 7.

27. Kerner distinguished *waldlos* (forestless) from *entwaldet* (deforested), ibid., 7.

28. Kerner, *Wald und Alpenwirtschaft*, 24–25.

29. Kerner, *Pflanzenleben der Donauländer*, 86.

30. Ibid., 89.

31. Kerner, "Österreichs waldlose Gebiete," 10.

32. Kerner, *Pflanzenleben der Donauländer*, 28.

33. Cf. Michael Gubser, *Time's Visible Surface: Alois Riegl and the Discourse on History and Temporality in Fin-de-Siècle Vienna* (Detroit: Wayne State University Press, 2006).

34. On late nineteenth-century efforts to aestheticize the landscape of the Russian steppes,

see Christopher Ely, *This Meager Nature: Landscape and National Identity in Imperial Russia* (De Kalb: Northern Illinois University Press, 2002).

35. Kerner, *Pflanzenleben der Donauländer*, 27.

36. Folders "Phaenologische Notizen, Ofen-Pest, 1856," "Ung. Tiefland. Verschiedene Notizen," "Höhen aus dem ungar. Tieflande: Notizen zur orografische hydograf. u. geologische Schilderung zu meteorolog[ischen Zwecken]," "Ung. Tiefen Geologie u. Orografie," and "Obere Grenzen," Box 305, 315, AK.

37. Quoted in Kronfeld, *Anton Kerner von Marilaun*, 310.

38. Larry Wolff, *Inventing Eastern Europe: The Map of Civilization on the Mind of the Enlightenment* (Stanford, CA: Stanford University Press, 1994).

39. A. Kerner, "Reiseskizzen aus dem ungarisch-siebenbürgischen Grenzgebirge," pt. 4, in subfolder "Wandern u. Wiener Zeitung," 131.33.5.2, AK.

40. Simon Schama, *Landscape and Memory* (New York: Vintage, 1996), pt. 1; Jane Costlow, *Heart-Pine Russia: Walking and Writing the Nineteenth-Century Forest* (Ithaca, NY: Cornell University Press, 2013).

41. Quoted in Kronfeld, *Kerner von Marilaun*, 191.

42. Kerner von Marilaun, "Goethes Verhältnis zur Pflanzenwelt," reprinted in Kronfeld, *Kerner von Marilaun*, 240–43.

43. Quoted in Kronfeld, *Kerner von Marilaun*, 193.

44. Perhaps inspired by Kerner, other Habsburg researchers (Čelakovský, August Neireich) subsequently published on *Stipa pennata* elsewhere in the Monarchy (Bohemia and Lower Austria, respectively).

45. Kerner, "Das ungarische 'Waisenmädchenhaar,'" *Die Gartenlaube* 10 (1862): 44–46, reprinted in Kronfeld, *Kerner von Marilaun*, 203–10, on 205, 206, 207, 210.

46. Ibid., 206.

47. Kerner, *Pflanzen der Donauländer*, 90, 20, 25.

48. László Kürti, trans., *The Remote Borderland: Transylvania in the Hungarian Imagination* (Albany: SUNY Press, 2001), 84.

49. Janet Browne, *The Secular Ark: Studies in the History of Biogeography* (New Haven, CT: Yale University Press, 1983), 175.

50. Kerner, "Gute und schlechte Arten," pt. 1, *Öst. Bot. Z.* 15 (1865): 6–8, on 7.

51. Kronfeld, *Kerner von Marilaun*, 98.

52. Endersby, *Imperial Nature*.

53. Kerner, "Gute und schlechte Arten," pt. 8, *Öst. Bot. Z.* 16 (1866): 51–57, on 51, 54.

54. Kerner, *Pflanzenleben der Donauländer*, 239.

55. Frank N. Egerton, "History of Ecological Sciences, Part 54: Succession, Community, and Continuum," *Bulletin of the Ecological Society of America* 96 (2015): 426–74, on 441.

56. Kerner, *Pflanzenleben der Donauländer*, 244.

57. Ibid., 247–49.

58. Ibid., 5–6.

59. Kronfeld, *Anton Kerner von Marilaun*, 121; *Botanik und Zoologie in Österreich in den Jahren 1850 bis 1900* (Vienna: Hölder, 1901).

60. Richard Wettstein, quoted in Kronfeld, *Anton Kerner von Marilaun*, 121.

61. Kronfeld, *Anton Kerner von Marilaun*, 82.

62. Kerner, *Pflanzenleben der Donauländer*, 4.

63. Kerner, "Beiträge zur Geschichte der Pflanzenwanderung," pt. 1, *Öst. Bot. Z.* 29 (1879): 174–82, on 176.

64. Kerner von Marilaun, *Das Pflanzenleben* (Leipzig and Vienna: Bibliographisches Institut, 1891), 2:4.

65. August Grisebach, *Die Vegetation der Erde nach ihrer Klimatischen Anordnung*, 2nd ed. (1884), explained plant distribution with reference to "centers of creation."

66. Nils Güttler, *Das Kosmoskop: Karten und ihre Benutzer in der Pflanzengeographie des 19. Jahrhunderts* (Göttingen: Wallstein, 2014).

67. Sander Gliboff, "Evolution, Revolution, and Reform in Vienna: Franz Unger's Ideas on Descent and Their Post-1848 Reception," *Journal of the History of Biology* 31 (1998): 179–209, on 185.

68. Franz Unger, *Versuch einer Geschichte der Pflanzenwelt* (Vienna: Braumüller, 1852), 254.

69. Ibid., 5, 347–49.

70. Martin J. Rudwick, trans., *Scenes from Deep Time: Early Pictorial Representations of the Prehistoric World* (Chicago: University of Chicago Press, 1995), 101.

71. Marianne Klemun, "Franz Unger and Sebastian Brunner on Evolution and the Visualization of Earth History; A Debate between Liberal and Conservative Catholics," in *Geology and Religion: A History of Harmony and Hostility*, ed. M. Kölbl-Ebert (London: Geological Society, 2009), 259–67.

72. Edward Forbes, "On the Connexion between the Distribution of the Existing Fauna and Flora of the British Isles, and the Geological Changes Which Have Affected Their Area," *Memoirs of the Geological Survey of England and Wales* 1 (1846): 336–432.

73. Ibid., 397.

74. A. Grisebach, "Der gegenwärtige Stand der Geographie der Pflanzen," *Geographisches Jahrbuch* 1 (1866): 373–402, esp. 379–91; Nicolaas Rupke, "Neither Creation nor Evolution," *Annals of the History and Philosophy of Biology* 10 (2005): 143–72.

75. Lorenz to Purkyně, 20 September 1878, EP.

76. Kronfeld, *Anton Kerner von Marilaun*, 358, 89.

77. Anton Kerner, "Chronik der Pflanzenwanderungen," *Öst. Bot. Z.* 21 (1871): 335–40, on 335, 336.

78. Kerner, "Chronik der Pflanzenwanderungen," 336.

79. Fritsch, "Kerner von Marilaun," 11.

80. Ibid., 20.

81. "Ein vaterländisches wissenschaftliches Unternehmen," *Neue Freie Presse*, 23 July 1886, 4.

82. Endersby, *Imperial Nature*; Güttler, *Das Kosmoskop*.

83. Kerner von Marilaun, *Das Pflanzenleben*, 1:18.

84. Kerner, *Das Pflanzenleben der Donauländer*, 197.

85. "Diluvialesfestland," 131.33.5.8, AK; cf. *Pflanzenleben der Donauländer*, 194.

86. Browne, *Secular Ark*, 200.

87. Lynn Nyhart, "Emigrants and Pioneers: Moritz Wagner's 'Law of Migration' in Context," in *Knowing Global Environments: New Historical Perspectives in the Field Sciences*, ed. Jeremy Vetter (New Brunswick, NJ: Rutgers University Press, 2010), 39–58.

88. Anton Kerner, "Können aus Bastarten Arten werden?," *Öst. Bot. Z.* 21 (1871): 34–41.

89. Anton Kerner, "Abhängigkeit der Pflanzengestalt vom Klima und Boden," in *Festschrift der 43. Versammlung Deutscher Naturforscher und Ärzte*, 1–38 (Innsbruck: Wagner, 1869), 30.

90. Ibid., 48.

91. Darwin, *Origin of Species*, chapter 12; Browne, *Secular Ark*, 199.

92. Kerner, "Beiträge zur Geschichte der Pflanzenwanderungen," 110; Anton Kerner, "Der Einfluß der Winde auf die Verbreitung der Samen im Hochgebirge," *Z. d. ö. AV* 2 (1871): 144–72, on 151.

93. Alphonse de Candolle, introductory note to "Expériences sur les graines de diverses espèces plongées dans de l'eau de mer," *Archives des sciences physiques et naturelles* 47 (1873): 177–79.

94. Quoted in Kronfeld, *Kerner von Marilaun*, 278.

95. Quoted in ibid., 255.

96. "Ein Instrument zur Messung des Thauniederschlages," *Centralblatt für das gesamte Forstwesen* 19 (1893): 185–86, on 186.

97. Kerner, "Einfluß der Winde," 144, 159–60.

98. Kerner, "Studien über die Flora der Diluvialzeit in den östlichen Alpen," *Wiener Berichte* II 97 (1888): 7–39, on 15.

99. Kerner, "Einfluß der Winde," 162.

100. Ibid., 162–65.

101. Ibid., 165.

102. Christian Körner, *Alpine Plant Life: Functional Plant Ecology of High Mountain Ecosystems* (Berlin: Springer, 1999), 275.

103. Kerner, "Einfluß der Winde," 171–72.

104. Kerner, "Flora der Diluvialzeit."

105. Ibid., 33; Eduard Brückner, "Entwicklungsgeschichte des kaspischen Meeres und seiner Bewohner," *Humboldt* 7 (1889): 209–14.

106. Kerner, "Flora der Diluvialzeit," 33.

107. Kerner, "Beiträge zur Geschichte der Pflanzenwanderungen," 181.

108. Kerner, "Flora der Diluvialzeit," 12.

109. Kerner, *Pflanzenleben*, 17–18.

110. Hanns Kerschner et al., "Paleoclimatic Interpretation of the Early Late-Glacial Glacier in the Gschnitz Valley, Central Alps, Austria," *Annals of Glaciology* 28 (1999): 135–40.

111. E. Schwienbacher et al., "Seed Dormancy in Alpine Species," *Flora* 206 (2011): 845–56. Auricula was not part of this study.

112. Quoted in Kronfeld, *Kerner von Marilaun*, 200–202.

113. Kerner, *Pflanzenleben*, 18.

114. Kerner, "Chronik der Pflanzenwanderungen," 336.

CHAPTER ELEVEN

1. Stifter, *Nachsommer*, 1:338.

2. Julius Hann, Diary A, 4–6, JH.

3. Alois Topitz, "Julius Hann, ein großer Oberösterreicher, zu seinem 50. Todestag,"

Oberösterreichische Heimatblätter 3 (1971): 126–29; Alois Topitz, "Der Meteorologe Julius Hann," *Historisches Jahrbuch der Stadt Linz* (1959): 431–44.

4. Diary A, 70, JH.
5. Diary A, 97, JH.
6. Topitz, "Der Meteorologe Julius Hann," 432.
7. N. Pärr, "P. Gabriel Strasser," in *Österreichisches Biographisches Lexicon, 1815–1950*, vol. 13 (Vienna: Österreichische Akademie der Wissenschaften, 1954), 362.
8. Hann, Diary A, 73, JH.
9. Hann, Diary A, 124, JH.
10. Hann, Diary A, 76, JH.
11. Carl Gustav Carus, *Nine Letters on Landscape Painting*, trans. David Britt (Los Angeles: Getty, 2002), 115.
12. E.g., Diary A, 136, JH.
13. Felix Exner, "Julius von Hann," *MZ* 38 (1921): 321–27, on 326.
14. Hann, Diary B, 54, JH.
15. Hann, Diary A, 89, 93, JH.
16. Andreas von Baumgartner, *Die Stellung der Astronomie im Reiche der Menschheit* (Brno: Carl Winiker, 1850), 6 (quoting Jean Paul).
17. Humboldt, *Kosmos*, 2:8.
18. Hann, Diary B, 30, 31, and Diary A, 86, JH.
19. Carl Ritter, *Einleitung zur allgemeinen vergleichenden Geographie* (Berlin: Reimer, 1852), 186; Hann, Diary B, 5b.
20. Hann, Diary B, 29, JH.
21. Hann, Diary A, 85, 89, 113, JH.
22. Hann, Diary A, 113, JH.
23. Hann, Diary C, 105, JH.
24. Hann, Diary A, 33, JH.
25. "Ich blick' in mein Herz und ich blick' in die Welt, / Bis vom Auge die brennende Träne mir fällt, / Wol leuchtet die Ferne mit goldenem Licht, / Doch hält mich der Nord—ich erreiche sie nicht. / O die Schranken so eng, und die Welt so weit, / Und so flüchtig die Zeit!"
26. Hann, Diary A, 50, JH.
27. Hann, Diary A, 116, JH.
28. Hann, Diary A, 128, JH.
29. Hann, Diary A, 58–59, JH.
30. Hann, Diary A, 120, JH.
31. Hann, Diary B, 50, JH.
32. Hann, Diary B, 90, JH.
33. Hann, Diary C, 40, JH.
34. Hann, Diary A, 130, JH.
35. Hann, Diary A, 132, JH.
36. Hann, Diary B, 68, 74, JH.
37. Hann, Diary B, 83, JH.
38. Hann, Diary C, 109, JH.

39. Cornelia Lüdecke, "East Meets West: Meteorological Observations of the Moravians in Greenland and Labrador since the 18th Century," *History of Meteorology* 2 (2005): 123–32.

40. See Andreas Hense and Rita Glowienka-Hense, "Comments On: On the Weather History of North Greenland, West Coast by Julius Hann," *MZ* 19 (2010): 207–11; Hew Davies, "Vienna and the Founding of Dynamical Meteorology," in Hammerl, *Zentralanstalt*, 301–12.

41. Hann, "Der Pulsschlag der Atmosphäre," *MZ* 23 (1906): 82–86, on 82.

42. Hann, Diary C, 47, JH.

43. Hann, Diary C, 67, JH.

44. Hann, Diary C, 69, JH.

45. Hann, Diary C, 85, 110, JH.

46. Topitz, "Hann, ein großer Oberösterreicher," 129.

47. Johann Heiss and Johannes Feichtinger, "Distant Neighbors: Uses of Orientalism in the Late Nineteenth-Century Austro-Hungarian Empire," in *Deploying Orientalism in Culture and History: From Germany to Central and Eastern Europe*, ed. James Hodkinson and John Walker, 148–65 (Rochester: Camden House, 2013).

48. E.g., Moritz Deutsch, *Die Neurasthenie beim Manne* (Berlin: H. Steinitz, 1907), 168; A. Eulenberg, "Die Balneologie in der Nervenheilkunde," *Berliner klinische Wochenschrift* 42 (1905): 589–93; on theories of climate and sexual function more generally, see Cheryl A. Logan, *Hormones, Heredity, and Race: Spectacular Failure in Interwar Vienna* (New Brunswick, NJ: Rutgers University Press, 2013), chapter 4.

49. Freud, *Three Essays on the Theory of Sexuality*, trans. James Strachey (New York: Basic, 1962), 5.

50. Richard Burton, *The Sotadic Zone* (New York: Panurge, ca. 1934), 18, 23.

51. Ellsworth Huntington, *Civilization and Climate* (New Haven, CT: Yale University Press, 1915), 46.

52. Leopold von Sacher-Masoch, *Venus im Pelz* (Berlin: Globus, 1910), 7, 40, 34.

53. Allan Janik and Stephen Toulmin, *Wittgenstein's Vienna* (New York: Simon and Schuster, 1973).

54. Laurence Cole, *Für Gott, Kaiser, und Vaterland: Nationale Identität der deutschsprachigen Bevölkerung Tirols, 1860–1914* (Frankfurt: Campus Verlag, 2000).

55. L. Ficker to C. Dallago, 26 April 1910, in Ludwig Ficker, *Briefwechsel*, vol. 1, ed. Ignaz Zangerle (Salzburg: O. Müller, 1986), 26.

56. C. Dallago to L. Ficker, 9 April 1910, in Ficker, *Briefwechsel*, 1:24.

57. Richard Huldschiner to L. Ficker, 6 May 1910, in Ficker, *Briefwechsel*, 1:27.

58. Ficker, *Klimatographie von Tirol*, 116.

59. Heinrich von Ficker, "Die Erforschung der Föhnerscheinungen in den Alpen," *Z. d. ö. AV* 43 (1912): 53–77, on 53.

60. Jim Doss and Werner Schmitt, trans., http://www.literaturnische.de/Trakl/english/ged-e.htm.

61. Ficker, "Erforschung der Föhnerscheinungen," 54.

62. Otto Marschalek, *Österreichische Forscher: Ein Beitrag zur Völker- und Länderkunde* (Mödling bei Wien: St. Gabriel, 1949), 124.

63. H. von Ficker, "Östliche Geschichte," 84–85, F1f 1909, LD.

64. Ibid., 75.

65. Afsaneh Najmabadi, *Women with Mustaches and Men without Beards: Gender and Sexual Anxieties of Iranian Modernity* (Berkeley: University of California Press, 2005), 34.

66. Ficker, "Östliche Geschichte," 77, F1f 1909, LD.

67. Ibid., 83.

68. Ficker, "Zur Meteorologie von West-Turkestan," *Denk. Akad. Wiss. math-nat.* 81 (1908): 533–59, on 558.

69. Ficker, "Untersuchungen über die meteorologischen Verhältnisse der Pamirgebiete," *Denk. Akad. Wiss. math-nat.* 97 (1921): 151–255, on 246.

70. Willi Rickmers, *Alai! Alai! Arbeiten und Erlebnisse der Deutsch-Russischen Alai-Pamir-Expedition* (Leipzig: Brockhaus, 1930), 240.

71. Willi Rickmers, "Vorläufiger Bericht über die Pamirexpedition des Deutschen und Österreichischen Alpenvereins," *Z. d. ö. AV* 45 (1914): 1–51, on 27.

72. Deborah R. Coen, "Imperial Climatographies from Tyrol to Turkestan," *Osiris* 26, *Klima* (2011): 45–65.

73. Sverker Sörlin, "Narratives and Counter-Narratives of Climate Change: North Atlantic Glaciology and Meteorology, c. 1930–1955," *Journal of Historical Geography* 35 (2009): 237–55.

74. Rickmers, "Vorläufiger Bericht," 51.

75. Carl Dallago, "Nietzsche und der Philister," *Der Brenner* 1 (1910): 26.

76. Ludwig Wittgenstein, "Tractatus Logico-Philosophicus," *Annalen der Naturphilosophie* 14 (1921): 185–262, on 262.

77. H. von Ficker, "Östliche Geschichte," 79–81, F1f 1909, LD.

CONCLUSION

1. Petra Svatek, "Hugo Hassinger und Südosteuropa: Raumwissenschaftliche Forschungen in Wien (1931–1945)," in *"Mitteleuropa" und "Südosteuropa" als Planungsraum*, ed. Carola Sachse, 290–311 (Göttingen: Wallstein, 2010).

2. Hugo Hassinger, *Österreichs Wesen und Schicksal, verwurzelt in seiner geographischen Lage* (Vienna: Freytag-Berndt, 1949), 10.

3. Ibid., 7, original emphasis.

4. Ludwig von Mises, "Vom Ziel der Handelspolitik," *Archiv für Sozialwissenschaft und Sozialpolitik* 42 (1916): 561–85, e.g., 562–63.

5. Oszkár Jászi, *The Dissolution of the Habsburg Monarchy* (Chicago: University of Chicago Press, 1929), 185.

6. Ibid., 188.

7. Robert Sieger, *Die geographischen Grundlagen der österreichisch-ungarischen Monarchie und ihrer Außenpolitik* (Leipzig: Teubner, 1915), 3; Robert Sieger, *Der österreichische Staatsgedanke und seine geographischen Grundlagen* (Vienna: C. Fromme, 1918), 5; Hans-Dietrich Schulze, "Deutschlands natürliche Grenzen: Mittellage und Mitteleuropa in der Diskussion der Geographen seit dem Beginn des 19. Jahrhunderts," *Geschichte und Gesellschaft* 15 (1989): 248–81, on 263.

8. Norbert Krebs, *Länderkunde der österreichischen Alpen* (Stuttgart: Engelhorn, 1913), 3.

9. Sieger, *Geographische Grundlagen*, 22, 44.

10. Richard von Coudenhove-Kalergi, *Apologie der Technik* (Leipzig: P. Reinhold, 1922), 8, 41.

11. Katiana Orluc, "A Wilhelmine Legacy? Coudenhove-Kalergi's Pan-Europe and the Crisis of European Modernity, 1922–1932," in *Wilhelminism and Its Legacies*, ed. Geoff Eley and James Retallack, 291–34 (New York: Berghahn, 2003); Marco Duranti, "European Integration, Human Rights, and Romantic Internationalism," in *The Oxford Handbook of European History, 1914–1945*, ed. Nicholas Doumanis (Oxford: Oxford University Press, 2016), 440–58.

12. On Hanslik, see Norman Henniges, "'Naturgesetze der Kultur': Die Wiener Geographen und die Ursprünge der Volks- und Kulturbodentheorie," *ACME* 14 (2015): 1309–51.

13. Jeremy King, *Budweisers into Czechs and Germans* (Princeton, NJ: Princeton University Press, 2002).

14. Erwin Hanslik, "Kulturgeographie der deutsch-slawischen Sprachgrenze," *Vierteljahrschrift für Sozial- und Wirtschaftsgeschichte* 8 (1910): 103–27, 445–75, on 470.

15. Erwin Hanslik, *Oesterreich als Naturförderung* (Vienna: Institut für Kulturforschung, 1917), 36.

16. Hanslik, "Deutsch-slawischen Sprachgrenze," 117.

17. E.g., Hanslik, *Österreich, Erde und Geist* (Vienna: Institut für Kulturforschung, 1917).

18. Erwin Hanslik, "Die Karpathen," in *Mein Österreich, Mein Heimatland*, vol. 1, ed. Siegmund Schneider and Benno Immendörfer, 76–82 (Vienna: Verlag für vaterländische Literatur, 1915).

19. Sieger, *Geographische Grundlage*, 40, 24n1; Wolff, *Inventing Eastern Europe*.

20. Max Bergholz, "Sudden Nationhood: The Microdynamics of Intercommunal Relations in Bosnia-Herzegovina after World War II," *AHR* 118 (2013): 679–707, on 684.

21. Krebs to Hettner, 3 November 1915 and 4 December 1919, D II 73, AH.

22. Sieger to W. M. Davis, 11 November 1919, folder 438; Brückner to W. M. Davis, 17 September 1922, folder 73, WMD.

23. For a defense of epistemic pluralism in the sciences, see Hasok Chang, *Is Water H_2O? Evidence, Realism and Pluralism* (Boston: Springer, 2012), 253–301.

24. As reported in *Jb. ZAMG* (1919): 4, twenty-eight in Bohemia, eighteen in Moravia, seven in Silesia, two in Galicia, three in Carniola, and two in Dalmatia. Cf. Coen, *Earthquake Observers*, chapter 7.

25. It is in the context of these disputes over ownership of instruments and recorded data that we need to understand the ZAMG's claim that its results were the work of "German scholars": Fasz. 681/Sig. 4A/Nr. 1277: 23 November 1918; Fasz. 686/Sig. 4A/Nr. 1340: 6 December 1918, SAU.

26. Michael Gordin, *Scientific Babel: How Science Was Done Before and After Global English* (Chicago: University of Chicago Press, 2015), and Surman, *Biography of Habsburg Universities*.

27. Jiří Martínek, "Radost z poznání nemusí vést k uznání. Julie Moschelesová," in Martínek, *Cesty k samostatnosti: Portréty žen v éře modernizace* (Prague: Historický ústav, 2010), 176–89.

28. Julie Moscheles to William Morris Davis, 15 August 1919, 15 November (no year), folder 336, WMD.

29. Fasz. 682/Sig. 4A/Nr. 20375: 23 June 1934, SAU, my emphasis.

30. Pollak, "Über die Verwendung des Lochkartenverfahrens in der Klimatologie," *Zeitschrift für Instrumentenkunde* 47 (1927): 528–32.

31. Helmut Landsberg, quoted in F. W. Kistermann, "Leo Wenzel Pollak (1888–1964): Czechoslovakian pioneer in Scientific Data Processing," *IEEE Annals of the History of Computing* 2 (1999): 62–68, on 65.

32. Edwards, *Vast Machine*, 99.

33. Robert Sieger to William Morris Davis, 26 January 1920, folder 438, WMD.

34. Heinrich Ficker, "Wo findet man in den deutsch-österreichischen Alpen einen Ersatz für Davos?," *MZ* 38 (1921): 307–9, on 309.

35. Alois Gregor, "Moderní klimatologie," *Spirála* 1 (1936): 449–75, on 466.

36. Klimatische Beobachtungsstationen 1930/Nr. 51584; Kurorte 1927/Nr. 21913, VG.

37. Ernst Brezina and Wilhelm Schmidt, *Das künstliche Klima in der Umgebung des Menschen* (Stuttgart: Enke, 1937), 207.

38. Alois Gregor, "Problémy velkoměstské klimatologie," *Sborník IV. sjezdu československých Geografů v Olomouci 1937* (Brno: Československá společnost zeměpisné, 1938), 82–85, on 82.

39. Steinhauser, "Großstadttrübung und Strahlungsklima," *Biokl. Beibl.* 3 (1934): 105–11, on 105.

40. Brezina and Schmidt, *Das künstliche Klima*, 3.

41. Gregor, "Problémy velkoměstské klimatologie," 84.

42. Brezina and Schmidt, *Das künstliche Klima*, 207.

43. Franz Linke, "Zur Einführung der 'Bioklimatischen Beiblätter der Meteorologischen Zeitschrift,'" *Biokl. Beibl.* 1 (1934): 1–2.

44. Report from the forty-third Balneological Congress in Baden, Kurorte 1928/Nr. 17591; reports from the 1932 meeting of the radiation committee of the International Meteorological Organization, Kl. Beob. St. 1933; Walter Hausmann, "Grundlagen und Organisation der lichtklimatischen Forschung in ihrer Beziehung zur öffentlichen Gesundheitspflege," *Mitteilungen des Volksgesundheitsamtes* (1932): 1–20.

45. Wilhelm Schmidt, "Das Bioklima als Kleinklima und Mikroklima," *Biokl. Beibl.* 1 (1934): 3–6.

46. Geiger and Schmidt, "Einheitliche Bezeichnungen in kleinklimatischer und mikroklimatischer Forschung," *Biokl. Beibl.* 4 (1934): 153–56.

47. Gregor, "Moderní klimatologie," 466.

48. Schmidt, "Kleinklimatische Beobachtungen in Österreich," *Geographischer Jahresbericht aus Österreich* 16 (1933): 42–72, on 43.

49. Bohuslav Hrudička, "Má dynamická klimatologie význam i pro geografický výklad?," *Sborník IV. sjezdu československých Geografů v Olomouci 1937* (Brno: Československá společnost zeměpisné, 1938), 90–92, on 90–91.

50. David Luft, ed. and trans., *Hugo von Hofmannsthal and the Austrian Idea: Selected Essays and Addresses, 1906–1927* (West Lafayette, IN: Purdue University Press, 2011), 99–102.

51. "Der Spätzünder," *Der Spiegel* 23 (1957): 53–58, on 57.

52. Heimito von Doderer, *Die Strudlhofstiege, oder Melzer und die Tiefe der Jahre* (Munich: C. H. Beck, 1995), 104.

53. Rudolf Brunngraber, *Karl und das 20. Jahrhundert* (Göttingen: Steidl, 1999), 162, 227; Rudolf Brunngraber, *Karl und das 20. Jahrhundert* (Kronberg: Scriptor, 1978), 66.

54. Robert Musil, *The Man without Qualities*, vol. 1, trans. Sophie Wilkins (New York: Vintage, 1996), 3.

55. Robert Musil, "The 'Nation' as Ideal and as Reality," in Musil, *Precision and Soul: Essays*

and Addresses, ed. and trans. Burton Pike and David S. Luft, 101–16 (Chicago: University of Chicago Press, 1990), 103 and 111.

56. Quoted in Stephen Walsh, "Between the Arctic and the Adriatic" (PhD diss., Harvard University, 2014), 221.

57. Emanuel Herrmann, *Cultur und Natur: Studien im Gebiete der Wirthschaft* (Berlin: Allgemeiner Verein für Deutsche Literatur, 1887), 320.

58. Purkyně: see chapter 9; Ficker: see chapter 11; J. Moscheles, "Logická soustava zeměpisu člověka," *Sborník Československé společnosti zeměpisné* 31 (1925): 247–56, on 252; Supan, *Die territoriale Entwicklung der europäischen Kolonien* (Gotha: Perthes, 1906), 313.

59. Supan, *Territoriale Entwicklung*, 322.

60. Claudia Ho-Lem et al., "Who Participates in the Intergovernmental Panel on Climate Change and Why," *Global Environmental Change* 21 (2011) 1308–17; "Activities," http://www.ipcc.ch/activities/activities.shtml, accessed 24 May 2017.

61. M. Hulme and M. Mahony, "What Do We Know about the IPCC?," *Prog. Phys. Geogr.* 34 (2010): 705–18; Thaddeus R. Miller et al., "Epistemological Pluralism: Reorganizing Interdisciplinary Research," *Ecology and Society* 13 (2008): art. 46.

62. Elisabeth Nemeth and Friedrich Stadler, eds., *Encyclopedia and Utopia: The Life and Work of Otto Neurath* (Dordrecht: Kluwer, 1996), 334.

63. Hann, "Thatsachen und Bemerkungen über einige schädliche Folgen der Zerstörung des natürlichen Pflankleides," *Zs. Ö. G. Meteo.* 4 (1869): 18–22, on 22.

64. Martin Bressani, *Architecture and the Historical Imagination: Eugène-Emmanuel Viollet-le-Duc, 1814–1879* (New York: Routledge, 2016), 481.

65. Eugène-Emmanuel Viollet-le-Duc, *Le Massif du Mont Blanc* (Paris: J. Baudry, 1876), 254. Cf. George Perkins Marsh, *Man and Nature, or Physical Geography as Modified by Human Action* (1864), 127.

66. Prosper Demontzey, *Studien über die Arbeiten der Wiederbewaldung und Berasung der Gebirge* (Vienna: C. Gerold, 1880), i; Ferdinand Wang, "Über Wildbachverbauung und Wiederbewaldung der Gebirge," *Österreichische Vierteljahresschrift für Forstwesen* 9 (1891): 219–37, on 227.

67. Dale Jamieson, *Reason in a Dark Time: Why the Struggle against Climate Change Failed—and What It Means for Our Future* (Oxford: Oxford University Press, 2014), 103; for other examples, see Brace and Geoghegan, "Human Geographies of Climate Change: Landscape, Temporality, and Lay Knowledges," *Progress in Human Geography* 35 (2010): 284–302, on 292; Birgit Schneider and Thomas Nocke, "Introduction," in *Image Politics of Climate Change: Visualizations, Imaginations, Documentations*, ed. Schneider and Nocke, 9–25 (Bielefeld: transcript, 2014), 13. For psychology, see Scott Slovic and Paul Slovic, *Numbers and Nerves: Information, Emotion, and Meaning in a World of Data* (Corvallis: Oregon State University Press, 2015).

SELECTED BIBLIOGRAPHY

All works consulted are cited in the endnotes, in full at first mention. See below for archival collections, frequently cited periodicals, and the most essential monographs.

ARCHIVAL COLLECTIONS

Correspondence of Emanuel Purkyně and George Engelmann, 1875–81, Biodiversity Heritage Library (EP-GE)

Nachlass Albrecht Penck, 871/3, Archiv für Geographie, Leibniz-Institut für Länderkunde, Leipzig (AP)

Nachlass Alfred Hettner, Heid. Hs. 3929, Universitätsbibliothek Heidelberg (AH)

Nachlass Anton Kerner, Sig. 131.33, Archive of the University of Vienna (AK)

Nachlass Julius Hann, Oberösterreichisches Landesarchiv, Linz (JH)

Nachlass Ludwig Ficker, Brenner-Archiv, University of Innsbruck (LF)

Nachlass Ludwig Prandtl, Archiv der Max-Planck-Gesellschaft, III. Abt., Rep. 61 (LP)

Nachlass Wladimir Köppen, Ms. 2054, Universitätsbibliothek Graz (WK)

Österreichisches Staatsarchiv, Allgemeine Verwaltungsarchiv, Ministerium für Cultus und Unterricht: Meteorologische Zentralanstalt (MCU)

Österreichisches Staatsarchiv, Archiv der Republik, Bundesministerium für soziale Verwaltung: Volksgesundheit (VG)

Österreichisches Staatsarchiv, Archiv der Republik, Deutsch-österreichisches Staatsamt für Unterricht: Meteorologische Zentralanstalt (SAU)

Písemná pozůtalost Emanuel Purkyně, Literární archiv PNP, Prague (EP)

Sammlung Ludwig Darmstaedter, Staatsbibliothek zu Berlin, Handschriftenabteilung (LD)

Teilnachlass Friedrich Simony, Geographisches Institut, University of Vienna (FS)

William Morris Davis Papers, Ms. Am. 1798, Houghton Library, Cambridge, MA (WMD)

FREQUENTLY CITED PERIODICALS

American Historical Review (*AHR*)
Annalen der Hydrographie und maritimen Meteorologie (*Ann. Hyd.*)
Austrian History Yearbook (*AHY*)
Bioklimatische Beiblätter (*Biokl. Beibl.*)
British Journal for the History of Science (*BJHS*)
Bulletin of the American Geographical Society (*BAGS*)
Bulletin of the American Meteorological Society (*BAMS*)
Denkschriften der kaiserlichen Akademie der Wissenschaften, mathemathisch-naturwissenschaftliche Klasse (*Denk. Akad. Wiss. math-nat.*)
Historical Studies in the Natural Sciences (*HSNS*)
Jahrbuch der k.k. Central-Anstalt für Meteorologie und Erdmagnetismus (Geophysik) (*Jb. ZAMG*)
Meteorologische Zeitschrift (*MZ*)
Mittheilungen der Geographischen Gesellschaft zu Wien (*Mitt. Geog. Ges.*)
Monthly Weather Review (*MWR*)
Österreichische Botanische Zeitschrift (*Öst. Bot. Z.*)
Schriften des Vereines zur Verbreitung naturwissenschaftlicher Kenntnisse in Wien (*Schr. d. Ver. z. Verbr. naturw. Kenntn.*)
Sitzungsberichte der kaiserlichen Akademie der Wissenschaften zu Wien, mathematisch-naturwissenschaftliche Klasse (*Wiener Berichte* II/IIa)
Studies in History and Philosophy of Biological and Biomedical Sciences (*SHPBBS*)
Verhandlungen des Zoologisch-Botanischen Vereins in Wien (*Verh. Zool.-Bot. Ver.*)
Zeitschrift der Österreichischen Gesellschaft für Meteorologie (*Zs. Ö. G. Meteo.*)
Zeitschrift des deutschen und österreichischen Alpenvereins (*Z. d. ö. AV*)

SELECTED MONOGRAPHS

Primary Sources

Andrássy, Julius. *Ungarns Ausgleich mit Österreich vom Jahre 1867*. Leipzig: Duncker & Humblot, 1897.

Blodget, Lorin. *Climatology of the United States*. Philadelphia: J. B. Lippincott and Co., 1857.

Brezina, Ernst, and Wilhelm Schmidt. *Das künstliche Klima in der Umgebung des Menschen*. Stuttgart: Enke, 1937.

Brückner, Eduard. *Klimaschwankungen seit 1700, nebst Bemerkungen über die Klimaschwankungen der Diluvialzeit*. Vienna: Hölzel, 1890.

Charmatz, Richard. *Minister Freiherr von Bruck, der Vorkämpfer Mitteleuropas: Sein Lebensgang und seine Denkschriften*. Leipzig: S. Hirzel, 1916.

Chavanne, Josef, ed. *Physikalisch-statistischer Handatlas von Österreich-Ungarn*. Vienna: E. Hölzel, 1887.

Chavanne, Josef. *Die Temperatur-Verhältnisse von Österreich-Ungarn dargestellt durch Isothermen*. Vienna: Gerold's Sohn, 1871.

Ficker, Heinrich von. *Die Zentralanstalt für Meteorologie und Geodynamik in Wien, 1851–1951*. Vienna: Österreichische Akademie der Wissenschaften, 1951.

Habsburg, Rudolf von, et al. *Die österreichisch-ungarische Monarchie in Wort und Bild*. 24 vols. Vienna: k.k. Hof- und Staatsdruckerei, 1886–1902.

Hann, Julius. *Atlas der Meteorologie*. Gotha: Justus Perthes, 1887.

Hann, Julius. *Handbuch der Klimatologie*. Stuttgart: Engelhorn, 1883.

Hann, Julius. *Klimatographie von Niederösterreich*. Vienna: Braumüller, 1904.

Hann, Julius. *Lehrbuch der Meteorologie*. 3rd ed. Leipzig: Tauchnitz, 1915.

Hann, Julius. *Die Vertheilung des Luftdruckes über Mittel- und Süd-Europa*. Vienna: Hölzel, 1887.

Hann, Julius von, et al. *Klimatographie von Österreich*. 11 vols. Vienna: Braumüller, 1904–30.

Hassinger, Hugo. *Österreichs Wesen und Schicksal, verwurzelt in seiner geographischen Lage*. Vienna: Freytag-Berndt, 1949.

Herrmann, Emanuel. *Cultur und Natur: Studien im Gebiete der Wirthschaft*. Berlin: Allgemeiner Verein für Deutsche Literatur, 1887.

Herrmann, Emanuel. *Miniaturbilder aus dem Gebiete der Wirthschaft*. Halle: L. Nebert, 1872.

Hettner, Alfred. *Vergleichende Länderkunde*, vol. 3, *Die Gewässer des Festlandes: Die Klimate der Erde*. Leipzig: Teubner, 1934.

Humboldt, Alexander von. *Cosmos*. Translated by E. C. Otte. New York: Harper and Brothers, 1858.

Kerner, Anton. *Die Botanischen Gärten, ihre Aufgabe in der Vergangenheit, Gegenwart und Zukunft*. Innsbruck: Verlag der Wagnerschen Universitätsbuchhandlung, 1874.

Kerner, Anton. *Das Pflanzenleben der Donauländer*. Innsbruck: Wagner, 1863.

Kerner von Marilaun, Anton. *Das Pflanzenleben*. 2 vols. Leipzig and Vienna: Bibliographisches Institut, 1888–91.

Kisch, Enoch. *Klimatotherapie*. Berlin: Urban and Schwarzenberg, 1898.

Kreil, Karl. *Die Klimatologie von Böhmen*. Vienna: Gerold's Sohn, 1865.

Lorenz, Josef Roman, and Carl Rothe. *Lehrbuch der Klimatologie mit besonderer Rücksicht auf Land- und Forstwirthschaft*. Vienna: Braumüller, 1874.

Lorenz von Liburnau, Josef Roman. *Wald, Klima, und Wasser*. Munich: R. Oldenbourg, 1878.

Penck, Albrecht. *Friedrich Simony: Leben und Wirken eines Alpenforschers*. Vienna: Hölzel, 1898.

Schmidt, Wilhelm. *Der Massenaustausch in freier Luft und verwandte Erscheinungen*. Hamburg: Henri Grand, 1925.

Sieger, Robert. *Die geographischen Grundlagen der österreichisch-ungarischen Monarchie und ihrer Außenpolitik*. Leipzig: Teubner, 1915.

Stifter, Adalbert. *Bunte Steine*. 4th ed. Pest: Hackenast, 1870.

Stifter, Adalbert. *Der Nachsommer: Eine Erzählung*. 2 vols. Pest: Heckenast, 1865.

Stifter, Adalbert. *Wien und die Wiener in Bildern aus dem Leben*. Edited by Elisabeth Buxbaum. Vienna: LIT, 2005.

Suess, Eduard. *Erinnerungen*. Leipzig: Hirzel, 1916.

Supan, Alexander. *Grundzüge der physischen Erdkunde*. Leipzig: Veit, 1911.

Supan, Alexander. *Statistik der unteren Luftströmungen*. Leipzig: Duncker & Humblot, 1881.

Secondary Sources

Anderson, Katharine. *Predicting the Weather: Victorians and the Science of Meteorology*. Chicago: University of Chicago Press, 2005.

Ash, Mitchell, and Jan Surman, eds., *The Nationalization of Scientific Knowledge in the Habsburg Empire, 1848–1918*. New York: Palgrave, 2012.

Bachl-Hofmann, Christina, ed. *Die Geologische Bundesanstalt in Wien: 150 Jahre Geologie im Dienste Österreichs*. Vienna: Böhlau, 1999.

Cooper, Alix. *Inventing the Indigenous: Local Knowledge and Natural History in Early Modern Europe*. Cambridge: Cambridge University Press, 2007.

Cordileone, Diana Reynolds. *Alois Riegl in Vienna, 1875–1905: An Institutional Biography*. Burlington, VT: Ashgate, 2014.

Darrigol, Olivier. *Worlds of Flow: A History of Hydrodynamics from the Bernoullis to Prandtl*. Oxford: Oxford University Press, 2005.

Dörflinger, Johannes. *Descriptio Austriae: Osterreich und seine Nachbarn im Kartenbild von der Spatantike bis ins 19. Jahrhundert*. Vienna: Edition Tusch, 1977.

Eckert, Max. *Die Kartenwissenschaft: Forschungen und Grundlagen zu einer Kartographie als Wissenschaft*. Berlin: De Gruyter, 1921.

Edwards, Paul N. *A Vast Machine: Computer Models, Climate Data, and the Politics of Global Warming*. Cambridge, MA: MIT Press, 2010.

Fichtner, Paula Sutter. *Emperor Maximilian II*. New Haven, CT: Yale University Press, 2001.

Fleming, James R. *Historical Perspectives on Climate Change*. Oxford: Oxford University Press, 1998.

Friedman, Robert Marc. *Appropriating the Weather: Vilhelm Bjerknes and the Construction of a Modern Meteorology*. Ithaca, NY: Cornell University Press, 1989.

Good, David F. *The Economic Rise of the Habsburg Empire, 1750–1914*. Berkeley: University of California Press, 1984.

Grove, Richard. *Green Imperialism: Colonial Expansion, Tropical Island Edens and the Origins of Environmentalism*. Cambridge: Cambridge University Press, 1995.

Hammerl, Christa, et al., eds. *Die Zentralanstalt für Meteorologie und Geodynamik, 1851–2001*. Graz: Leykam, 2001.

Hanik, Jan. *Dzieje meteorologii i obserwacji meteorologicznych w Galicji od XVIII do XX wieku*. Wrocław: Zakład Narodowy im. Ossolińskich, 1972.

Imbrie, John, and Katherine Palmer Imbrie. *Ice Ages: Solving the Mystery*. Cambridge, MA: Harvard University Press, 1979.

Janko, Jan, and Soňa Štrbáňová. *Věda Purkyňovy doby*. Prague: Academia, 1988.

Judson, Pieter. *The Habsburg Empire: A New History*. Cambridge, MA: Harvard University Press, 2016.

Kaufmann, Thomas DaCosta. *The Mastery of Nature: Aspects of Art, Science, and Humanism in the Renaissance*. Princeton, NJ: Princeton University Press, 1993.

Khrgian, A. Kh. *Meteorology: A Historical Survey*. Edited by Kh. P. Pogosyan. Jerusalem: Israel Program for Scientific Translations, 1970.

Klemm, Fritz. *Die Entwicklung der meteorologischen Beobachtungen in Österreich einschließlich Böhmen und Mähren bis zum Jahr 1700*. *Annalen der Meteorologie* 21. Offenbach am Main: Deutscher Wetterdienst, 1983.

Komlosy, Andrea. *Grenze und ungleiche regionale Entwicklung: Binnenmarkt und Migration in der Habsburgermonarchie*. Vienna: Promedia, 2003.

Kronfeld, E. M. *Anton Kerner von Marilaun*. Leipzig: Tauchnitz, 1908.

Krška, Karel, and Ferdinand Šamaj. *Dějiny meteorologie v českých zemích a na Slovensku*. Prague: Karolinium, 2001.

Krueger, Rita. *Czech, German, and Noble: Status and National Identity in Habsburg Bohemia*. Oxford: Oxford University Press, 2009.

Kutzbach, Gisela. *The Thermal Theory of Cyclones: A History of Meteorological Thought in the Nineteenth Century*. Boston: American Meteorological Society, 1979.

Martin, Craig. *Renaissance Meteorology: Pomponazzi to Descartes*. Baltimore: Johns Hopkins Press, 2011.

Moon, David. *The Plough That Broke the Steppes: Agriculture and Environment on Russia's Grasslands, 1700–1914*. Oxford: Oxford University Press, 2013.

Phillips, Denise, and Sharon Kingsland, eds. *New Perspectives on the History of Life Sciences and Agriculture*. New York: Springer, 2015.

Przybylak, Rajmund, et al., eds. *The Polish Climate in the European Context: An Historical Overview*. Dordrecht: Springer, 2010.

Rácz, Lajos. *The Steppe to Europe: An Environmental History of Hungary in the Traditional Age*. Cambridge: White Horse Press, 2013.

Raffler, Marlies. *Museum—Spiegel der Nation? Zugänge zur Historischen Museologie am Beispiel der Genese von Landes- und Nationalmuseen in der Habsburgermonarchie*. Vienna: Böhlau, 2008.

Rampley, Matthew. *The Vienna School of Art History: Empire and the Politics of Scholarship, 1847–1918*. University Park: Penn State Press, 2013.

Singh, Simron Jit, et al., eds. *Long Term Socio-Ecological Research: Studies in Society-Nature Interactions across Spatial and Temporal Scales*. Dordrecht: Springer, 2013.

Surman, Jan. *Biography of Habsburg Universities, 1848–1918*. West Lafayette, IN: Purdue University Press, forthcoming.

Telesko, Werner. *Geschichtsraum Österreich: Die Habsburger und ihre Geschichte in der bildenden Kunst des 19. Jahrhunderts*. Vienna: Böhlau, 2006.

Telesko, Werner. *Kulturraum Österreich: Die Identität der Regionen in der bildenden Kunst des 19. Jahrhunderts*. Vienna: Böhlau, 2008.

Wawrik, Franz, and Elisabeth Zeilinger, eds. *Austria Picta: Österreich auf alten Karten und Ansichten*. Graz: Akademische Druck- und Verlagsanstalt, 1989.

Wolff, Larry. *Inventing Eastern Europe: The Map of Civilization on the Mind of the Enlightenment*. Stanford, CA: Stanford University Press, 1994.

INDEX